Advances in Materials Science for Environmental and Energy Technologies II

Advances in Materials Science for Environmental and Energy Technologies II

Ceramic Transactions, Volume 241

Edited by
Josef Matyáš
Tatsuki Ohji
Xingbo Liu
M. Parans Paranthaman
Ram Devanathan
Kevin Fox
Mrityunjay Singh
Winnie Wong-Ng

WILEY

Published by John Wiley & Sons, Inc., Hoboken, New Jersey.
Published simultaneously in Canada.

Library of Congress Cataloging-in-Publication Data is available.

ISBN: 978-1-118-75104-6
ISSN: 1042-1122

Printed in the United States of America.

10 9 8 7 6 5 4 3 2 1

Contents

MATERIALS AND SYSTEMS FOR ENERGY APPLICATIONS

Preface

The Materials Science and Technology 2012 Conference and Exhibition (MS&T'12) was held October 7–11, 2012, in Pittsburgh, Pennsylvania. One of the major themes of the conference was Environmental and Energy Issues. Papers from five of the symposia held under that theme are included in this volume. These symposia included Materials Issues in Nuclear Waste Management for the 21st Century; Green Technologies for Materials Manufacturing and Processing IV; Energy Storage: Materials, Systems and Applications; Energy Conversion—Photovoltaic, Concentrating Solar Power and Thermoelectric; and Materials Development for Nuclear Applications and Extreme Environments.

The success of these symposia and the publication of the proceedings could not have been possible without the support of The American Ceramic Society and other organizers of the program. The program organizers for the above symposia are appreciated. Their assistance, along with that of the session chairs, was invaluable in ensuring the creation of this volume.

Josef Matyáš, Pacific Northwest National Laboratory, USA
Tatsuki Ohji, AIST, JAPAN
Xingbo Liu, West Virginia University, USA
M. Parans Paranthaman, Oak Ridge National Laboratory, USA
Ram Devanathan, Pacific Northwest National Laboratory, USA
Kevin Fox, Savannah River National Laboratory, USA
Mrityunjay Singh, NASA Glenn Research Center, USA
Winnie Wong-Ng, NIST, USA

Materials for Nuclear Waste Disposal and Environmental Cleanup

BORON AND LEAD BASED CHEMICALLY BONDED PHOSPHATES CERAMICS FOR NUCLEAR WASTE AND RADIATION SHIELDING APPLICATIONS

Henry A. Colorado[1, 2*], Jason Pleitt[3], Jenn-M. Yang[1], Carlos H. Castano[3]
[1]Materials Science and Engineering, University of California, Los Angeles.
[2]Universidad de Antioquia, Mechanical Engineering Department. Medellin-Colombia.
[3]Missouri University of Science and Technology, Nuclear Engineering Department.

*Corresponding author:
Email: henryacl@gmail.com. Address: MSE 3111 Engineering V, 410 Westwood Plaza, Los Angeles, CA 90095, USA.

ABSTRACT

We present research about the development of a fast setting ceramic materials fabricated with boron oxide, and wollastonite powders with potential use for nuclear waste transport, disposal, stabilization, and radiation shielding applications. The boron oxide is mixed with calcium silicate and phosphoric acid in order to consolidate a ceramic by an acid base-reaction. These materials have high compression strength when compared with cements and with other available solutions for the treatment of nuclear wastes, and are suitable as fire resistant shielding materials. Nuclear attenuation experiments, compressive strength, x-ray diffraction, and scanning electron microscopy were used to characterize these materials. Our work indicates that a balance between lead oxide and boron oxide is needed to obtain a suitable hybrid material with good mechanical properties that can attenuate gammas and neutrons.

INTRODUCTION

Chemically Bonded Phosphate Ceramics (CBPCs) are formed in low temperature conditions by mixing a base powder with phosphoric acid. The compound that forms from the cement like processing has the properties of an engineering ceramic with the advantage of being able to be formed with relative ease and low expense[1, 2, 3]. An additional advantage is that the compound can be formulated with additives that can enhance the properties of the ceramic without causing a major degradation of the structural properties. The high melting temperatures and the good thermal resistance allow this ceramic to be used as a firewall to prevent transformer explosions[4]. Based upon these principles the ceramic was examined with several different additives to determine the improvement of the linear attenuation coefficient for dry cask spent fuel storage and/or transportation cask for spent nuclear fuel.

Dry cask spent fuel storage involves storing spent nuclear fuel above ground and/or transporting it in casks that are lined with gamma and neutron shields[5, 6]. One issue that arises with these casks is that while they are designed to withstand high temperatures for short periods of time, there have been examples of traffic accidents[7,8] that have resulted in temperatures and times exceeding the current safety design parameters. This research examines CBPCs with regards to neutron shielding in order to obtain preliminary data for an improved design of nuclear casks that can survive a traffic accident of similar nature to the new scenarios considered.

EXPERIMENTAL PROCEDURE

Manufacturing

CBPC samples were fabricated by mixing an aqueous phosphoric acid formulation and natural wollastonite powder ($CaSiO_3$) M200 (from Minera Nyco; see Table 1). Also, boron oxide (from Alfa Aesar; see Table 2) was added to the mixture. For all samples, the 1.2 ratio liquid (phosphoric acid formulation) to powders (wollastonite + B_2O_3) was maintained constant, see Table 3. The mixing process of the components was conducted in a Planetary Centrifugal Mixer (Thinky Mixer® AR-250, TM). First, wollastonite was mixed with the acidic liquid for 1 min, then, boron oxide was added and mixed for 1 min. All samples were dried for 2 days at 100°C in order to stabilize the weight.

Table 1. Chemical composition of wollastonite powder.

Composition	CaO	SiO_2	Fe_2O_3	Al_2O_3	MnO	MgO	TiO_2	K_2O
Percentage	46.25	52.00	0.25	0.40	0.025	0.50	0.025	0.15

Table 2. Chemical composition of B_2O_3 powder.

Composition	B_2O_3	SO_4	Al_2O_3	Cl	H_2O (insoluble)
Percentage	95.00	0.7	0.1	0.2	0.02

Table 3. Raw materials amounts in the samples fabricated.

Sample composition	Wollastonite ($CaSiO_3$) (g)	B_2O_3 (g)	Total powders (g)	Phosphoric acid formulation (g)
CBPC	100	0	100	120
CBPC-2 B_2O_3	98	2	100	120
CBPC-10 B_2O_3	90	10	100	120
CBPC-15 B_2O_3	85	15	100	120
CBPC-20 B_2O_3	80	20	100	120

Characterization

For compression tests, samples were fabricated using glass molds of 12.7 mm diameter and 100 mm long. A diamond saw was used to cut cylinders of 28 mm length. Samples were ground (with silicon carbide papers of grit ANSI 400) using a metallic mold until flat, parallel, and smooth surfaces were obtained. The final length was 25.4 mm. Compression tests were conducted in an Instron® machine 3382. A set of 5 samples was tested for each composition of Table 3. The crosshead speed was 1 mm/min. For Scanning Electron Microscopy (SEM) examination (sample sections were ground using silicon carbide papers grit ANSI 240, 400 and 1200 progressively), samples were mounted on an aluminum stub and sputtered in a Hummer 6.2 system (15mA AC for 30 sec) creating approximately a 1nm thick film of Au. The SEM used was a JEOL JSM 6700R in high vacuum mode. X-Ray Diffraction (XRD) experiments were conducted using an X'Pert PRO (Cu Kα radiation, λ=1.5406 Å), at 45KV and scanning between 10° and 80°. M200, M400 and M1250 wollastonite samples (before and after the drying process) were ground in an alumina mortar and XRD tests were done at room temperature. The above characterization was conducted for the compositions summarized in Table 3. A similar set of experiments for PbO additions (instead of B_2O_3), was presented before[6].

In order to determine the shielding effect of CBPCs a method for measuring the attenuation of neutrons was implemented. In order to allow for quick measuring of the attenuation coefficients a He-3 neutron detector was used. The He-3 detector gives a count of thermal neutrons over a set time period. This detector was connected to a lower level discriminator to reduce noise and a scaler to obtain counts. A PuBe neutron source was used at the Missouri University of Science and Technology Research Reactor (MSTR). The production of neutrons from the PuBe sources occurs when the alpha particle released from the plutonium reacts with beryllium in the source producing neutrons according to the reaction shown in equation 1.

$$ {}_{2}^{4}He + {}_{4}^{9}Be \rightarrow {}_{6}^{12}C + {}_{0}^{1}n \tag{1} $$

This produces neutrons with a wide energy range; the neutron energy spectrum for PuBe can be seen in Figure 1a.

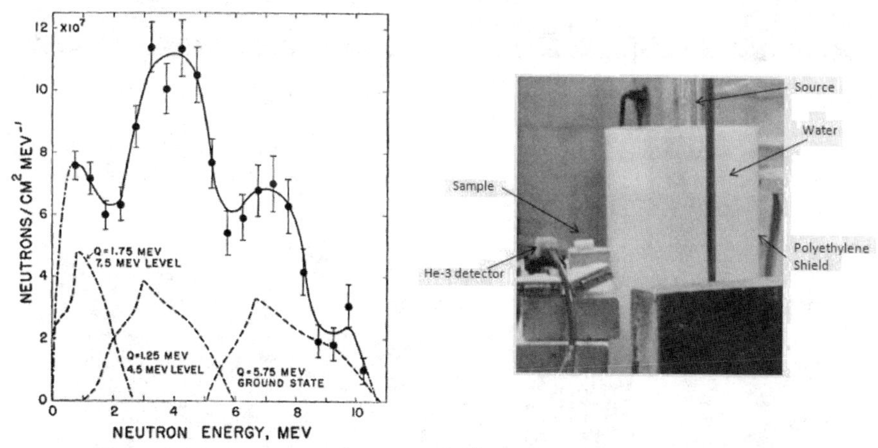

Figure 1 a) PuBe neutron energy spectrum[9], and b) neutron attenuation setup

Since the spectrum of the PuBe source contains more fast neutrons then thermal neutrons a method for thermalizing the neutrons was used. The average distance for thermalizing a fast neutron in water is about 12.7 cm (estimated from the neutron age for fast neutrons ~27 cm^2) [10]. The source was suspended in water in a container of low density polyethylene (borosilicate glass must be avoided) with a distance to the wall of 9 cm. The sample was placed on a platform with the edge of the sample 1 to 2 cm away from the polyethylene wall. The samples were then counted using a He-3 detector with 10 minute count rates with and without a 0.6 mm cadmium shield placed over it. When present, the cadmium shield absorbs all the thermalized neutrons traveling through the sample, and the difference measured with and without the Cd shield gives the neutron difference used to calculate the attenuation for the samples. A picture of the setup is shown in Figure 1b.

RESULTS AND ANALYSIS

Figure 2 shows boron oxide (B_2O_3) and wollastonite ($CaSiO_3$) powders respectively. Wollastonite grains are needle like shaped while B_2O_3 grains do not have specific shape.

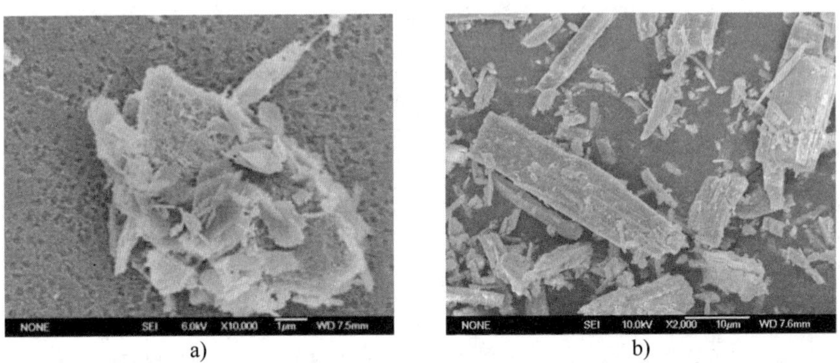

a) b)

Figure 2 SEM images of the raw powders used to fabricate the CBPC, a) B_2O_3, and b) $CaSiO_3$.

Figure 3 shows SEM images for the CBPCs fabricated. $CaSiO_3$, $CaHPO_4.2H_2O$, BPO_4 and B_2O_3 have been identified in the images. These results were confirmed by the XRD data as shown in Figure 4.

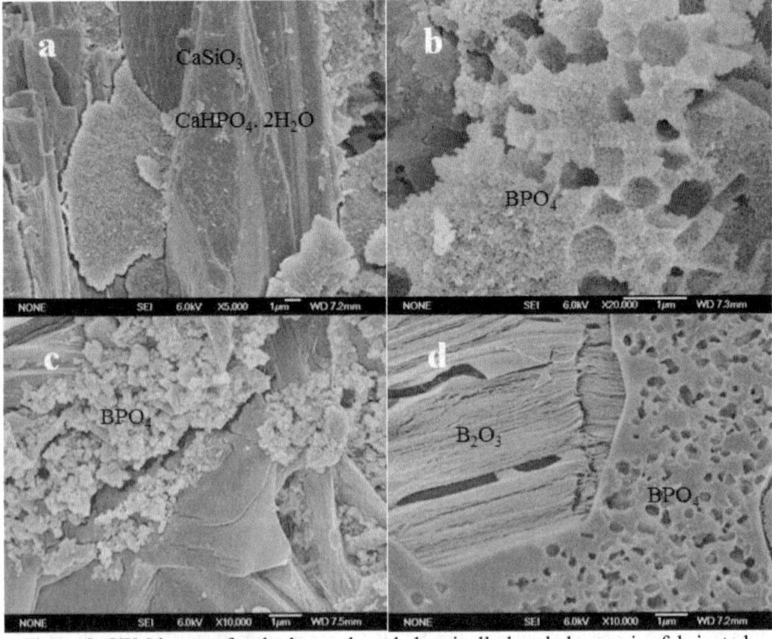

Figure 3 SEM images for the boron-based chemically bonded ceramics fabricated.

Figure 4 shows the XRD for the materials used (raw powders) and fabricated in this research (see Table 3). $CaSiO_3$, $CaHPO_4.2H_2O$, BPO_4, and B_2O_3 are shown. B_2O_3 is mostly amorphous. Brusite ($CaHPO_4.2H_2O$) is the result of the reaction between $CaSiO_3$ and H_3PO_4. Brusite and boron phosphate are the binding phases.

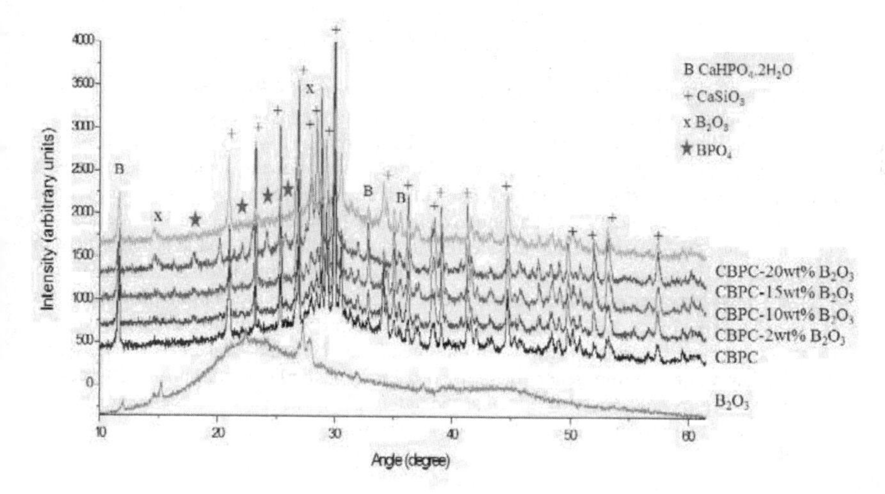

Figure 4 XRD for the raw powders and ceramics fabricated.

Figure 5 shows the compressive strength for the CBPCs with B_2O_3 contents. The standard deviation was bigger for the sample with 10% B_2O_3. As the B_2O_3 contents increases, the compressive strength decreases.

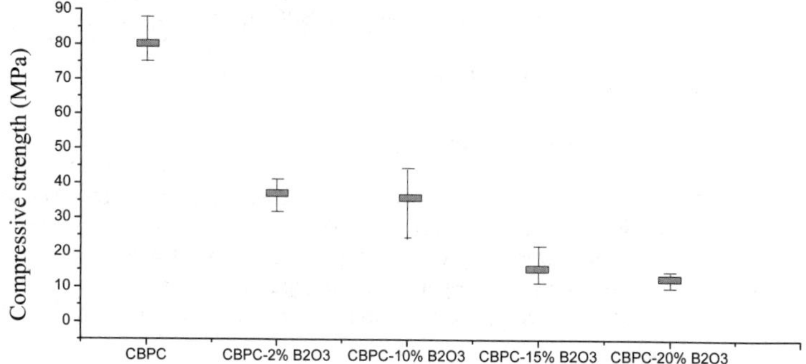

Figure 5 Compressive strength for the boron-based chemically bonded ceramics fabricated.

Results for the neutron attenuation experiment for several different CBPCs are summarized in Table 4.

Table 4: Thermal Neutron Attenuation Values for CBPCs

	Thickness (cm)	Linear Attenuation (cm^{-1})	Error
20% Boron	2.774	0.089	0.017
0%	1.664	0.060	0.018
50% PbO	1.284	0.045	0.023
Boron oxide	1.308	0.068	-

From the results the addition of boron improved the attenuation of CBPCs by 48% for thermal neutrons. However the addition of PbO caused a decrease in attenuation by 25%. The CBPCs compared well to boron oxide during the testing being only 11% lower. The CBPC also compares favorably when it was combined with boron oxide showing a definite increase in thermal neutron attenuation for both materials. However, these results still have large error values and further testing on neutron attenuation with a more intense source is recommended.
One of the reasons for the relatively large errors is due to the continued thermalization of neutrons within the sample, i.e. the sample is not only blocking neutrons but is still thermalizing them. Another possible error is that the source location has small variations in its position. This happens because the source had to be repositioned several times over the course of the experiment to guarantee a safe working environment for the experimenter (i.e. ALARA). Missouri S&T has a new neutron generator in the process of being licensed. Utilizing a neutron generator as a source of neutrons will alleviate the safety problems, since the generator can be started and stopped at will as opposed to the PuBe neutron source.

DISCUSSION

The reaction generating brushite is shown in equation 2. $CaHPO_4 . 2H_2O$ is a calcium phosphate that under time and temperature can lose the bonded water molecule and transform to a more stable phase, monetite, $CaHPO_4$.
In the acidic formulation used, boron oxide is dissolved and boric acid (H_3BO_3) is generated. Then, the reaction of boric acid and phosphoric acid can generate boron phosphate (BPO_4) as found experimentally.We suggest the reaction generating BPO_4 in equation 3,

$$CaSiO_3 + H_3PO_4 + 2H_2O = SiO_2.H_2O + CaHPO_4 . 2H_2O \quad (2)$$

$$H_3PO_4 + H_3BO_3 = BPO_4 + 3\ H_2O \quad (3)$$

As B_2O_3 presence increases, the compressive strength is decreased. This can be associated with two factors. The first one is likely due to a weak reaction of saturation of boron ions in the acidic solution. The second one is that when the residual boron oxide content increases, there is more possibility to have agglomeration of these particles. The agglomeration produces a poor liquid impregnation, which decreases the compressive strength.

CONCLUSION

The addition of boron oxide to CBPC shows significant improvement in neutron attenuation for the ceramic. A previously prepared sample with PbO however, demonstrated a significant drop in neutron attenuation. For applications of this ceramic as a shielding material, a balance between boron oxide and lead oxide needs to be determined in order to function as an effective radiation shield. In this regard further testing on the ceramic with more variation in regards to boron content will be performed.

ACKNOWLEDGMENTS
The authors desire to express their gratitude to Colciencias the Administrative Department for Science Technology and Innovation of the Republic of Colombia for their support of Henry A. Colorado to pursue this research, also NRC Grant PPR-NRC-38-10-966 for partially funding J. Pleitt.

REFERENCES

[1] Della M. Roy. New Strong Cement Materials: Chemically Bonded Ceramics. *Science*, **235**, 651-658 (1987).

[2] Henry Colorado, Clem Hiel, H. Thomas Hahn and Jenn-Ming Yang Chemically Bonded Phosphate Ceramic Composites, Metal, Ceramic and Polymeric Composites for Various Uses, John Cuppoletti (Ed.), ISBN: 978-953-307-353-8, InTech, 265-282 (2011).

[3] H.A. Colorado, H.T. Hahn, C. Hiel, Pultruded glass fiber-and pultruded carbon fiber-reinforced chemically bonded phosphate ceramics. Journal of Composite Materials, **45**(23), 2391–2399 (2011).

[4] H. A. Colorado, C. Hiel and H. T. Hahn. Chemically bonded phosphate ceramic composites under thermal shock and high temperature conditions. Society for the Advancement of Material and Process Engineering (SAMPE), May 17-20, 2010. Seattle, Washington USA (2010).

[5] Jason Pleitt, Henry A. Colorado, Carlos H Castano. Materials Science & Technology 2011 Conference. Materials for Nuclear Waste Disposal and Environmental Cleanup Symposium. Radiation Shielding Simulation for Wollastonite-Based Chemically Bonded Phosphate Ceramics. Columbus, Ohio (2011).

[6] H.A. Colorado, J. Pleitt, C. Hiel, J.M. Yang, H.T. Hahn, C.H. Castano. Wollastonite based-Chemically Bonded Phosphate Ceramics with lead oxide contents under gamma irradiation. *Journal of Nuclear Materials*, **425**, 197–204 (2012).

[7] H.E. Adkins, B.J. Koeppel, J.M. Cuta, A.D. Guzman, C. S. Bajwa. Spent Fuel Transportation Package Response to the Caldecott Tunnel Fire Scenario. NUREG/CR-6894, Rev. 1. PNNL-15346 (2006).

[8] R. J. Halstead and F. Dilger. Implications of the Baltimore rail tunnel fire for full-scale testing of shipping casks. Waste Management Conference 2003. February 23-27, Tucson, AZ (2003).

[9] Stewart, L. Neutron Spectrum and Absolute Yield of a Plutonium-Beryllium Source. *Physical Review*, 740-743 (1954).

[10] John R. Lamarsh, A. J. *Introduction to Nuclear Engineering*. New Jersey: Prentice-Hall, Inc. (2001).

ADVANCED CERAMIC WASTEFORMS FOR THE IMMOBILISATION OF RADWASTES

M.C. Stennett,* L.D. Casey, C.L. Corkhill, C.L. Freeman, A.S. Gandy, P.G. Heath, I.J. Pinnock, D.P. Reid, and N.C. Hyatt*
Department of Materials Science & Engineering, The University of Sheffield, Mappin Street, Sheffield, S1 3JD. UK.

ABSTRACT

Recent progress in the synthesis, characterisation and radiation damage behaviour of advanced ceramic wasteforms for the immobilisation of actinides and halide radionuclides is reviewed. A systematic methodology is described to probe the structure and evolution of the radiation damaged structure of model wasteform materials, combining ex-situ ion beam irradiation of bulk ceramics with X-ray Absorption Spectroscopy (XAS), to quantify damage induced changes in element speciation. The defect chemistry and crystal structure of cerium brannerite, $Ce_{0.975}Ti_2O_{5.95}$, was clarified by Rietveld analysis and defect energy calculations, combined with careful investigation of the phase diagram. Formation of oxygen vacancies at the O1 site, charge compensated by Ce vacancies, relived considerable coulombic repulsion and structural strain associated with short O1-O1 contacts forming the shared edge of neighbouring TiO_6 polyhedra. The rapid synthesis of $Pb_5(VO_4)_3I$, a potential immobilisation host for iodine radioisotopes, was achieved in an open container by microwave dielectric heating of a mixture of PbO, PbI_2, and V_2O_5 at a power of 800 W for 180 s (at 2.45 GHz). The resulting ceramic bodies exhibited a zoned microstructure, differentiated by inter-granular porosity and phase assemblage, as a consequence of the inverse temperature gradient characteristic of microwave dielectric heating.

INTRODUCTION

This paper presents progress in three key areas of research related to the management and geological disposal of radioactive wastes in the UK, with particular reference to the immobilisation and disposal of actinide and halide radionuclides in ceramic wasteforms.

- The application of XAS to the characterisation of surface-amorphised $Gd_2Ti_2O_7$, produced by ion beam implantation, as an analogue for metamict minerals and actinide wasteforms amorphised through natural α-decay.
- The understanding of defect chemistry and crystal structure of titanate brannerites, ideally MTi_2O_6 (M = U, Th, Ce), which are of interest as actinide host phases in multi-component ceramics, such as the Synroc phase assemblage.
- The development of a novel method for the immobilisation of iodine radioisotopes in $Pb_5(VO_4)_3I$ by rapid microwave dielectric heating, in an open container, without significant volatilisation of iodine.

The background to each of these studies is discussed in more detail in later sections of this paper.

EXPERIMENTAL

$Gd_2Ti_2O_7$ was synthesised by conventional solid state reaction using oxide (Gd_2O_3 and TiO_2) precursors. 20 mm diameter pellets were produced by sintering at 1600 °C for 4 hours and irradiated with 2 MeV Kr^+ ions at a fluence of 5 x 10^{15} ions cm^{-2} at the Ion Beam Centre (IBC), University of Surrey, Guildford, UK. Displacement profiles were calculated using the software package SRIM [1] which predicted the peak in the damage profile to occur at ~800 nm. Ti K-edge XAS measurements were conducted on beamline X23A2 of the National Synchrotron Light Source (NSLS), Brookhaven National Laboratory (BNL), USA. Transmission measurements were collected on a finely ground specimen of pristine $Gd_2Ti_2O_7$ dispersed in PEG to achieve a thickness of one absorption length. Fluorescence measurements were collected on pristine (pristine here indicates sintered pellets which were not irradiated) and irradiated pellets oriented such that the X-ray beam grazed the sample surface at a shallow angle. Glancing angles were selected such as to maintain a path length of at least two absorption lengths within the 800 nm amorphised surface layer, across the range of incident X-ray energy. Two absorption lengths corresponds to least 95% of detected fluorescence photons arising from the surface amorphised layer. A major drawback with fluorescence measurements of concentrated absorbers is that a significant fraction of fluorescence X-rays may be reabsorbed by the absorber atoms in the sample. This causes an attenuation of the fluorescence signal and systematic errors in the determination of nearest neighbour co-ordination numbers. The samples were mounted on a stage which could translate in all three directions, to allow accurate positioning of the sample with respect to the beam, and tilted in relation to the plane of the incoming beam to within an accuracy of 0.1°. All data analysis was performed using Athena and Artemis [2].

Cerium brannerite samples, $Ce_{1-x}Ti_2O_{6-2x}$ with x = 0.000, 0.025 and 0.050, were synthesised by solid state reaction of CeO_2 and TiO_2 in alumina crucibles at 1350 °C for a total of 96 hours with intermediate grinding. An Oxford Instruments INCA X-sight energy dispersive X-ray spectrometer was used for quantitative elemental analysis. Atomic number, absorption and fluorescence (ZAF) correction factors, for each element, were calculated by measuring standards of known composition. The sample measurement position was fixed and the beam current was allowed to stabilise before being calibrated using a cobalt standard mounted flush with the sample in the holder as described elsewhere [3]. Ce L_{III}-edge XAS data from $Ce_{0.975}Ti_2O_{5.95}$, CeO_2, and $CePO_4$ (with the monazite structure) were acquired on beamline X23A2 of the NSLS, BNL, USA. Data were acquired in transmission mode and analysis was performed using the program Athena [2]. Neutron diffraction data were acquired at room temperature on the HRPT diffractometer of the SINQ spallation neutron source at the Paul Scherrer Institute (PSI), Switzerland [4]. Rietveld analysis of neutron powder diffraction data was undertaken using the GSAS suite of programs [5]. Defect calculations were performed with the General Utility Lattice Program [6] using the Mott-Littleton Method [7] and the potential set developed by Gilbert and Harding to describe the structure [8]. The polarisability of the ions was accounted for used the shell model [9].

A phase assemblage of 90% $Pb_5(VO_4)_3I$ and 10% $Pb_3(VO_4)_2$ was targeted for the lead iodovanadate phase to provide a buffer against iodine substoichiometry [10]. Stochiometric quantities of PbO, PbI_2 and V_2O_5 were homogenized by ball milling in a polyethylene mill pot with yttria-stabilised milling media and isopropanol as a carrier fluid. Synthesis was conducted in a modified domestic microwave oven (DMO) operated at 2.45 GHz with a maximum power of 800 W. The glass turntable and support mechanism were removed from the DMO and replaced by a fire-brick approximately 50 x 50 x 60 mm with a core-drilled cavity, approximately 25 mm diameter and 25 mm deep. 1.00 gram batches of the milled reaction mixture were pressed at 50 MPa in a 13 mm diameter hardened stainless steel die and placed in a thick walled alumina crucible in the cavity of the alumina block. The whole assembly was positioned in the centre of the microwave oven on small alumina supports. Temperature measurement during irradiation was made by inserting a shielded K-type thermocouple into a pre-drilled hole within the alumina block to make contact with the bottom of the alumina crucible. The thermocouple was connected using a TC08 interface to a computer and samples were microwaved continuously at 800 W for a pre-determined time period between 120 – 500 s.

RESULTS AND DISCUSSION

Study of radiation damaged ceramics using grazing angle XAS

Assessment of the long-term behaviour of synthetic mineral wasteforms designed for actinide disposition is critical to underpinning the scientific case for their utilisation as disposal matrices. The known retention of actinide species in certain mineral systems over geological timescales guides the choice and design of synthetic materials for the disposition of actinide species arising from civilian and military nuclear activities. Whilst natural analogues give confidence in the ability of such systems to retain actinide species over the desired time-scales, the effect of α-decay induced amorphisation on the crystalline lattice and other key material properties are not well understood.

Several methods exist for simulating radiation damage occurring over disposal timescales. Damage rates can be accelerated by the incorporation of short half-life actinide species, generating fully amorphous bulk samples over timescales of several years [11, 12]. *In-situ* ion-beam irradiation on thin specimens coupled with inspection of selected area diffraction patterns allows the determination of the critical ion fluence (ions cm^{-2}) required to cause amorphisation under a heavy ion beam flux, at a given temperature [13-15]. In this study, irradiation of the surface of $Gd_2Ti_2O_7$ samples with heavy ions has been combined with *ex-situ* analysis of the specimens by GA-XAS.

The Ti K-edge XANES spectra obtained from both pristine and irradiated specimens all exhibit a distinctive pre-edge feature. Systematic studies have been conducted by Farges [16] and Waychunas [17] which demonstrated that the intensity and position of this pre-edge feature are related to the Ti cation co-ordination and geometry. Figure 1 shows the pre-edge features for $Gd_2Ti_2O_7$ powder measured in transmission and the pristine and irradiated monolithic samples measured by fluorescence in glancing angle geometry. The glancing angle fluorescence data was corrected for self-absorption using the Troger algorithm [18]. To ensure the robustness of the self-absorption correction the corrected pristine fluorescence data set was compared to the pristine powder sample measured in transmission. This self-absorption correction was then applied to the irradiated fluorescence data set.

Pre-edge peak height and absolute energy positions were extracted according to the methodology proposed by Waychunas [17] and plotted in Figure 2 along with model Ti containing compounds from Farges [19]. Three distinct regions are indicated on the diagram associated with Ti in 6-, 5- and 4-fold co-ordination environments. Overlaid on this are the extracted pre-edge height and absolute energy positions for the pristine and irradiated samples measure in glancing angle geometry and also parameters from powder samples of two 5-fold coordination standards measured in transmission. The two 5-fold co-ordination standards are Dy_2TiO_5 where the Ti cations are located in trigonal bipyramidal (TBP) sites and Gd_2TiO_5 where the Ti are located in square pyramidal (SP) sites. The pre-edge features for the two 5-fold geometries both have similar absolute energy position but the variation in height indicates differences in site symmetry. As expected the pristine sample parameters are consistent with 6-fold co-ordinated Ti whilst the irradiated sample parameters are consistent with 5-fold co-ordinated Ti in mixed TBP and SP geometry. The increase in the pre-edge peak height indicates that the Ti co-ordination environment in the irradiated sample is non-centrosymmetric, with respect to the pristine material.

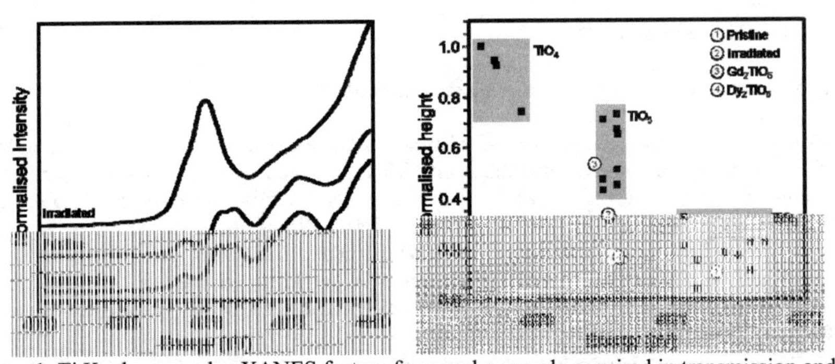

Figure 1. Ti K edge pre-edge XANES feature for powder sample acquired in transmission and pristine and irradiated $Gd_2Ti_2O_7$ samples acquired in glancing angle geometry.
Figure 2. Correlation between Ti co-ordination environment and normalized height and energy position of pre-edge feature in Ti K edge XANES spectra. Solid squares are data from Ti bearing model compounds reported by Farges [19]. Open circles are data from this study.

This work has demonstrated the feasibility of *quantitative* XAS measurement of thin ion beam amorphised surface layers on ceramic specimens. Quantitative XANES analysis has shown, for the first time, that the co-ordination number of Ti is reduced from 6 in pristine $Gd_2Ti_2O_7$ to ~5 in the material amorphised by heavy ion implantation. Comparison of post edge XANES data shows damping of the near edge structure in the spectra of ion beam irradiated specimens, compared to data acquired from pristine $Gd_2Ti_2O_7$. This damping is understood to arise from random phase decoherence of single and multiple scattering paths as a consequence of atomic disorder in aperiodic materials [16]. As discussed previously, the relative intensity and precise energy position of the pre-edge are sensitive to the local environment of Ti which may be identified by careful fitting of the XANES spectral features [17]. This is demonstrated clearly by Figure 2, elaborated from the analysis of Farges et al [19], in which the height and energy position of the pre-edge feature are shown to be well correlated with co-ordination number in well characterised titanate materials. A similar analysis has also revealed significant changes in the Gd co-ordination environment. Work is on-going to apply this methodology to establishing structure-property relations in radiation amorphised ceramics for actinide immobilization.

Defect chemistry of cerium brannerite

Synthetic titanate brannerites, ideally MTi_2O_6 (M = U, Th, Ce), are of interest as actinide host phases in multi-component ceramics, such as the Synroc phase assemblage and special purpose pyrochlore formulations, for the immobilisation of radioactive wastes and excess weapons plutonium [20-25]. Natural brannerite, ideally UTi_2O_6, is an important accessory phase in uranium mineral ores and is mined as the principal ore mineral in some localities and are generally found to be X-ray amorphous due to accumulated damage from the recoil daughters of U and Th α-decay [26, 27]. This observation is consistent with systematic ion beam irradiation studies of synthetic titanate brannerites which were found to be susceptible to radiation induced amorphisation [28]. The archetype brannerite structure was determined by Syzmanski and Scott in 1982 from a synthetic single crystal specimen of UTi_2O_6 [26]. The structure adopts monoclinic symmetry, space group C2/m, and comprises layers of anatase-like edge sharing TiO_6 octahedra which are linked by UO_6 octahedra (Figure 3). The objective of our investigation was to determine an accurate structural model for $CeTi_2O_6$ using high resolution neutron powder diffraction which is more sensitive to the position of O in the presence of Ce (neutron scattering lengths of 5.80 and 4.84 barns, respectively [29]). Using combined powder diffraction, micro-chemical analysis, and atomistic modelling calculations we showed that $CeTi_2O_6$ is in fact non-stochiometric and more properly expressed as $Ce_{1-x}Ti_2O_{6-2x}$, with $x \approx 0.025$.

Figure 4 shows selected X-ray diffraction (XRD) data, backscattered electron micrographs and corresponding energy dispersive X-ray (EDX) spectra acquired for $Ce_{1-x}Ti_2O_{6-2x}$ brannerite compositions with x = 0.000, 0.025 and 0.050. The XRD data range is selected to highlight the major reflections from brannerite and secondary phases of CeO_2 and TiO_2 (rutile). Inspection of both XRD and SEM / EDX data show the presence of reflections and secondary phase inclusions of CeO_2 and TiO_2 for compositions with x = 0.000 (Figures 4b) and x = 0.050 (Figures 4e), respectively. The presence of CeO_2 impurity in nominally stoichiometric $CeTi_2O_6$ is in agreement with the previous investigation of Helean et al [30]. XRD and SEM / EDX data provided no evidence of secondary phases for the composition $Ce_{0.975}Ti_2O_{5.95}$ (x = 0.025), quantitative EDX analysis confirmed the stoichiometry to be $Ce_{0.978(6)}Ti_{1.996(6)}O_{5.95}$, consistent with the nominal composition (based on analysis of ten separate grains).

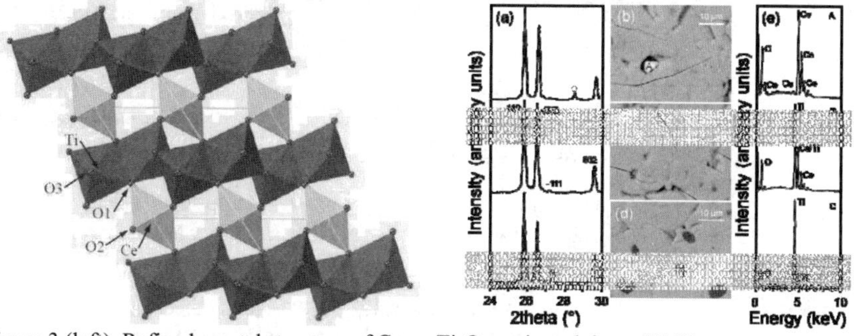

Figure 3 (left). Refined crystal structure of $Ce_{0.975}Ti_2O_{5.95}$ viewed down [010].
Figure 4 (right). (a) XRD patterns for $Ce_{1-x}Ti_2O_{6-2x}$ where x = 0.000 (top), x = 0.025 (middle) and x =0.050 (bottom). (b-d) backscattered electron micrographs for x = 0.000, x = 0.025 and x = 0.500, respectively. (e) Typical EDX spectra acquired from points A (top), B (middle), and C = bottom indicated on backscattered electron micrographs.

Figure 5 shows Ce L_{III} edge XANES of $Ce_{0.975}Ti_2O_{5.95}$, together with the data acquired for selected well characterised standard compounds which permit straightforward fingerprinting of Ce oxidation state. From comparison of the spectra it is apparent that the XANES data of

$Ce_{0.975}Ti_2O_{5.95}$ are very similar to those of $SrCeO_3$, demonstrating that Ce^{4+} is the dominant species. Notably, feature C is prominent and well resolved in the XANES of $SrCeO_3$ and $Ce_{0.975}Ti_2O_{5.95}$ which contain Ce^{4+} in six-fold co-ordination, however, this feature is less well resolved in the XANES of CeO_2, which contains Ce^{4+} in 8-fold co-ordination. In summary, the Ce L_{III} edge XANES of $Ce_{0.975}Ti_2O_{5.95}$ confirm the presence of octahedrally co-ordinated Ce^{4+} as the dominant species, consistent with our structure refinement discussed below. For detailed interpretation of the initial and final states giving rise to the observed features in the XANES spectra, the reader is referred to the work of Bianconi and Soldatov [31-32].

Structure refinement initially assumed the ideal stoichiometry of $CeTi_2O_6$, with anisotropic thermal parameters, in the higher symmetry space group C2/m, consistent with the published structures of UTi_2O_6 and related compounds [26, 33-35]. Refinements quickly converged to a satisfactory fit with χ^2 = 2.74, R_{wp} = 5.69 %, R_p = 4.55 %, for 41 variables including 28 structural parameters. However, the equivalent isotropic thermal parameter of the Ce cation (U_{iso} = 0.0165(5) $Å^2$) was significantly larger than that of Ti cation (U_{iso} = 0.0085(3) $Å^2$), suggesting the presence of Ce vacancies (charge compensated by O vacancies), in agreement with the analysis of XRD and SEM / EDX data. Refinement of Ce and O vacancies over all available sites, under charge balance constraints, confirmed the presence of Ce vacancies charge balanced by O vacancies at the O1 site only. The final structural model allowed for Ce and O1 vacancies, subject to electroneutrality constraints, provided a substantially improved fit to the data with χ^2 = 2.50, R_{wp} = 5.57, and R_p = 4.37; for 42 variables with 29 structural parameters. The final refined structural parameters are given in Table I, the final profile fit is shown in Figure 6.

Table I: Refined structural parameters for $Ce_{0.975}Ti_2O_{5.95}$ determined from Rietveld analysis of neutron powder diffraction data; [a] denotes equivalent isotropic thermal parameter, [b] occupancies constrained to be equal for electroneutrality.

Space group: C2/m			a = 9.8320(1) Å		b = 3.75287(6)Å		c = 6.8852(1) Å		β = 119.230(1)°	
Atom	Site	x	y	z	U x 100 ($Å^2$)					Frac.
					U_{iso}	U_{11}	U_{22}	U_{33}	U_{13}	
Ce	2a	0	0	0	1.22[a]	2.10(15)	1.26(13)	0.97(15)	0.95(14)	0.976(3)[b]
Ti	4i	0.8246(3)	0	0.3907(4)	0.80[a]	1.26(13)	1.14(11)	0.66(10)	0.82(1)	1
O1	4i	0.9775(1)	0	0.3077(2)	1.36[a]	0.37(10)	2.60(9)	0.91(10)	0.18(7)	0.976(3)[b]
O2	4i	0.6519(2)	0	0.1026(3)	1.85[a]	1.46(9)	1.53(9)	1.92(9)	0.23(6)	1
O3	4i	0.2793(2)	0	0.4032(2)	1.53[a]	2.33(9)	0.82(7)	2.71(9)	2.13(6)	1
Powder statistics:			χ^2 = 2.50		R_{wp} = 5.57 %		R_p = 4.37 %			

The individual point defect energies determined for the Ce and O sites in $CeTi_2O_6$ were: Ce 86.901 eV; O1 17.944 eV; O2 18.167 eV; and O3 18.832 eV. The total defect energy of the non-stochiometry (i.e. one isolated V_{Ce} and two isolated V_O) is thus 122.69 eV. The high local charge of defect sites frequently causes defects of opposite charge to cluster within a material as the Coulombic interaction favourably attracts the defects. Defect energies were calculated for the charge neutral cluster "CeO_2" with the six different permutations of O site vacancies. The binding energy is the energy gain or loss associated with bringing the isolated defects together as a cluster within the crystal. It is calculated by subtracting the defect energy of isolated defects from the defect energy of the cluster; a negative binding energy implies the defects are attracted and the cluster is energetically more favourable than isolated defects. All six permutations of combined Ce and O vacancies were found to have negative binding energies, with the lowest energy cluster determined as the O1-Ce-O1 defect with the most negative binding energy.

Phase diagram investigation, neutron diffraction data and defect energy calculations provide evidence for the presence of vacancies at the Ce and O1 sites in $Ce_{0.975}Ti_2O_{5.95}$. The lower defect energy of the O1 site can be understood from consideration of the distorted nature of the TiO_6 octahedra (Figure 3). The O1, atoms which form the edge shared between the adjacent TiO_6 octahedra, are brought into close proximity (2.464 Å) leading to Coulombic repulsion compared to the other O sites for which the oxygen-oxygen contact distances are greater. This distortion is partially caused by the regular CeO_6 octahedra which effectively compact the O1-O1 separation. It has been demonstrated that similar compacted oxygen-oxygen separations play a key role in determining the defect properties of hexagonal $BaTiO_3$ [36] and garnets [37]. Removal of both O1 and the Ce from the same CeO_6 octahedra allows the remaining four O2 anions of the CeO_6 octahedra to relax towards their respective Ti cations reducing the distortion within each of these TiO_6 octahedra. Separately removing the Ce cation to the O1 anions as isolated defects will cause an even greater distortion in the octahedra incorporating the O1 anions as these are pulled closer towards the Ti cations. Therefore, the binding energy for the O1-Ce-O1 cluster is the most favourable. The large distortion that occurs in the TiO_6 octahedra of all synthetic brannerites is likely to generate the same characteristic defect properties. The high stress of the O1 site should facilitate the formation of M and O vacancies leading to similar non-stoichiometry in α-$ThTi_2O_6$, UTi_2O_6 and related compound.

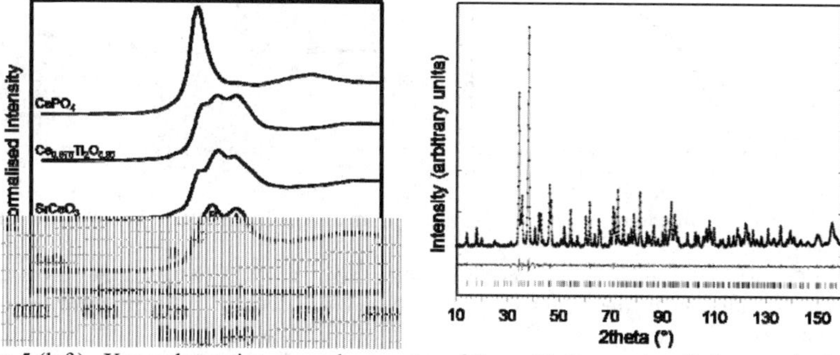

Figure 5 (left). X-ray absorption near edge spectra of $Ce_{0.975}Ti_2O_{5.95}$ and well characterised reference compounds collected at the Ce L_{III} absorption edge with $E_0 \approx 5723$ eV.
Figure 6 (right). Showing fit (solid line) to high resolution neutron diffraction data (points) for $Ce_{0.975}Ti_2O_{5.95}$ at 25 °C in space group C2/m; tick marks show allowed reflections, the difference profile (lower solid line) demonstrates an excellent fit to the data.

Rapid synthesis of $Pb_5(VO_4)_3I$ by microwave dielectric heating

^{129}I constitutes an important fraction of the fission product inventory in spent nuclear fuel. ^{129}I is a β and γ emitter and as a result of its long half life (15.7 x 10^6 y), high mobility in the environment, and bioactivity poses a potential long term dose risk [38]. Historically ^{129}I has been discharged into the marine environment but future regulatory changes will probably require it, and other radioactive halides, to be immobilised in a safe, robust wasteform. Audubert *et al.*

proposed a lead iodovanadate phase, $Pb_5(VO_4)_3I$, as a potential matrix for the immobilisation of ^{129}I [39]. $Pb_5(VO_4)_3I$ adopts the apatite structure incorporating the large iodine anions in one dimensional tunnels formed by corner sharing VO_4 polyhedra. Due to the volatility of iodine, solid state synthesis of $Pb_5(VO_4)_3I$ is only possible in a closed system e.g. by isostatically pressing a mixture of $Pb_3(VO_4)_2$ and PbI_2 in a sealed metal can. Typical reaction conditions are 500 – 700 °C and 9 – 200 MPa pressure [39-41]. In an attempt to develop a rapid and direct synthesis route for $Pb_5(VO_4)_3I$, under ambient pressure conditions, mixtures of PbO, V_2O_5 and PbI_2 were reacted by microwave dielectric heating in a modified domestic microwave oven (DMO). V_2O_5 is known to couple extremely efficiently to 2.45 GHz radiation and the rapid volumetric heating and inverse temperature profile of microwave heating were anticipated to be advantageous in reducing iodine volatilization [42-46].

The temperature profile of the reaction mixture as a function of irradiation time was initially investigated to assist optimization of the processing conditions. Following an induction period, the temperature of the sample increased as shown in Figure 7. After 180 seconds of irradiation the sample temperature had reached ~500°C.

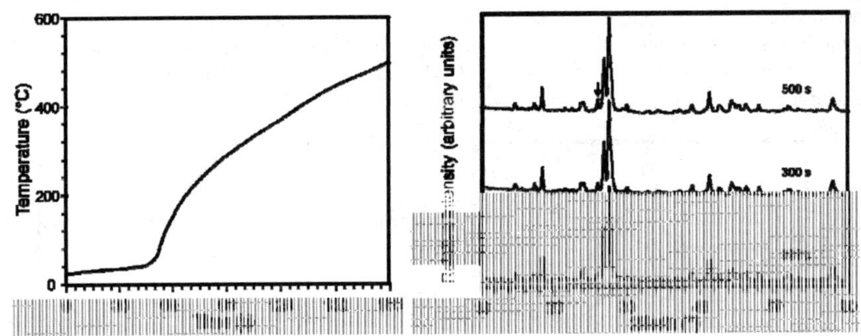

Figure 7 (left). Temperature of $Pb_5(VO_4)_3I$ reaction mixture as a function of irradiation time in modified DMO.
Figure 8 (right). Typical X-ray powder diffraction patterns of $Pb_5(VO_4)_3I$ samples prepared by microwave irradiation in the DMO for 180, 300 and 500 seconds. Arrow indicates major reflection from $Pb_3(VO_4)_2$.

Samples were prepared in duplicate and irradiated for 180, 300 and 500 seconds. Powder X-ray diffraction (XRD) was used to confirm the phase assemblage in each sample. Surprisingly, the XRD data confirmed the formation of $Pb_5(VO_4)_3I$ after only 180 seconds of irradiation at 800 W. Bluish purple vapour was observed (assumed to be volatilised iodine) when samples were heated for periods longer than 180 seconds concomitant with an increase in the relative $Pb_3(VO_4)_2$ content as observed in the XRD patterns (see Figure 8). Thermo-gravimetric analysis (TGA) confirmed that $Pb_5(VO_4)_3I$ decomposed above 500 °C which is consistent with the observations of increased $Pb_3(VO_4)_2$ content in samples heated for long time periods.

The microstructure of a typical $Pb_5(VO_4)I$ ceramic processed by microwave irradiation in the DMO at 800W for 180s is shown in Figure 9. Zone A consisted of partially sintered ceramic with large, mainly acicular, voids approximately 10 – 20 µm in diameter arising from mechanical polishing as a consequence of the friable nature of the specimen. At high magnification in backscattered electron mode, inset in Figure 9, a major phase of uniform light grey contrast was observed, with inclusions of a second phase of dark grey contrast a few microns in size. Chemical analysis of the light grey phase by quantitative Energy Dispersive X-ray Spectroscopy (EDX), indicated an average chemical composition consistent with the formation of stoichiometric $Pb_5(VO_4)_3I$. The analysed composition, based on 20 independent point analyses, was: Pb, 23.3(2) at. % (expected 23.8 at. %); V 14.0(1) % (expected 14.3 at. %), and I 4.5(1) at. % (expected: 4.8 at.%). The analysed composition was therefore found to be within three standard deviations of the ideal stoichiometric composition, with no evidence for Pb or I non-stoichiometry. The dark grey inclusions apparent in Figure 9 were shown to contain negligible amounts of I and it is considered that these inclusions are the $Pb_3(VO_4)_2$ phase, confirmed to be present by XRD. A representative high magnification back scattered electron image of Zone B, is shown as an inset in Figure 9. This zone exhibited percolating inter-granular porosity characteristic of a poorly sintered ceramic. The uniform grey contrast and elemental X-ray maps demonstrate a homogeneous chemical composition at the micron scale. Quantitative EDX analysis revealed this phase to stoichiometric within experimental precision. The microstructure of Zone C, inset in Figure 9, was found to comprise a poorly sintered assemblage of several phases with considerable inter-granular porosity. The major matrix phase was found to have a composition consistent with $Pb_5(VO_4)_3I$, based on elemental X-ray maps and EDX point analyses. Further analysis suggested the presence of: $Pb_3(VO_4)_2$ (mid-grey); V_2O_5, (dark grey); and two bright contrast phases, PbI_2 and PbO.

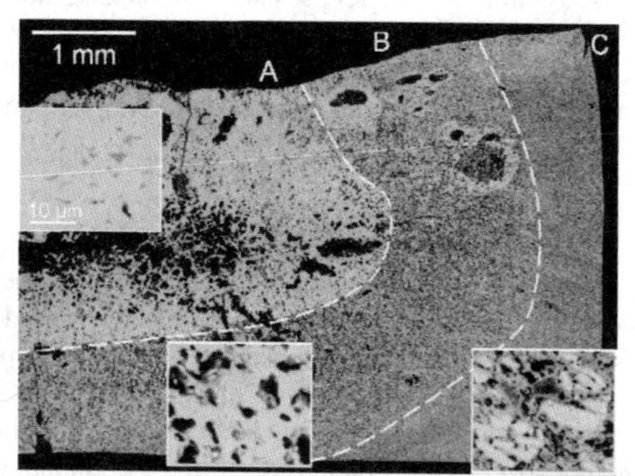

Figure 9. Backscattered electron micrograph showing microstructure of $Pb_5(VO_4)I$ ceramic formed by irradiation for 180 seconds in the DMO. Scale bars are the same for all insets.

The combination of volumetric heating and inverted temperature profile is considered critical for the synthesis of $Pb_5(VO_4)_3I$ without the use of a sealed container or over-pressure. Conventional radiant heat transfer relies on absorption of infra-red frequencies in the outer few microns of the sample, resulting in a normal temperature gradient where the surface temperature exceeds that of the interior. Consequently, ceramic $Pb_5(VO_4)_3I$ cannot be prepared by conventional solid state synthesis without the use of a sealed reaction vessel, due to the long heating times promoting iodine volatilisation. In the case of microwave irradiation, volumetric heating of the specimen occurs, in a short time scale, leading to quantitative retention of the iodine inventory. A further key point is that the microwave susceptibility of $Pb_5(VO_4)_3I$ at 2.45 GHz is rather poor, even at high temperature. This was demonstrated by failure to re-heat a specimen of $Pb_5(VO_4)_3I$ by microwave irradiation after cooling *in situ* to ~300 °C in the absence of the microwave field. Therefore, as V_2O_5 is consumed and $Pb_5(VO_4)_3I$ is formed, the effective microwave susceptibility of the phase assemblage is reduced and the sample does not reach sufficiently high temperature to induce complete decomposition of the target phase. It should be noted that, as a consequence of the inverse temperature profile, microwave synthesis of $Pb_5(VO_4)_3I$ is sensitive to sample size. In the case of large samples (> 3g), reaction temperatures commonly exceeded 650 °C and complete melting of the sample was observed. This effect could be mitigated by use of a tuneable microwave power source and temperature feedback loop in a purpose designed microwave oven. The lower inter-granular porosity observed in the interior of microwave processed $Pb_5(VO_4)_3I$ samples, Zone A in Figure 9 suggests that densification may be assisted by formation of a liquid phase. The reaction temperature of ~440 °C attained at 180s was measured by a thermocouple located near to the centre of the lower face of the sample and is hence likely to somewhat underestimate the temperature at the interior centre of the sample. From the melting points of PbI_2 (402 °C), V_2O_5 (696 °C) and PbO (886 °C) [47], we infer the presence of molten PbI_2 at the interior centre of the pellet which may assist rapid formation of $Pb_5(VO_4)_3I$ (from component oxides or intermediates) by enhanced ionic diffusion. It should also be noted that the PbI_2 – PbO phase diagram exhibits a PbI_2 rich eutectic with a melting point of 375 °C, which could also assist phase formation [48]. Since decomposition of $Pb_5(VO_4)_3I$ is rapid above 520 °C, demonstrated by TGA analysis, the formation of this phase within Zone A indicates that the maximum temperature must be below 520 °C: the presence of molten V_2O_5 and PbO is therefore considered unlikely. The presence of substantial residual inter-granular porosity in Zones B and C, and minor relics of the reagent materials observed in Zone C, suggest that if molten PbI_2 assists formation of $Pb_5(VO_4)_3I$ in these zones, then volume of liquid phase must be significantly smaller than at the interior centre of the sample, consistent with the inverse temperature gradient.

CONCLUSIONS

In this study, application of hard XAS in grazing angle mode was demonstrated to be essential for determination of cation speciation in ion beam amorphised surface layer of polycrystalline ceramics. Analysis of Ti K edge XAS data demonstrate that the TiO_6 polyhedra in crystalline $Gd_2Ti_2O_7$ are unstable with respect to ion beam amorphisation, with quantitative conversion to TiO_5 species. This conclusion is supported by quantified changes in the height and energy of the pre-edge feature which are consistent TiO_5 polyhedra being the dominant species.

Combined powder diffraction, backscattered electron microscopy, and defect energy calculations demonstrated that $CeTi_2O_6$ is in fact non-stochiometric and more properly expressed as $Ce_{1-x}Ti_2O_{6-2x}$, with $x \approx 0.025$. Rietveld analysis of high resolution neutron powder diffraction data with a model permitting cerium vacancies and oxygen vacancies at the O1 site, afforded an improved fit to the diffraction data in comparison with a model assuming the stoichiometric composition. Defect energy calculations confirmed the favourable formation of cerium and oxygen vacancies and indicated a preference for oxygen vacancies at the O1 site. The preferential formation of oxygen vacancies at the O1 site was rationalised by consideration of the short O1-O1 contact distance forming the shared edge of neighbouring TiO_6 octahedra.

Synthesis of $Pb_5(VO_4)_3I$ has been shown to be possible without the use of a sealed container, by microwave irradiation in a modified domestic microwave oven. The success of the techniques is due to the inverse temperature profile, rapid volumetric heating and poor microwave susceptibility of the product phase $Pb_5(VO_4)_3I$, thus limiting the reaction temperature attained. It has also been shown that analogous halide containing phases can be synthesised using a commercially available microwave muffle furnace. The uniform nature of the ceramic monoliths produced suggests that this may be a promising route for forming dense single phase wasteforms over relatively short timescales without loss of the volatile halide components.

ACKNOWLEDGEMENTS

The authors gratefully acknowledge part support from the Engineering and Physical Sciences Research Council under grant numbers EP/I012214/1, EP/G037140/1, EP/F055412/1, EP/I001514/1, EP/I016589/1 and EP/G005001/1. Experiments at the SINQ neutron source were supported by a beamtime allocation from the Paul Scherrer Institute. Use of the National Synchrotron Light Source, Brookhaven National Laboratory, was supported by the US Department of Energy, Office of Science, Office of Basic Energy Sciences, under Contract no. DE-AC02- 98CH10886. We are grateful to Dr Denis Sheptyakov and Dr Bruce Ravel for assistance and advice in undertaking neutron diffraction and XAS experiments, respectively. NCH is grateful to the Royal Academy of Engineering and Nuclear Decommissioning Authority for funding support.

REFERENCES

[1] J. F. Zeigler, Nuclear Instruments and Methods in Physics Research Section B: Beam Interactions with Materials and Atoms, **219-220** (2004) 1027.
[2] B. Ravel and M. Newville, Journal of Synchrotron Radiation, **12** (2005) 537.
[3] E. Lifshin and R. Gauvin, Microscopy and Microanalysis, **7** (2001) 168-177.
[4] P. Fischer, G. Frey, M. Koch, M. Könnecke, V. Pomjakushin, J. Schefer, R. Thut, N. Schlumpf, R. Bürge, U. Greuter, S. Bondt, and E. Berruyer, Physica B, **146** (2000), 276–278.
[5] A.C. Larson and R.B. von Dreele, General Structure Analysis System (GSAS), Los Alamos National Laboratory, New Mexico, (2004).
[6] J.D. Gale and A. Rohl, Molecular Simulation, **29** (2003) 291-341.
[7] N. Mott and M. Littleton, Transactions of the Faraday Society, **34** (1938) 0485-0499.
[8] M. Gilbert and J.H. Harding, Physical Chemistry Chemical Physics, **13** (2011) 13021-13025.
[9] B.G. Dick and A.W. Overhauser, Physical Review, **112** (1958) 90-103.
[10] F. Audubert, J.M. Savariault and J.L. Lacout, Acta Crystallographica, C55, (1999) 271.
[11] F.W. Clinard Jr., D.L. Rohr and R.B. Roof. J Nucl. Mater. 126, 245 (1984).
[12] W.J. Weber, L.W. Wald and Hj. Matzke. J. Nucl. Mater. 138, 196 (1986).
[13] R.C. Ewing and L.M. Wang. Nucl. Inst. Meth. Phys. Res. B65, 319 (1992).
[14] K.L. Smith, N.J. Zaluzec and G.R. Lumpkin. J. Nucl. Mater. 250, 36 (1997).
[15] S.X. Wang, G.R. Lumpkin, L.M. Wang and R.C. Ewing. Nucl. Inst. Meth. Phys. Res. B166-167, 293 (2000).
[16] F. Farges, G.E. Brown and J.J. Rehr, Phys. Rev. B, 56, 1809 (1997).
[17] G.A. Waychunas. Am. Mineral. 72, 89 (1987).
[18] L. Troger, D. Arvanitis, K. Baberschke, H. Michaelis, U. Grimm and E. Zshech. Phys. Rev. B. 46, 3283 (1992).
[19] F. Farges. Am. Mineral. 82, 44 (1997).
[20] B.B. Ebbinghaus, R.A. Van Konynenburg, F.J. Ryerson, E.R. Vance, M.W.A. Stewart, A. Jostsons, J.S. Allender, T. Rankin, and J. Congdon, “Ceramic formulation for the immobilization

of plutonium", in Proceedings of Waste Management 98, Tucson, Arizona, March 1–5, 1998. Published on CD ROM. Tucson, Arizona: Waste Management Symposia, Inc., 1998.
[21] E.R. Vance, A. Jostsons, S. Moricca, M.W.A. Stewart, R.A. Day, B. Begg, M.J. Hambley, K.P. Hart, and B.B. Ebbinghaus, Ceramic Transactions, **93** (1999) 323-329.
[22] A.E. Ringwood, S.E. Kesson, K.D. Reeve, D.M. Levins, and E.J. Ramm, "Synroc" in: W. Lutze and R.C. Ewing (Eds.), Radioactive Waste Forms for the Future, (1988).
[23] M.L. Carter, H. Li, Y. Zhang, E.R. Vance, and D.R.G. Mitchell, Journal of Nuclear Materials, **384** (2009) 322-326.
[24] D.P. Reid, N.C. Hyatt, M.C. Stennett, and E.R. Maddrell, Materials World, **19** (2011) 26-28.
[25] W.E. Lee, M.I. Ojovan, M.C. Stennett, and N.C. Hyatt, Advances in Applied Ceramics, **105** (2006) 3-12.
[26] J.T. Szymanski and J.D. Scott, Canadian Mineralogist, **20** (1982) 271-279.
[27] J.W. Anthony, R.A. Bideaux, K.W. Bladh, and M.C. Nichols, Eds., Handbook of Mineralogy, Mineralogical Society of America, Chantilly, VA 20151-1110, USA.
[28] J. Lian, L.M. Wang , G.R. Lumpkin, and R.C. Ewing , Nuclear Instruments and Methods in Physics Research B, **191** (2002) 565–57.
[29] V.F. Sears, Neutron News, **3** (1992) 29-37.
[30] K.B. Helean, A. Navrotsky, G.R. Lumpkin, M. Colella, J. Lian, R.C. Ewing, B. Ebbinghaus, and J.G. Catalano, Journal of Nuclear Materials **320** (2003) 231–244.
[31] A. Bianconi, A. Marcelli, H. Dexpert, R. Karnatak, A. Kotani, T. Jo, and J. Petiau, Physical Review B, **35** (1987) 806-812.
[32] A. Soldatov, T. Ivanchenko, S. Dellalonga, A. Kotani, Y. Iwamoto, and A. Bianconi, Physical Review B, **50** (1994) 5074-5080.
[33] M. James and J. N. Watson, Journal of Solid State Chemistry, **165** (2002) 261–265.
[34] M. James, M.L. Carter, and J.N. Watson, Journal of Solid State Chemistry **174** (2003) 329–333.
[35] R. Ruh and A.D.Wadsley, Acta Crystallographica, **21** (1966) 974-978.
[36] J.D. Dawson, C.L. Freeman, L.B. Ben, J.H. Harding, and D.C. Sinclair, Journal of Applied Physics, **109** (2011) 084102.
[37] C.L. Freeman, M.Y. Lavrentiev, N.L. Allan, J.A. Purton, and W. van Westrenen, Journal of Molecular Structure: Theochem, **727** (2005) 199-204.
[38] P.D. Wilson, The Nuclear Fuel Cycle From Ore To Waste, Oxford University Press (1996).
[39] F. Audubert, J. Carpena, J.L. Lacout and F. Tetard, Solid State Ionics, 95 (1997) 113.
[40] M. Uno, M. Shinohara, K. Kurosaki and S. Yamanaka, Journal of Nuclear Materials, 294 (2001) 119.
[41] E.R. Maddrell and P.K. Abraitis, Materials Research Society Symposium Proceedings, 807 (2004) 261.
[42] D.R. Baghurst and D.M.P. Mingos, Chemical Communications, 12 (1988) 829-830.
[43] D.E. Clark and W.H. Sutton, Annual Reviews of Materials Science, 26 (1996) 299-331.
[44] J.D. Katz, Annual Review of Materials Science, 22 (1992) 153-170
[45] Microwave Processing of Materials, National Research Council, Publication NMAB-473, National Academy Press (1994).
[46] K.J. Rao, B. Vaidhyanathan, M. Ganguli, and P. A. Ramakrishnan, Chemistry of Materials, 11 (1999) 882–895.
[47] D.R. Lide (Ed.), Handbook of Chemistry and Physics, 75th ed., CRC Press, 1994.
[48] W. Rolls, E.A. Secco, U.V. Varadaraju, Materials Science and Engineering, 65 (1984) L5-8.

MIGRATION OF IODINE SOLIDIFIED IN ETTRINGITE INTO COMPACTED BENTONITE

Kazuya Idemitsu[1], Yoshihiko Matsuki[1], Masanao Kishimoto[1], Yaohiro Inagaki[1], Tatsumi Arima[1], Yoshiko Haruguchi[2], Yu Yamashita[2], Michitaka Sasoh[2]
[1] Kyushu University, Fukuoka, Japan
[2] Toshiba, Kawasaki, Japan

ABSTRACT

Nuclear reprocessing plants produce materials containing radioactive iodine-129. Cement can be used to immobilize and solidify radioactive iodine by fixing it in the form of the iodate ion in the mineral ettringite. Because the half-life of ^{129}I is 15.7 million years, radioactive wastes that contain ^{129}I require disposal by burial deep underground. In the disposal of such wastes, compacted bentonite is used as a buffer material as well as the disposal of high-level radioactive wastes. Because there is a concern that radioactive iodine could leak from disposed wastes by diffusion through compacted bentonite over a long period, the release behavior of iodine from ettringite was examined by means of electromigration studies in compacted bentonite. Most of the calcium and iodine is retained within the structure of iodate-containing ettringite (IO_3-AFt), even if the chemical form of iodine changes from iodate to iodide ion as a result of reaction with ferrous ions. However, acid produced by precipitation of ferric hydroxide might destroy the structure of IO_3-AFt in cementitious materials and cause release of iodine. It might therefore be necessary to adopt measures to prevent intrusion of acid or ferrous ions into cementitious wastes containing radioactive iodine.

INTRODUCTION

Radioactive iodine-bearing materials, such as spent silver adsorbent, are produced in nuclear reprocessing plants. In Japan, radioactive wastes that contain a certain quantity of iodine-129 are classified as Transuranic Waste Group 1 (TRU 1) for spent silver adsorbent or as Group 3 for bitumen-solidified waste[1] and they should be disposed of by burial deep underground. Because the half-life of ^{129}I is 15.7 million years, it would be difficult to prevent release of ^{129}I from the wastes into the surrounding environment over such a prolonged time. Moreover, because iodine in its ionic forms is soluble and not readily adsorbed, its migration is not retarded significantly in engineered or natural barriers. Therefore the release of ^{129}I from nuclear wastes needs to be restricted to permit reliable safety assessment; this technique is called "controlled release."

Several techniques for immobilization of iodine have been developed for this purpose.[2-8] These techniques are classified into three types: the leaching model, the distribution equilibrium model, and the solubility-equilibrium model.[9] In the leaching model, ^{129}I is physically sealed into intergranular solids from which it is released by diffusion and from which it will also leach by surface dissolution of the solid. In the distribution equilibrium model, the release of iodine is dependent on its sorption/desorption properties of solid hydrates. In the solubility equilibrium model, iodine is fixed in an insoluble mineral as a component element.

Immobilization of ^{129}I in cementitious material is one of the techniques for controlled release, and it is categorized as a distribution equilibrium model. Iodide ions (I^-) and iodate ions (IO_3^-) are sorbed by certain hydrates present in cementitious materials, such as calcium silicate gel, tetracalcium aluminate monosulfate dodecahydrate (AFm: monosulfate, $3CaO{\bullet}Al_2O_3{\bullet}CaSO_4{\bullet}12H_2O$) and calcium aluminate trisulfate dotriacontahydrate (AFt: ettringite, $3CaO{\bullet}Al_2O_3{\bullet}3CaSO_4{\bullet}32H_2O$).[10] Distribution coefficients of iodate ion on AFm and AFt are two or three orders of magnitude larger than those of iodide ion, and AFm has slightly larger distribution coefficient for iodate ion (2.5 m^3/kg) compared with that for AFt (0.97 m^3/kg).[11] In terms of both iodine sorption and the mechanical strength of the waste, the most favorable formulation for an adsorbent would be a mixture of alumina cement and calcium sulfate dihydrate ($CaSO_4 \cdot 2H_2O$) in a SO_4/Ca mole ratio of 0.16, corresponding to a weight ratio of 100:15.5.[8] As a pretreatment for the immobilization process, iodine is separated from spent silver adsorbent in alkaline solution and is then converted into iodate ions by treatment with ozone. The iodate ions are immobilized in a cementitious material by kneading and solidification. As a result of the pretreatment in alkaline solution, transition of iodine into the gas phase is rare, and the rate of recovery of iodine is extremely high (over 99.96%). Immersion tests with the cementitious material for one year confirm that it is possible to maintain distribution coefficients of over 0.9 m^3/kg and that iodine is immobilized in AFt.[12]

When the wastes are subjected to disposal, compacted bentonite would be used as a buffer material as well as disposal of high-level radioactive wastes.[1] There is a concern, however, that ^{129}I might leak from disposed waste through the compacted bentonite during a prolonged period of time. Bentonite is a type of clay product and consists mainly of clay minerals such as montmorillonite. Clay minerals have a layered structure, and the interlayer surfaces carry negative charges. Therefore, the interlayers contain exchangeable cations, and it would be expected that migration of radioactive cations should be delayed as a result of ion exchange. However, migration of radioactive anions would not be expected to be delayed in this manner, although they would be affected by anion exclusion.[13]

Our aim in the research reported in this paper was to investigate the release behavior of iodine from ettringite under reducing conditions by using the electromigration method[14] and by diffusion experiments.

EXPERIMENTAL

Cementitious Materials

We prepared three types of iodine-bearing cementitious materials: IO_3-cement, IO_3-AFt and IO_3-AFm.

The first material, IO_3-cement, was fabricated by the same method proposed for use in a private reprocessing plant in Japan, as illustrated in Figure 1. Iodine is separated from iodine-bearing silver adsorbent in alkaline solution by treatment with Na_2S and is then converted into iodate ions by treatment with ozone. The iodate ions are immobilized through kneading and

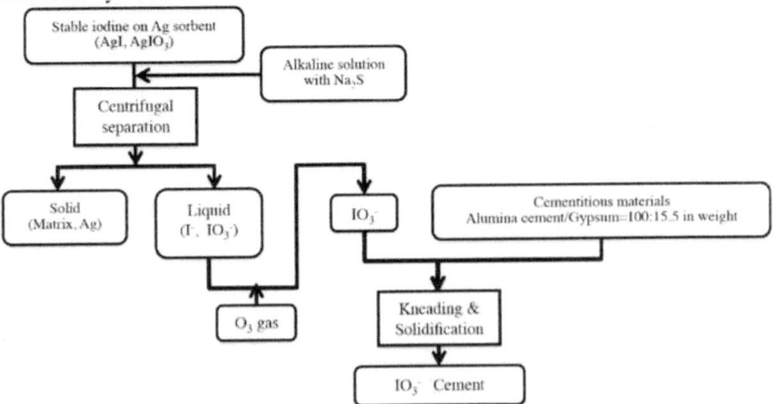

Figure 1. Process for immobilization of iodine from a silver sorbent into a cementitious material

solidification with alumina cement and gypsum in a ratio of 100:15.5 (wt/wt). The iodine content in this cement was estimated to be about 2%.

The second material, IO_3-AFt, was synthesized by the following procedure. Sodium aluminate, calcium hydroxide, and sodium iodate were dissolved in 1 dm^3 of distilled water in amounts of 3.3, 20, and 20 mmole, respectively. Addition of sucrose accelerated the formation of AFt. The slurry was cured for more than seven days under an atmosphere of argon gas then rinsed and dried. The resulting IO_3-AFt powder was analyzed by X-ray diffractometry (XRD: Rigaku; RINT2000), scanning electron microscopy–energy dispersive x-ray spectroscopy (SEM-EDS: JEOL; JSM-6360LA and SII; SPS3000). The XRD pattern (Figure 2) confirmed the presence of ettringite (AFt). Some images of IO_3-AFt by SEM are shown in Figure 3. Acicular crystals, typical of ettringite, were observed. Component analysis of the acicular crystals by EDS showed that the mole ratio of Ca, Al, and I was 3:1:1. The Ca/Al/S ratio for normal ettringite is 6:2:3, so the Ca/Al/I ratio of IO_3-AFt should be 3:1:3 if each sulfate ion is replaced by two iodate ions. The model crystal structure of ettringite {AFt: $(Ca_6[Al(OH)_6]_2(SO_4)_3{\bullet}26H_2O)$} consists of four columns of calcium and aluminum polyhedrons with sulfate tetrahedrons and rigid water molecules.[15] The structure of ettringite is shown in Figure 4. The dotted-line box corresponds to a unit cell of AFt. Because there are only four spaces and six negative charges in total for anions in a unit cell, one of the three of sulfate ions in AFt might be replaced by two iodate ions, so that the structural formula of IO_3-AFt could be $Ca_6[Al(OH)_6]_2(SO_4)_2(IO_3)_2{\bullet}26H_2O$. The Ca/Al/I ratio is consequently 3:1:1, as shown in Figure 4. The iodine content of this IO_3-AFt is estimated to be 16.8 wt%. However, because there was no sulfate in the raw material, each sulfate ion could be replaced by two hydroxy ions, giving a structural formula for IO_3-AFt of $Ca_6[Al(OH)_6]_2(OH)_4(IO_3)_2{\bullet}26H_2O$, the iodine content of which is estimated to be 18.3 wt%.

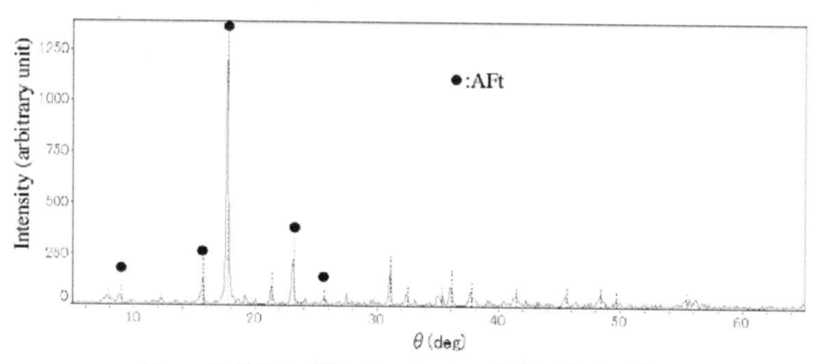

Figure 2. X-Ray diffraction pattern of IO_3-AFt powder

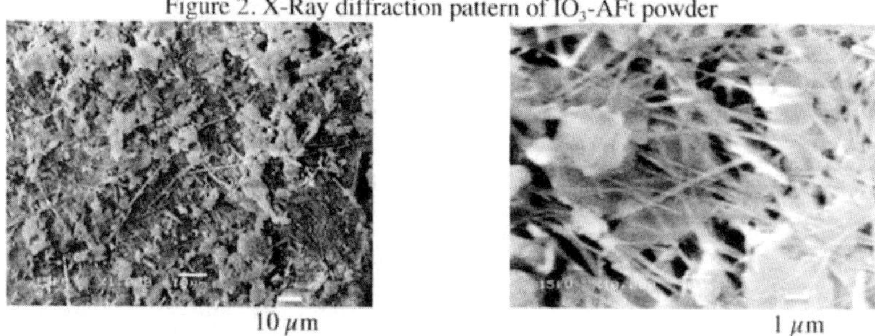

Figure 3. Scanning electron micrographs of IO_3-AFt

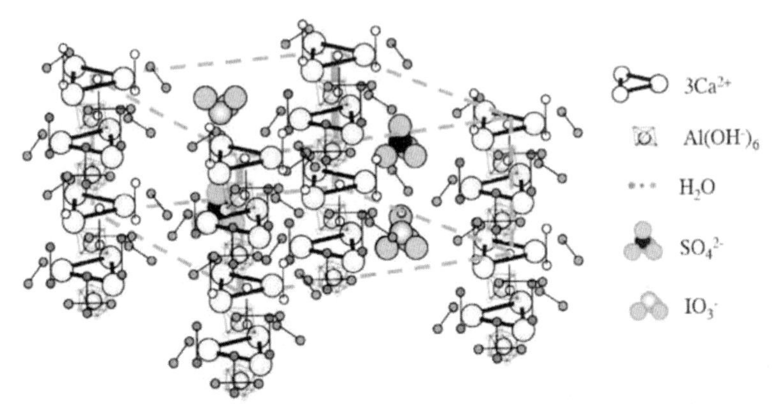

Figure 4. Proposed structure of IO_3-AFt. The dotted-line box corresponds to a unit cell of ettringite. There are four spaces and six negative charges in total for oxoanions in each unit cell. The structural formula of IO_3-AFt is $Ca_6[Al(OH)_6]_2(SO_4)_2(IO_3)_2•26H_2O$ or $Ca_6[Al(OH)_6]_2(OH)_4(IO_3)_2•26H_2O$ when each sulfate ion is replaced by two hydroxy ions.

The last material, IO_3-AFm, was synthesized as follows. Calcium iodate and tricalcium aluminate (C3A) were mixed in a 1:1 molar ratio and the mixture was ground. Water at a water-to-solids (W/S) ratio of 1 or 10 was added and the mixture was kneaded for four or ten days under argon, then dried in vacuum. The resulting IO_3-AFm powders were analyzed by XRD and SEM-EDS as described above. The XRD patterns are shown in Figure 5 for the powder with W/S = 1 after four days of kneading and for W/S = 10 after ten days of kneading. Some peaks for C3A were observed in the case of W/S = 1 after four days of kneading, but these disappeared in the case of W/S = 10 after ten days of kneading. Because no clear peak for AFm was observed in the XRD, even for W/S = 10 after ten days of kneading, AFm may be in an amorphous phase.

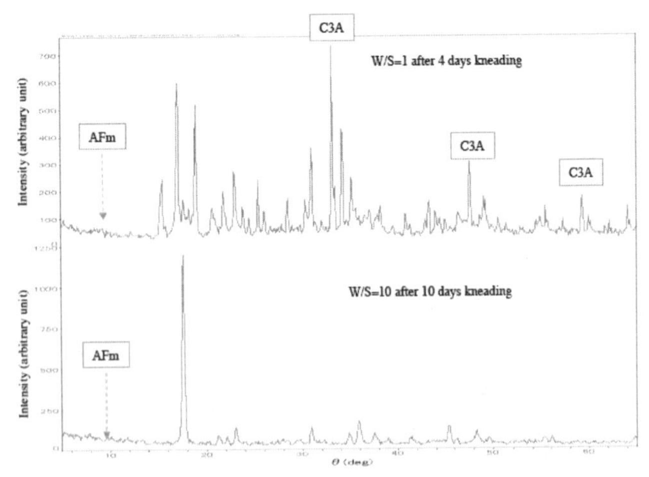

Figure 5. X-Ray diffraction patterns of IO_3-AFm powder

Figure 6 shows SEM images of IO_3-AFm. Planar crystals can be observed in both cases. Component analysis of the planar crystals by EDS showed that the mole ratio of Ca, Al, and I was 2:1:1. The elemental ratio, Ca/Al/S, of the ordinary monosulfate $\{Ca_4[Al(OH)_6]_2SO_4 \bullet 6H_2O\}$ is 4:2:1, so the Ca/Al/I ratio of IO_3-AFm should be 2:1:1 if each sulfate ion is replaced by two iodate ions. Because the estimated Ca/Al/I mole ratio agrees with the observed ratio, we propose that the structural formula of IO_3-AFm is $Ca_2Al(OH)_6IO_3 \bullet 3H_2O$; the estimated iodine content of this IO_3-AFm is 29 wt%. The model crystal structure for the monosulfate consists of sheets of calcium and aluminum polyhedrons with positive charges, interspersed by sulfate or iodate tetrahedrons and water.[16]

5 μm

(a) W/S =1 after four days of kneading.

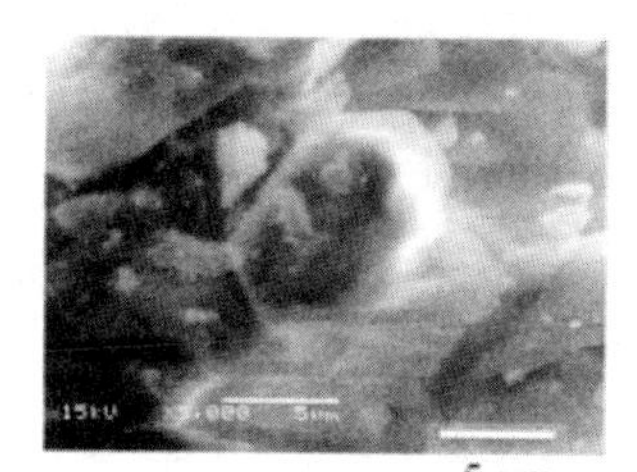

5 μm

(b) W/S =10 after ten days of kneading

Figure 6. Scanning electron micrographs of IO_3-AFm

Bentonite Material

A typical Japanese purified sodium bentonite (Kunipia-F; Kunimine Industries Co. Ltd.) was used in the experiments. Kunipia-F contains approximately 95 wt% of montmorillonite and its estimated chemical formula is $(Na_{0.3}Ca_{0.03}K_{0.004})(Al_{1.6}Mg_{0.3}Fe_{0.1})Si_4O_{10}(OH)_2$.

X-ray Absorption Near Edge Structure Measurements

Iodine L_{III}-edge X-ray absorption near edge structure (XANES) measurements were performed at beam-line BL-11 (SAGA Light Source; Tosu, Japan). A Si(111) double-crystal monochromator was used, and the beam was focused by using bent conical mirrors coated with rhodium. The width of the beam at the measurement position was 1 to 5 mm (controlled by a slit) and its height was 1 mm. The beam density was about 10^9 photons/s. The sample was sealed in vinyl plastic and examined under an ordinary atmosphere.

Spectra of XANES were measured in the fluorescence mode by using a silicon drift detector (SDD). In this mode, the electrical signal of iodine L_α X-rays from the SDD was selected by means of a single-channel analyzer at the L_{III}-edge. X-ray fluorescence spectra were also obtained by using the SDD coupled with a multichannel analyzer. Data analysis, including background subtraction, normalization, and linear combination fitting of XANES spectra, was performed with REX2000 Version 2.5 software (Rigaku).

Electromigration Experiments

The electromigration method was used to study the release of iodine from ettringite under reducing condition at the presence of ferrous ions.[14] Equal weights of IO_3-AFt and bentonite powder were mixed and compacted into thin pellets with a diameter of 10 mm and a thickness of 1 mm with a dry density of 1.4 Mg/m^3. One of these pellets was saturated with 0.01 M aqueous NaCl for more than one month [Figure 7(a)]. After saturation, a carbon-steel coupon was attached and a reference electrode of Ag/AgCl and a counter electrode of platinum foil were both inserted into the upper part of the apparatus, which was filled with 0.01 M aqueous NaCl [Figure 7(b)]. The carbon-steel coupon was connected to a potentiostat to serve as a working electrode and it was supplied with an electrical potential of +300, 0, −300, or −500 mV versus the Ag/AgCl electrode at 25 °C for 1–7 days. In all cases, the working electrode had an electrical potential that was 0.5 to 1.5 V higher than that of the counter electrode. After saturation with water and subsequently supplying +300 mV for up to seven days, the sample was optionally kept without electrical potential for further three days to permit further diffusion of iodine. The bentonite specimen was expelled from the column and sliced in steps of 0.5 to 2 mm. Half of each slice was submerged in 1.0% aqueous hexamethyltetraammonium hydroxide solution to extract iodine and the sample was then kept in an oven at 70 °C for about four hours. The liquid phase was separated by centrifugation and the supernatant was filtered with a 10,000-Da cut-off filter.[17] The supernatant was then analyzed by inductively coupled plasma-mass spectrometry (ICP-MS: Agilent; 7500C) for iodine measurement. The other half of each bentonite slice was submerged in 1 M aqueous HCl to extract cations such as sodium, calcium, and iron. Concentrations of sodium, iron, and calcium in the liquid extract were measured by means of atomic absorption spectrometry (AAS: Shimazu; AA-6300).

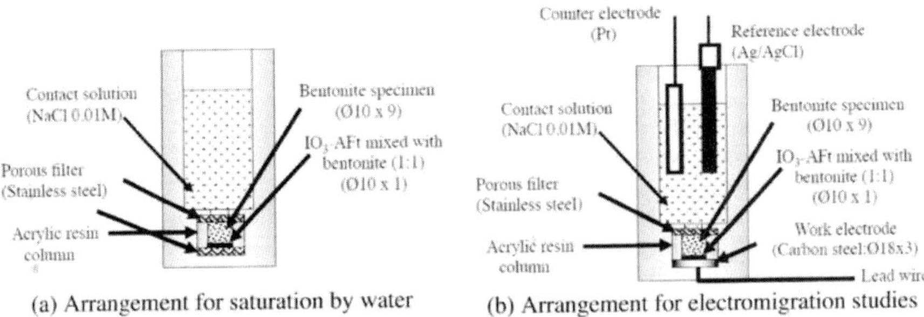

Figure 7. Schematic representations of electromigration experiments

RESULTS AND DISCUSSION

XANES Spectra

Results of iodine L_{III}-edge (4557 eV) XANES for IO_3-cement, IO_3-AFt, IO_3-AFm, and several reference materials, together with previous data from the literature,[18] are shown in Figure 8. The contribution of background was subtracted from the original spectra by extrapolation of the linear absorption or curve (as defined by the Victoreen equation) from the pre-edge region.

The characteristics of spectra reported in the literature differed among samples treated with KI, I^- solution, KIO_3, IO_3^- solution, I_2, and methyl iodide (CH_3I),[18] suggesting that the identification of iodine species in these materials is possible based on the simulation of XANES spectra [Fig. 8(a)]. The spectra of IO_3-cement, IO_3-AFt, and IO_3-AFm were slightly broader than that of $Ca(IO_3)_2$, but were quite different from that of CaI_2 [Figure 8(b)]. On the other hand, there were no clear differences among IO_3-cement, IO_3-AFt, and IO_3-AFm. The spectra of IO_3-cement, IO_3-AFt and IO_3-AFm were in perfect agreement with the spectrum of the IO_3^- solution [Figure 8(a)]. This means that iodine in these materials exists as the iodate ions with surrounding water molecules, as described in the structural formula discussed above. We wish to emphasize that the spectra were obtained under the conditions where calcium is present, despite the fact that the $L_{\alpha 1}$

and $L_{\alpha2}$ peaks for iodine (3938 and 3926 eV, respectively), which were used in our measurements, are close to the $K_{\beta1}$ peak for Ca (4013 eV). In the current system, calcium does not affect the XANES spectra of the iodine L_{III}-edge, even in the fluorescence mode using SDD.

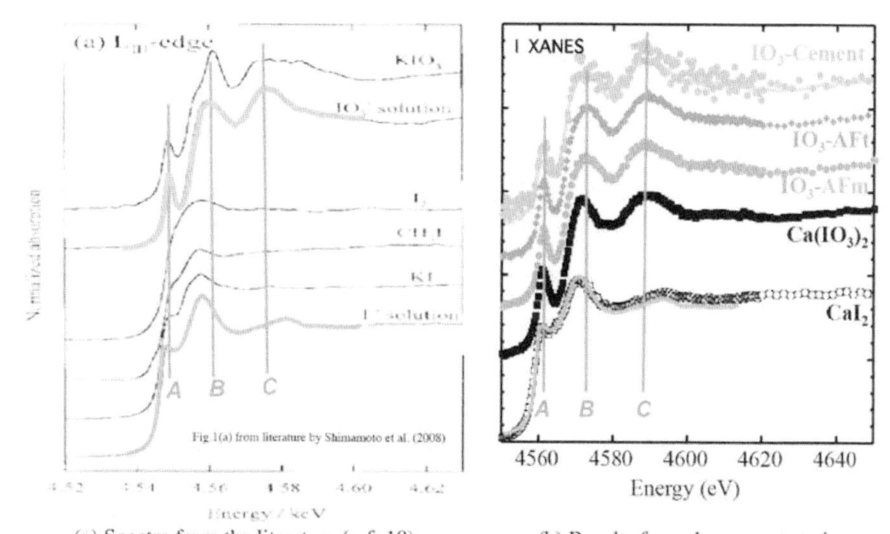

(a) Spectra from the literature (ref. 18) (b) Results from the current study

Figure 8. Iodine L_{III}-edge XANES spectra of IO_3-cement, IO_3-AFt, IO_3-AFm, and reference materials. Lines A, B and C in the figures correspond to IO_3^-.

Release of Iodine from IO_3-AFt During Saturation by Water for Up to Two Months

Figure 9 shows iodine profiles from IO_3-AFt during saturation by water. Compacted bentonite samples 10 mm thick usually became saturated with water within two weeks, so iodine released from IO_3-AFt could diffuse for two to six weeks. These two profiles show a typical diffusion patterns for the case of an instantaneous plane source according to the following equation:

$$C(x,t) = \frac{M}{\sqrt{\pi Dt}} exp\left(-\frac{x^2}{4Dt}\right), \quad (1)$$

where C is the concentration of the diffusing substance at distance x form the source; t is the diffusion period; D is the diffusion coefficient, and M is the total amount of diffusing substance in a unit cross-section.[19] The diffusion coefficients of iodine in the compacted bentonite, as estimated by using Eq. 1, are in the range 2–6 $\mu m^2/s$, but the precise diffusion period could not be measured. This diffusion coefficient is much smaller than the values given in the literature.[20] The total amounts of diffusing iodine were 1–2 μmole, as calculated by Eq. 1. Because the initial amount of iodine in the mixture of IO_3-AFt and bentonite was about 80 μmole, a few percent of the iodine present might be released instantaneously from the IO_3-AFt.

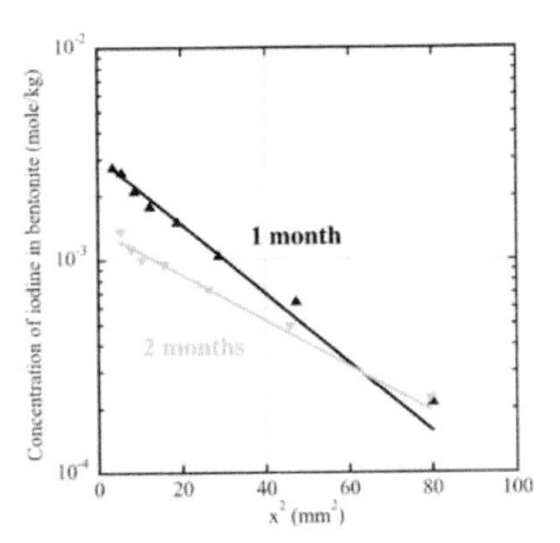

Figure 9. Iodine profile in compacted bentonite with a dry density of 1.4 Mg/m^3 during saturation by water. The x on the horizontal axis corresponds to the distance from the interface between bentonite and the mixture of IO_3-AFt and bentonite.

Electromigration Experiments

Figure 10(a) shows typical cation profiles in compacted bentonite after the electromigration experiment. Iron ions migrate as ferrous (Fe^{2+}) ions through the interlayer of montmorillonite, replacing exchangeable sodium ions in the interlayer.[14] Iodine was not observed in the bentonite region. This means that negatively charged iodine could be swept outwards toward the working electrode when an electrical potential was supplied. On the other hand, calcium levels showed a peak value [Figure 10(b)]. The lines in Figure 10(b) are model curves for electromigration. The movement of ions under the influence of an electric potential gradient (electromigration) can be described by means of the dispersion–convection equation:

$$\frac{\partial C}{\partial t} = D\frac{\partial^2 C}{\partial x^2} - V\frac{\partial C}{\partial x} \quad (2)$$

where D is the apparent dispersion coefficient, V is the apparent migration velocity (consisting mainly of electromigration of iron with negligible electroosmotic flow of water). If an instantaneous plane source at $x = 0$ and an initial distribution of calcium are assumed, the solution of Eq. 2 is given as follows:

$$C = \frac{M}{2A\sqrt{\pi D t}} exp\left(-\frac{(x-Vt)^2}{4Dt}\right) + \frac{C_0}{2} erfc\left(\frac{Vt-x}{2\sqrt{Dt}}\right), \quad (3)$$

where M is total amount of calcium in the plane source, A is the cross-sectional area, and C_0 is the initial concentration of calcium in the compacted bentonite. The first term on the right-hand side of Eq. 3 corresponds to the contribution of the plane source and the second term corresponds to the initial redistribution of calcium. The model curves agree well with the profiles shown in Figure 10(b). This means that peaks could arise from IO_3-AFt as an instantaneous plane source. The total amounts of diffusing calcium from the instantaneous plane source (M) were 30 to 60 μmole or 10 to 25% of the calcium initially present in the IO_3-AFt. Correspondingly, 90 to 75% of the calcium in IO_3-AFt was retained in the structure.

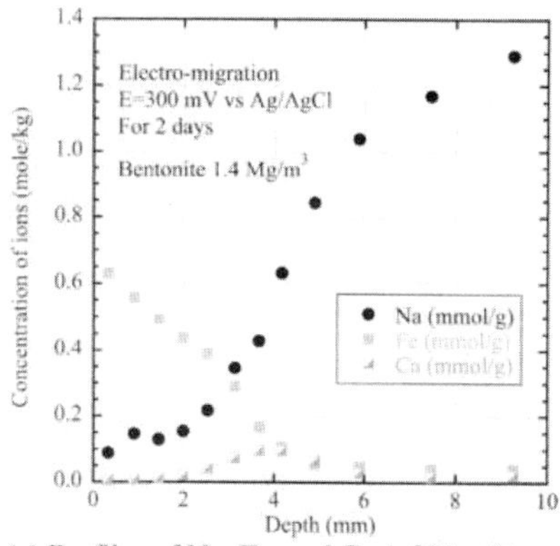

(a) Profiles of Na, Fe, and Ca (+300 mV vs. Ag/AgCl for two days)

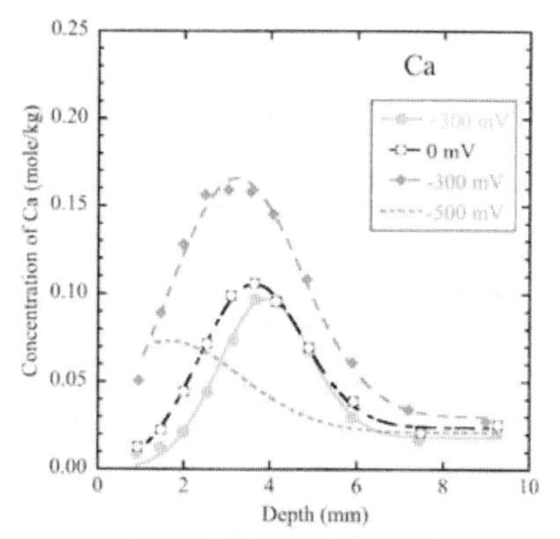

(b) Ca profiles (+300, 0, −300, or −500 mV vs. Ag/AgCl for two days)

Figure 10. Typical profiles of Na, Fe, and Ca in compacted bentonite (1.4 Mg/m^3) immediately after the electromigration experiment.

Chemical State of Iodine in IO_3-AFt After Injection of Ferrous Ions

The amount of iron penetrating into a bentonite specimen can be calculated from the corrosion current, if it is assumed that iron is present as ferrous ions.[14] In the presence of ferrous ions, iodate ions are reduced to iodide ions as follows:

$$IO_3^- + 6Fe^{2+} + 15H_2O \Rightarrow I^- + 6Fe(OH)_3 + 12H^+. \quad (4)$$

Measurements of XANES on IO_3-AFt after injection of ferrous ions were performed as described above. Figure 11 shows the iodine L_{III}-edge XANES spectra of IO_3-AFt subjected to an electric potential. The ratios of iodide ion increased with increasing duration of injection of ferrous ions (Figure 11). The ratios of iodate ions to iodide ions in the IO_3-AFt were plotted as a function of the amount of ferrous ion supplied (Fig. 12) and the results were found to be consistent with Eq. 4.

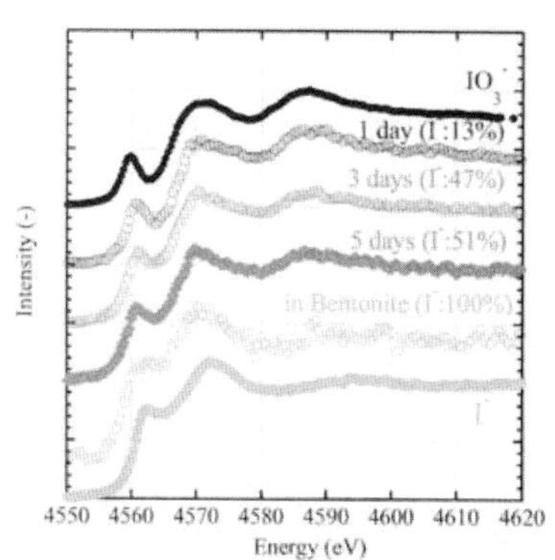

Figure 11. Iodine L_{III}-edge XANES spectra of IO_3-AFt after injection of ferrous ions

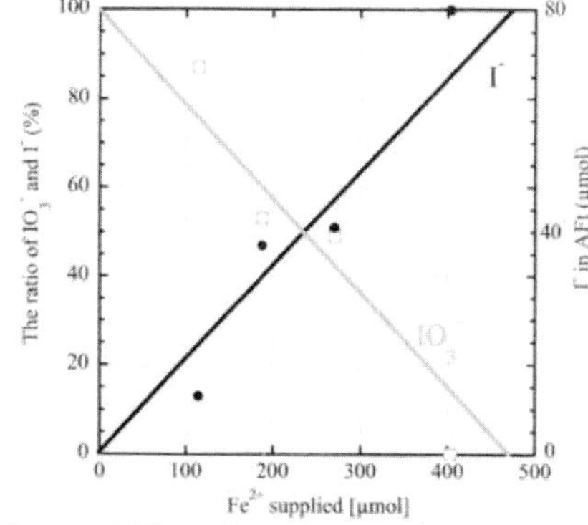

Figure 12. The ratio of IO_3^- and I^- as a function of the ferrous ion supplied

Diffusion of Iodine, Calcium, and Iron After Application of an Electric Potential

Although iodine did not migrate, its chemical state might have been changed by exposure to an electrical potential. We therefore performed additional experiments, such as diffusion experiments, after applying an electric potential for three days. Profiles of iodine, iron, and calcium recorded three days after applying an electric potential of +300 mV vs. Ag/AgCl to the working electrode for one to seven days are shown in Figure 13. Iodine collected at the working electrode during the application of the electric potential and it subsequently diffused into the compacted bentonite, as shown in Figure 13(a). The diffusion coefficients of iodine obtained form these profiles by using Eq. 1, are in the range 37–95 $\mu m^2/s$, and agree with the values given in the literature.[20] The amounts of diffusing iodide ion were 4 to 18 μmole or 5 to 23% of the initial amount of iodine present as iodate ions in IO_3-AFt. These ratios are consistent with the release ratios of calcium during application of the electric potential, as described above. This means that diffusing iodine, in the form of iodide ions, could be released from part of the IO_3-AFt, probably as a result of damage by the acid produced by ferric ion precipitation, as shown in Eq. 4.

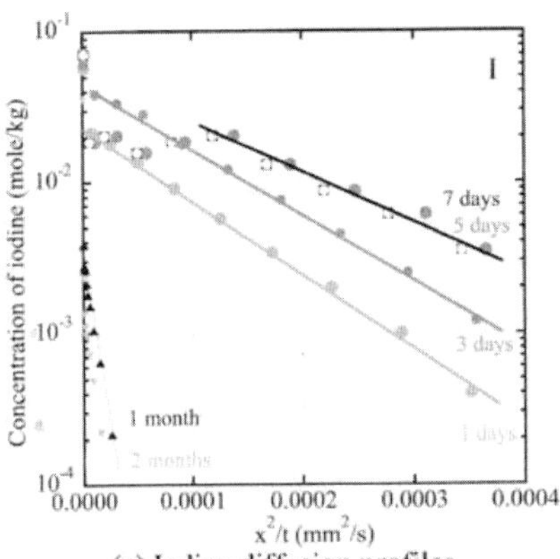

(a) Iodine diffusion profiles
Profiles recorded during water saturation are also shown in this figure (1 and 2 months).

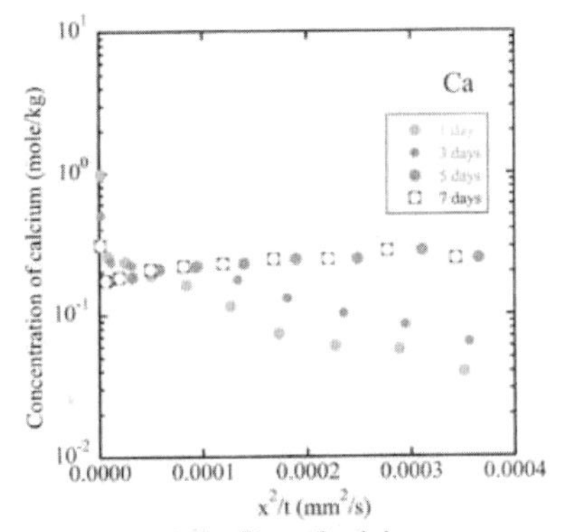

(b) Profiles of calcium
Calcium retained in each first slice decreased with increasing duration of application of electric potential.

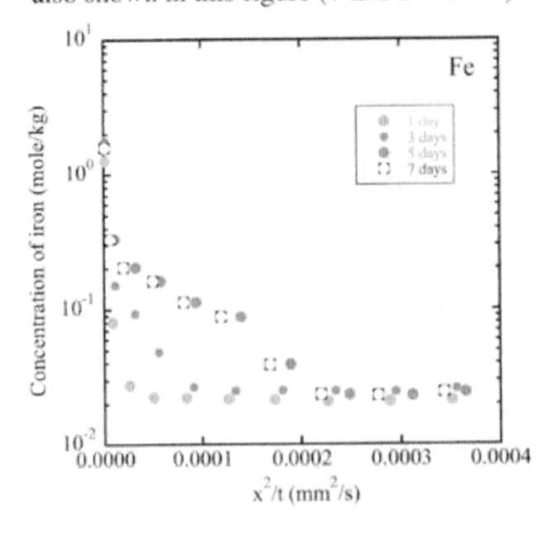

(c) Profiles of Iron
Concentrations at the first slice were much higher than those in other positions.

Figure 13. Diffusion profiles for three days of I, Fe, and Ca in compacted bentonite with a density of 1.4 Mg/m^3 after application of an electric potential for one to seven days.

The profiles for calcium are shown in Figure 13(b). Calcium retained in each first slice decreased with increasing duration of application of electric potential. Release of calcium from IO_3-AFt could be also caused by the acid produced by precipitation of ferric ion, as shown in Eq. 4. Profiles for iron are shown in Figure 13(c). The concentration of iron in each first slice was much higher than those in other positions. This indicates that the ferrous ions produced by anode corrosion were consumed in the reduction of iodate ions to iodide ions and that they then precipitated as ferric hydroxide, as shown in Eq. 4.

CONCLUSIONS

We prepared three types of iodine-bearing cementitious materials, IO_3-cement, IO_3-AFt and IO_3-AFm, and we characterized these materials by using XANES. The release behavior of iodine from ettringite was studied under reducing condition by using the electromigration method for iodide ion and iodate ion in compacted bentonite. We reached the following conclusions.

(1) The structural formula of IO_3-AFt is probably $Ca_6[Al(OH)_6]_2(SO_4)_2(IO_3)_2{\bullet}26H_2O$, so that the Ca/Al/I ratio is 3:1:1; the iodine content is then estimated to be 16.8 wt%. In the case where no sulfate is present in the raw materials, the sulfate ion might be replaced by two hydroxy ions, so that the structural formula of IO_3-AFt would then be $Ca_6[Al(OH)_6]_2(OH)_4(IO_3)_2{\bullet}26H_2O$ and the estimated iodine content in the IO_3-AFt would be 18.3 wt%.

(2) Iodine L_{III}-edge (4557 eV) XANES measurements in the fluorescence mode by using a silicon-drift detector and a multi-channel analyzer were capable of distinguishing chemical states such as iodide ion and iodate ion in cementitious materials, but were not capable of detecting any differences among IO_3-cement, IO_3-AFt, and IO_3-AFm.

(3) A few percent of the iodine fixed in IO_3-AFt can be released instantaneously from the IO_3-AFt to bentonite.

(4) On supplying ferrous ions, the iodate ions in IO_3-AFt are reduced to iodide ions as follows:

$$IO_3^- + 6Fe^{2+} + 15H_2O \Rightarrow I^- + 6Fe(OH)_3 + 12H^+.$$

(5) Most of the calcium and iodine present are retained within the structure of IO_3-AFt, even if the chemical form of iodine changes from iodate to iodide ion as a result of reaction with ferrous ions.

(6) However, acid produced by precipitation of ferric hydroxide might destroy the structure of IO_3-AFt and cause release of iodine from IO_3-AFt in cementitious materials. It is therefore necessary to take some measures to prevent intrusion of acid or ferrous ions into cementitious wastes containing radioactive iodine.

ACKNOWLEDGMENTS

Measurements of XANES were performed with the approval of SAGA Light Source (1012002AS).

REFERENCES

[1]JAEA, Second Progress Report on Research and Development for TRU Waste Disposal in Japan, *JAEA Rev.* **2007-010**, 2-43-44 (2007); available online at www.jaea.go.jp/04/be/docu/tru_eng/ tru-2e_index.htm (accessed Sept. 18th, 2012).

[2]R. Wada, T. Nishimura, Y. Kurimoto, and T. Imakita, HIP Rock Solidification Technology for Radioactive Iodine-Contaminated Waste: *Kobe Seiko Giho* **53**, 47–55 (2003); in Japanese.

[3]H. Fujihara, T. Murase, T. Nishi, K. Noshita, T. Yoshida, and M. Matsuda, Low Temperature

Vitrification of Radioiodine Using AgI–Ag_2O–P_2O_5 Glass System, *Mater. Res. Soc. Symp. Proc.*, **556**, 375–82 (1999).
[4]H. Kato, O. Kato, and H. Tanabe, Review of Immobilization Techniques of Radioactive Iodine for Geological Disposal, In: *Proceedings of the International Symposium NUCEF-2001*, 31st Oct.–2nd Nov 2001, JAERI, Tokai, Japan; 697–704 (2002).
[5]T. Advocat, C. Fillet, F. Bart *et al.*, New Conditionings for Enhanced Separation Long-lived Radionuclides, In: *Proceedings of the International Conference "Back-End of Fuel Cycle: From Research to Solution", Global'01*, Sept. 9th–13th, 2001, Paris, France.
[6]J. Izumi, I. Yanagisawa, K. Katsurai, *et al.*, Multi-layered Distributed Waste-Form of Iodine-129: Study on Iodine Fixation of Iodine Adsorbed Zeolite by Silica CVD, In: *Proceedings of the Symposium on Waste Management 2000*, Feb. 27th–March 2nd, 2000, Tucson, Arizona, USA.
[7]T. Nakazawa, H. Kato, K. Okada, S. Ueta, and M. Mihara, Iodine Immobilization by Sodalite Waste Form, *Mater. Res. Soc. Symp. Proc.*, **663**, 51–7 (2001).
[8]M. Toyohara , M. Kaneko , H. Ueda , N. Mitsutsuka, H. Fujiwara , T. Murase, and N. Saito, Iodine Sorption onto Mixed Solid Alumina Cement and Calcium Compounds, *J. Nucl. Sci. Technol. (Tokyo, Jpn.)*, **37**, 970–978 (2000).
[9]JAEA, Second Progress Report on Research and Development for TRU Waste Disposal in Japan, *JAEA Rev.* **2007-010**, 7-5-9 (2007); available online at www.jaea.go.jp/04/be/docu/tru_eng/ tru-2e_index.htm (accessed Sept. 18th, 2012).
[10]M. Atkins and F. P. Glasser, Encapsulation of Radioiodine in Cementitious Waste Forms, *Mater. Res. Soc. Symp. Proc.*, **176**, 15–22 (1989).
[11]F. Tomita, M. Wada, M. Mitsutsuka, M. Kaneko, M. Toyohara, T. Murase, and H. Fujihara, Development of Iodine Immobilization Process with Cementitious Materials, In: *Proceedings of the 7th International Conference on Radioactive Waste Management and Environmental Remediation* (1999), ASME: New York, NY (1999), CD-ROM.
[12]T. Nishimura, T. Sakuragi, Y. Nasu, H. Asano, and H. Tanabe, Development of Immobilization Techniques for Radioactive Iodine for Geological Disposal, In: *Proceedings of Workshop "Mobile Fission and Activation Products in Nuclear Waste Disposal"*, Jan. 16th–19th 2007, La Baule, France, NEA No. 6310, 221–34 (2009).
[13]J. I. Drever, *The Geochemistry of Natural Waters: Surface and Groundwater Environments* (3rd ed.), Prentice Hall: Upper Saddle River, NJ, 69–86 (1997)
[14]K. Idemitsu, S. A. Nessa, S. Yamazaki, H. Ikeuchi, Y. Inagaki, and T. Arima, Migration Behaviour of Ferrous Ion In Compacted Bentonite Under Reducing Conditions Controlled with Potentiostat, *Mater. Res. Soc. Symp. Proc.*, **1107**, 501–8 (2008).
[15]A. E. Moore and H. F. W. Taylor, Crystal Structure of Ettringite, *Acta Crystallogr., Sect. B: Struct. Crystallogr. Cryst. Chem.*, **26**, 386–93 (1970).
[16]J. P. Rapin, A. Walcarius, G. Lefevre. and M. François, A Double-Layered Hydroxide, $3CaO \cdot Al_2O_3 \cdot CaI_2 \cdot 10H_2O$, *Acta Crystallogr., Sect. C: Cryst. Struct. Commun.*, **55**, 1957–9 (1999).
[17]H. Yamada, T. Kiriyama, and K. Yonebayashi, Determination of Total Iodine in Soils by Inductively Coupled Plasma Mass Spectrometry, *Soil Sci. Plant Nutr. (Abingdon, U. K.)*, **42**, 859–66 (1996).
[18]Y. S. Shimamoto and Y. Takahashi, Superiority of K-edge XANES over L_{III}-edge XANES in the Speciation of Iodine in Natural Soils, *Anal. Sci.*, **24**, 405–9 (2008).
[19]J. Crank, *The Mathematics of Diffusion* (2nd ed.), Clarendon Press: Oxford, 11 (1975).
[20]Y. Matsuki, K. Idemitsu, D. Akiyama, Y. Inagaki, and T. Arima, In: *Proceedings of GLOBAL 2011*, Makuhari, Japan, Dec. 11th–16the, Paper No. 391339 (2011).

RADIOACTIVE DEMONSTRATIONS OF FLUIDIZED BED STEAM REFORMING (FBSR) WITH HANFORD LOW ACTIVITY WASTES

C.M. Jantzen, C.L. Crawford, P.R. Burket, C.J. Bannochie, W.G. Daniel, C.A. Nash, A.D. Cozzi and C.C. Herman
Savannah River National Laboratory
Aiken, South Carolina 29808

ABSTRACT

Several supplemental technologies for treating and immobilizing Hanford low activity waste (LAW) are being evaluated. One immobilization technology being considered is Fluidized Bed Steam Reforming (FBSR) which offers a low temperature (700-750°C) continuous method by which wastes high in organics, nitrates, sulfates/sulfides, or other aqueous components may be processed into a crystalline ceramic (mineral) waste form. The granular waste form produced by co-processing the waste with kaolin clay has been shown to be as durable as LAW glass in this study and other studies dating back to 2002. The granular waste form is composed of insoluble sodium aluminosilicate (NAS) minerals known as the feldspathoid family of minerals which includes sodalite, nosean, and nepheline. These minerals have aluminosilicate cages and rings that sequester radionuclides, halides, and Tc-99 and Cs-137. The FBSR mineral product has been produced at the industrial, engineering, pilot, and laboratory scales using simulants. Testing at SRNL with a bench-scale steam reformer (BSR) demonstrated the FBSR technology on radioactive LAW tank waste. Carefully controlled mass balance of the radioactive species demonstrated that Cs-137, Tc-99 and I-125/I-127/I-129 were all retained in the mineral waste form. Analyses demonstrated that the same mineralogy was obtained in the radioactive testing as in all previous testing on simulants. Durability testing using ASTM C1285 demonstrated the durability of the waste form in granular form. Subsequent monolithing of the granular product and testing also demonstrated comparable durability but is reported elsewhere.

INTRODUCTION

The Hanford Site in Washington State has 56 million gallons of radioactive and chemically hazardous wastes stored in 177 underground tanks.[1] The U.S. Department of Energy (DOE), Office of River Protection (ORP), through its contractors, is constructing a Waste Treatment Plant (WTP) to convert the radioactive and hazardous wastes into stable glass waste forms for disposal. Pretreated High Level Waste (HLW) will be sent to the HLW Vitrification Facility, and pretreated Low Activity Waste (LAW) will be sent to the LAW Vitrification Facility. The vitrification facilities will convert these process streams into borosilicate glass, which is poured directly into stainless steel canisters. The immobilized LAW canisters will be disposed of on the Hanford site in the Integrated Disposal Facility (IDF).

The projected throughput capacity of the WTP LAW Vitrification Facility is insufficient to complete the River Protection Program (RPP) mission in the time frame required by the Hanford Federal Facility Agreement and Consent Order, also known as the Tri-Party Agreement (TPA). Without additional LAW treatment capacity, the mission would extend an additional 40 years beyond December 31, 2047, the TPA milestone date for completing all tank waste treatment. Supplemental Treatment is, therefore, required to meet the TPA and to cost effectively complete the tank waste treatment mission. The radioactive durability testing results are documented in this study. More details about non-radioactive production, durability testing, and performance testing of FBSR products since 2002 are given elsewhere[2,3,4,5,6,7,8] and summarized in Table I.

Monolithing of the granular FBSR product is being investigated due to regulatory concerns, i.e. to prevent dispersion during transport or to meet burial/storage requirements of the IDF. Monolithing is not necessary to meet durability performance requirements because the granular mineral product is very durable: mineral waste forms degrade by the breaking of atomic bonds in the mineral structure in the same fashion that atomic bonds are broken in vitreous waste forms. Thus the long term performance of both glass and mineral waste forms are controlled by a rate drop that is affinity controlled. Considerable durability testing has already been performed by on the non-radioactive granular and monolith FBSR forms: see Table I and Reference 9.

MINERALIZATION OF LAW

Principal contaminants of concern (COC) contained in the LAW stream that are expected to impact disposal are Tc-99, I-129, U, Cr, and nitrate/nitrite.[16] During the FBSR process the nitrate and nitrites will be converted to N_2 which will exit the process as a gas. Any organics will be pyrolyzed into CO_2 and steam. The mineral waste form will sequester any halides, sulfates, sulfides, and radionuclides while an iron oxide (Fe_2O_3-FeO-Fe_3O_4) denitration catalyst used in the process will sequester the Cr as $FeCr_2O_4$ spinels. The spinel can also accommodate Ni, Pb, Mn and other transition metal species.

The Na-Al-Si (NAS) based minerals are all members of the feldspathoid family of minerals: nepheline (ideally $NaAlSiO_4$), sodalite (ideally $Na_8[AlSiO_4]_6(Cl)_2$), and nosean which has a sodalite structure (ideally $Na_8[AlSiO_4]_6SO_4$). The sodalites sequester the halides (including I-129), the sulfates/sulfides, and oxyanions such as ReO_4^-, TcO_4^- (ideally $Na_8Al_6Si_6O_{24}(ReO_4)_2$). Nepheline which has a ring structure sequesters Cs-137, K and Rb. The rhenium sodalite has been made phase pure (Table II) and the phase pure pertechnetate sodalite has been made in the Shielded Cells at the SRNL.[10]

The sodalite and nosean minerals have unique aluminosilicate cages that bond the halides and oxyanions atomically into the cage and ring structures. The sodalite minerals are classified[11] as "clathrasils" which are structures with large polyhedral cavities that the "windows" in the cavity are too small atomically to allow the encaged polyatomic ions and/or molecules pass once the structure is formed. They differ from zeolites in that the zeolites have tunnels or larger polyhedral cavities interconnected by windows large enough to allow ready diffusion of the guest species through the crystal.[11] The cavities or cages are formed of a mixture of Si and Al tetrahedra but the ratio of Si:Al can vary from the nominal 1:1. Therefore, Si deficient, Al deficient, and stoichiometric sodalites and nephelines are all treated as solid solutions with the same cavity or ring structures.[11]

The sodalite minerals are known to accommodate Be in place of Al and S_2 in the cage structure along with Fe, Mn, and Zn (Table II). These cage-structured sodalites were minor phases in HLW supercalcine waste forms (1973-1985) and were found to retain Cs, Sr, and Mo into the cage-like structure. In addition, sodalite structures are known to retain B and Ge in the cage like structures (Table II).

The mineral waste form is produced by co-processing waste with kaolin clay. The cations in the LAW waste; Na, Cs-137, Tc-99, etc, and other species such as Cl, F, I-129, and SO_4 are immediately available to react with the added clay because the clay dehydrates at the FBSR temperatures and the dehydration causes the aluminum atoms in the clay become charge imbalanced. Hence the clay becomes amorphous (loses its crystalline structure) and very reactive at the FBSR temperatures. The cations and other species in the waste react with the

Table I. Summary of All Previous Work for FBSR Granular/Monolith Product Durability Testing from 2002-Present

Pilot Scale Facility	Date	FBSR Diam.	Acidic and Basic Wastes	Granular PCT Testing	TCLP Granular Form	Granular SPFT Testing	Preliminary Performance Assessment	Product Tested	Coal	Particle Size Distrib.	Monolith & Monolith Testing)
Non-Radioactive Testing											
HRI/ TTT	December 2001	6”	LAW Env. C	Ref. 12	Ref. 12, 13	Ref. 14,15 (and PUF testing)	Ref. 16	Bed	Removed By Hand	Gaussian	No
	Ref. 13	6”	LAW Env. C	Ref.17,18,19		None	“Tie-back” Strategy	Fines	Removed by 525 °C Roasting		No
SAIC/ STAR	July 2003 Ref. 20,21	6”	SBW				None	Bed			Yes (20% LAW, 32 % SBW and 45% Startup Bed Ref 22,23
SAIC/ STAR	August 2004 Ref.24	6”	LAW (68 Tank Blend)			Ref. 19,25,26	Data from Ref. 19,25,26 “Tie-back” Strategy	Bed and Fines Separate			
SAIC/ STAR	July and Sept. 2004 Ref.27	6”	SBW			Ref. 19,25	None				
HRI/ TTT	12/06 28	15”	SBW	Ref. 29		None	None				No
HRI/ TTT	2008 Ref.30	15”	LAW (68 Tank Blend)	Ref. 31 and 32	Ref. 32	Ref. 4	“Tie-back” Strategy	Bed and Fines Together	Not removed	Bi-Modal	Yes Ref. 5,6,7, 31,32, 33
		15”	WTP-SW (recycle)			None	None				
Radioactive Testing (This manuscript and additional references that contain more detail)											
SRNL/ BSR	2010 2011	2.75”	WTP-SW (recycle)	Ref. 2,3,5,6		None	None	Bed and Fines Together	Not removed	Gaussian	Ref. 2,3,5,6
SRNL/ BSR	2010 2011	2.75”	LAW (68 Tank Blend), SX-105, AN-103	Ref. 2,5,7,8		PNNL	“Tie-back” Strategy				Ref. 2, 4,5,7,8

PCT – Product Consistency Test method (ASTM C1285-08); TCLP – Toxicity Characteristic Leaching Procedure; SPFT – Single Pass Flow-Through test (ASTM C1662); ANSI16.1/ASTM C1308/EPA 1315 – monolith emersion tests all similar with different leachate replenishment intervals; HRI/TTT – Hazen Research Inc/THOR Treatment Technologies; SAIC/STAR – Science Applications International Corporation/Science and Technology Applications Research; LAW Env. – Hanford low activity waste envelope A, B, and C; SBW – Idaho Sodium Bearing Waste; FY1 1 – Joint program between SRNL, PNNL, ORNL; N/A – not applicable

Table II. Substitution of Cations and Oxy-anions in Feldspathoid Mineral Structures

Nepheline – Kalsilite Structures[34, 41]	Sodalite Structures[35,36,37,38,39, 40,41]
$Na_xAl_ySi_zO_4$ *where x=1-1.33, y and z = 0.55-1.1*	$[Na_6Al_6Si_6O_{24}](NaX)_2$
Hosts Na, K, Cs, Rb, Ca, Sr, Ba and Y, La, Nd [41]; Iron, Ti^{3+}, Mn, Mg, Ba, Li, Rb, Sr, Zr, Ga, Cu, V, and Yb all substitute in trace amounts[42]	Hosts C., F, Cl, I, Br,Re, Mn, B, Be, Mo, SO_4, and S Higher valent anionic groups such as AsO_4^{3-} and CrO_4^{2-} form Na_2XO_4 groups in the cage structure where X= Cr, Se, W, P, V, and As

amorphous meta-kaolin to form new stable crystalline mineral structures allowing formation and structural templating at the nano-scale by the following types of reactions with NaOH or $NaNO_3/NaNO_2$ or Cs/K hydroxides, nitrates, or nitrites:

$$\underbrace{6NaOH + Na_2SO_4}_{waste} + \underbrace{3(Al_2O_3 \bullet 2SiO_2)}_{kaolin clay additive} \rightarrow \underbrace{Na_6Al_6Si_6O_{24}(Na_2SO_4)}_{Nosean\ product} + 3H_2O \uparrow \quad (1)$$

$$\underbrace{6NaOH + 2NaCl}_{waste} + \underbrace{3(Al_2O_3 \bullet 2SiO_2)}_{kaolin clay additive} \rightarrow \underbrace{Na_6Al_6Si_6O_{24}(2NaCl)}_{Sodalite\ product} + 3H_2O \uparrow \quad (2)$$

$$\underbrace{6NaOH + 2Na(Re,Tc)O_4^-}_{waste} + \underbrace{3(Al_2O_3 \bullet 2SiO_2)}_{kaolin clay additive} \rightarrow \underbrace{Na_6Al_6Si_6O_{24}(2Na(Re,Tc)O_4)}_{Sodalite\ product} + 3H_2O \uparrow \quad (3)$$

$$\underbrace{6NaOH}_{waste} + \underbrace{3(Al_2O_3 \bullet 2SiO_2)}_{kaolin clay additive} \rightarrow \underbrace{6NaAlSiO_4}_{Nepheline\ product} + 3H_2O \uparrow \quad (4)$$

RADIOACTIVE TESTING AND "TIE-BACK" STRATEGY TO NON-RADIOACTIVE TESTING

Bench-scale, pilot-scale, and engineering-scale tests using kaolin clay have all formed the mineral assemblages discussed above with a variety of legacy US DOE waste simulants. A summary of this testing is given in Table I along with a synopsis of the types of durability tests performed on each product and whether or not monolithic waste forms were fabricated and also tested.

The BSR was available at SRNL to treat actual radioactive wastes to confirm the findings of the non-radioactive FBSR pilot-scale tests performed in 2001, 2004, and the engineering-scale tests performed in 2008 (references given in Table I). Using this "tie-back" strategy, i.e. demonstrating the similarity of the radioactive mineral products and their durability to the non-radioactive tests, allows one to determine the suitability of the waste form for disposal at Hanford based on a 2003 Risk Assessment (RA) performed on non-radioactive FBSR products. Detailed discussions of the preliminary RA results are included in Mann et.al.[16]

Building correlations between work with radioactive samples and simulants is critical to being able to conduct future relevant simulant tests, which are more cost effective and environmentally sensitive than tests with radioactive wastes. Hanford's blended 68 tank average LAW known as the Rassat simulant[43] was used in the 2008 engineering-scale FBSR tests [30] and the 2004 pilot-scale FBSR tests.[24] This same recipe was used in the radioactive testing at SRNL in the BSR described in the next section. The Rassat simulant represents about 85% of the LAW chemistry in the single shell tanks.

RADIOACTIVE PRODUCTION OF FBSR PRODUCT

Radioactive testing at SRNL included four radioactive demonstrations which were designated as Modules A through D. For all radioactive tests a simulant was prepared and initial testing was performed on the simulants to determine the operational parameters for the radioactive BSR in the shielded cells.

Module A was a demonstration to convert Hanford's Waste Treatment Plant-Secondary Waste (WTP-SW which represented the WTP vitrification facility melter off-gas) into an FBSR product. Melter off-gas from the SRS HLW Defense Waste Processing Facility (DWPF) was shimmed with a mixture of I-125, I-129, and Tc-99 to chemically resemble the anticipated WTP-SW. Test results for this radioactive demonstration are given elsewhere.[3,6]

Module B testing used SRS LAW from Tank 50 chemically trimmed to resemble the Rassat formulation. The BSR Module B testing was shimmed with excess Resource Conservation and Recovery Act (RCRA) elements as was the 2008 engineering tests for comparison. The BSR was additionally shimmed with Tc-99, Re, I-125, and I-129. The Tank 50 waste had enough Cs that an additional shim was not necessary. In addition, 300 mg Tc-99 per kg of product was shimmed into the last 100 mL of feed processed in the BSR to facilitate the X-ray Absorption Spectroscopy (XAS) studies being performed at the National Synchrotron Light Source (NSLS) located at Brookhaven National Laboratory. The samples were simultaneously shimmed with Re to determine how good a surrogate Re is for Tc in the sodalite mineral structures. Discussion of the XAS and TCLP studies is documented elsewhere.[5] Approximately six hundred forty (640) grams of radioactive product were made for this extensive testing and comparative "tie-backs" to the data collected from non-radioactive pilot-and engineering-scale tests performed in 2004 and 2008.

Module C testing was performed on actual waste from Hanford Tank SX-105 which contained moderate concentrations of anions such as Cl and SO_4. No shims of excess RCRA components or radionuclides were added. Three hundred seventeen (317) grams of radioactive product were made for testing. In addition, 200 mg Tc-99 per kg of product was shimmed into the last 100 mL of feed processed in the BSR for XAS studies. The samples were simultaneously shimmed with Re to determine the effectiveness of Re as a surrogate for Tc in the sodalite mineral structures. Discussion of the XAS studies is documented elsewhere.[5]

Module D testing was performed on actual waste from Hanford Tank AN-103 which is a low anion, high sodium tank waste. Two hundred twenty four (224) grams of radioactive product were made for subsequent testing.

TESTING RESULTS AND DISCUSSION

Mass Balance

Determining the disposition of key contaminants within a treatment process is a critical consideration for any technology selection process. Previous FBSR engineering-scale tests with LAW simulants indicated that >99.99% of the nonradioactive surrogates for Tc-99 and Cs-137 and >94% of the I-129 surrogate were captured in the mineral product. For the radioactive BSR tests, mass balance data have been obtained for Tc-99, I-129, I-125, Cs-137 and rhenium. This includes analyzing the granular product, liquid condensate, off-gas filters, and rinse solutions from the post-test cleanout of the BSR apparatus.

Although mass balance does not relate directly to waste form performance, confirming the fate of Tc, Re and I from the actual waste tests is important to confirm prior data from tests with simulants. Reproducible mass balance results add confidence that the key contaminants of concern can be accurately accounted for within the limits of measurement accuracy and detection limits. Mass balance targets for previous non-radioactive demonstrations were to close within +/-10% for major constituents and +/- 30% for minor constituents.[30] In the BSR testing, Tc, Re, I were all present at levels considered minor constituents. The mass balances for Modules A, B, C, and D consisted of identifying key input and output streams and then analyzing these streams for key species.

Mass balance results from Modules A, B, C and D are given in Table III. Module A findings for off-gas are considered relevant for comparison and are also summarized in Table III and more details are given in reference 6. Good mass balance closure on Tc-99, Re, Cs-137, and I-129,-125,-127 was achieved in all BSR radioactive testing and this agreed with all previous non-radioactive testing.

Mineralogy

The mineralogy observed for the BSR non-radioactive and radioactive samples for Module B (Rassat simulant) are about the same as those of the 2008 engineering-scale bed products made with the Rassat simulant. The phases were primarily, nepheline (74 wt%), sodalite (7 wt%), and nosean (12 wt%) as calculated from the process control model (MINCALC™) used to target the engineering-scale and BSR campaigns.

For Module C, the mineralogy of the non-radioactive product from the BSR matched the mineralogy of the radioactive product from the BSR. The phases observed agree with the predicted mineralogy from MINCALC™ of ~ 80 wt% nepheline with ~5 wt% sodalite and 7 wt% nosean.

For Module D, the mineralogy of the non-radioactive product from the BSR matched the mineralogy of the radioactive product from the BSR and also agreed with MINCALC™. This low anion feed produced the highest nepheline (~93 wt%), sodalite (~5 wt%), and nosean (~1 wt%).

Waste Form Durability (Product Consistency Test; ASTM C1285)

In mineral waste forms, as in glass, the molecular structure controls dissolution (contaminant release) by establishing the distribution of ion exchange sites, hydrolysis sites, and the access of water to those sites.[44] During durability testing of the mineral albite and an albite composition glass, the author states, "the same mechanisms are operating with both glasses and minerals but at different rates."[45] Thus the long term performance of both glass and mineral waste forms are controlled by a rate drop that is affinity controlled.

Short term PCT tests were performed by SRNL and PNNL to compare the relative stability of the LAW BSR products (radioactive and non-radioactive) to the durability of the 2001 and 2004 LAW pilot scale tests and the 2008 LAW engineering scale tests on simulants.

Long term PCT tests were also performed (e.g. 1, 3, 6, and 12 month) at SRNL to confirm that the performance of the mineral waste form is affinity controlled like vitreous waste forms This data is not included in this study as not all of the analyses have been completed.

The short-term PCT data generated by SRNL and PNNL is in agreement with the data generated by SRNL in 2001 on AN-107 FBSR product, the 2004 pilot-scale facility FBSR products with the Rassat simulant, and the 2008 engineering-scale FBSR products made with the Rassat simulant.

The durability correlations shown in Figure 1 were generated with the 7 available PCT responses from the 2001 and 2004 testing of both the FBSR bed and the fines products. The results from the 2008 engineering-scale studies are overlain on the durability plots for comparison (see LAW samples P1B Product Receipt, PR, and High Temperature Filter, HTF, fines) which appear as "x" marks on the graphs in Figure 1. The 2008 engineering-scale studies for the WTP-SW are overlain (PR and HTF) as open diamonds in Figure 1. The BSR data for

Table III. Mass Balance Closure for Radioactive Testing of Modules A, B, C and D

Method	Specie	RAD A (DWPF Melter Recycle WTP Formulation)		RAD B (Tank 50 Rassat Formulation)		RAD C (SX-105)		Rad D (AN-103)	
		Total Recovery (%)	% in Solids[a]	Total Recovery (%)	% in Solids[b]	Total Recovery (%)	% in Solids[c]	Total Recovery (%)	% in Solids[d]
Radio-metric	Cs-137	94	99.32	124	99.0	Indeterminate		Indeterminate	
	I-125	93	98.23	84	95.12	Not Shimmed		Not Shimmed	
	I-129	98	98.04	69-87	94.50	74.6-88.7	98.33-98.59	100.26	99.58
	Tc-99	109	100	87	87.90	80.24	99.74	86.15	99.70
ICP-MS	Tc-99	Not Measured		Below Detection		82.51	99.70	82.85	100
	Re	102	99.49	98	97.90	70.73	99.53	87.69	99.59
	I-127	151	94.0	94	94.94	Not Shimmed		Not Shimmed	
ICP-AES	Al	100	99.94	110	100	105.35	99.99	98.35	100
	Cl	129	100	83	94.10	77.73	98.62	Indeterminate	
	Cr	181	99.94	120	99.90	107.75	100	Indeterminate	
	Na	151	99.72	104	99.50	103.82	99.95	101.70	99.97
	Si	110	99.91	110	100	108.52	99.98	105.00	99.98
IC	SO_4	Indeterminate		113	95.80	100.33	99.02	Indeterminate	

a solids include bed and fines; fines in condensate and crossbar ranged from zero to 0.04% of the solids
b. solids include bed and fines; fines in condensate and crossbar ranged from zero to 0.5% of the solids
c solids include bed and fines; fines in condensate were zero as a quartz wool plug had been added; crossbar solids ranged from 0- 2.79% except for I-129 which was 12.2% and Cl which was 4.98%
d solids include bed and fines; fines in condensate were zero as a quartz wool plug had been added; crossbar solids ranged from 0- 2.83% except for I-129 which was 30.54%

non-radioactive and radioactive Modules B and C are overlain with "doughnut" shaped circles in Figure 1.

For all the short term durability test data, the pH increases (becomes more caustic) as the surface area of the material is decreased (see Figure 1a). For glass waste forms, pH usually increases with increasing surface area. This is indicative that a buffering mechanism may be occurring. Based on the trend of alkali (Na) release, which is co-linear with Al release (Figure 1b), it was hypothesized that this was an aluminosilicate buffering mechanism as known to occur in nature during mineral weathering.[18,19]

All the remaining cations appear to be released as a function of the solution pH (Figure 1c, d and e) and this includes Si, S, and Re. It should be noted that all the releases in Figure 1 are below 2 g/m^2 which is the Hanford limit for release from LAW glass.

The Re release plot for the BSR products (radioactive and non-radioactive Modules B and C), appear in Figure 1e. It is noteworthy that the Re release from the Module B non-radioacive and radioactive PCT for Module B are the same. These Re concentrations as measured by SRNL are biased low compared to the Re release measured by PNNL for non-radioactive Module B PCT's. However, the non-radioactive Module B Re release measured by PNNL, tracks with the radioactive Tc-99 measured by SRNL. Likewise, for Module C, the SRNL analyses for Re in the non-radioactive and radioactive campaigns track each other and track the Tc-99 measured by SRNL. This demonstrates that Re is a good surrogate for Tc-99 during leaching experimentation for this type of waste form and that the current radioactive and non-radioactive BSR campaign products match the historic and engineering scale data. Thus the "tie-back" strategy is proven.

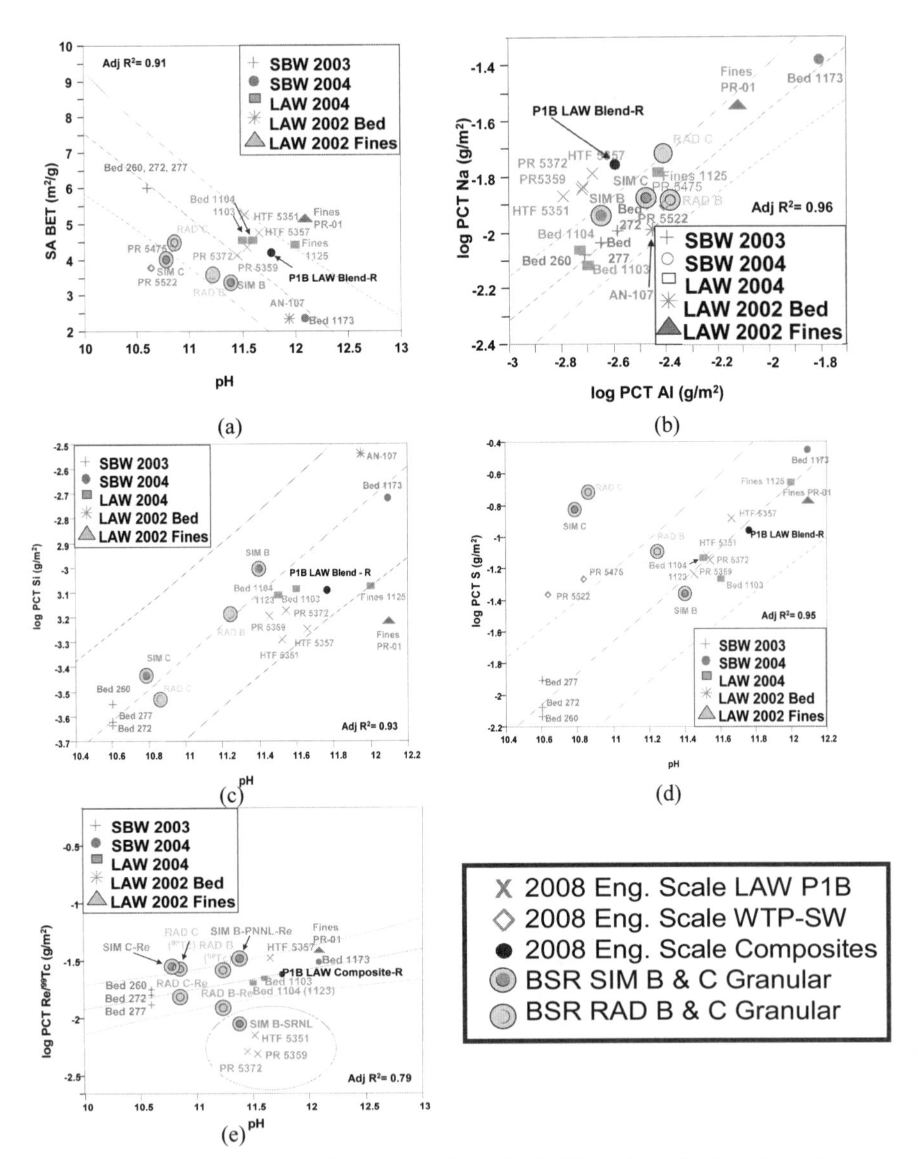

Figure 1. Comparison of the PCT response from the BSR products (radioactive and non-radioactive) to previous pilot and engineering-scale products tested.

CONCLUSIONS

The FBSR process appears to be a good technology for Hanford Supplementary Treatment. The mass balance data indicates that Tc-99, Re, Cs, and I (all isotopes) report to the mineral product and not to the off-gas. The Tc-99 and Re show similar behavior in partitioning between product and off-gas so for mass balance Re is an acceptable simulant for Tc-99.

The mineralogy testing indicates that the phases observed agree with the predicted mineralogy from MINCALC™ of ~ 90 wt% nepheline with ~10 wt% sodalite and nosean. The amount of sodalite is limited by the amount of halides in the waste and the amount of nosean is limited by the amount of sulfate/sulfide in the waste. Since the waste is predominately alkali (sodium and/or potassium) the predominant mineral is nepheline. All are feldspathoid minerals with cages and rings to sequester the COC. The mineralogy of all the radioactive campaigns and simulant products from the BSR and all the pilot and engineering scale tests with Hanford simulants and waste all produce the same mineral assemblages.

The conclusions from the short-term durability testing using ASTM C1285 are as follows: (1) short term ASTM C1285 testing is below 2 g/m^2 for the constituents of concern (COC), (2) Use of BET surface area to account for the surface roughness of the mineral granules demonstrates that the FBSR product is 2 orders of magnitude lower than the 2 g/m^2 benchmark durability for LAW glass, (3) Use of the geometric surface area, which ignores the surface roughness of the mineral granules compared to glass, gives an equivalent leach rate to vitreous waste forms, (4) Re is a good surrogate for Tc-99 during leaching experimentation for this type of waste form, (5) Durability of the radioactive and simulant BSR campaign products match the historic and engineering scale data.

ACKNOWLEDGEMENTS

The work performed at SRNL was supported under Contract No. DE-AC09-08SR22470 with the U.S. DOE. The Module B through D campaigns were funded by EM-31 through Washington River Protection Solutions (WRPS). The Module A campaigns were funded by a DOE Advanced Remediation Technologies (ART) Phase 2 Project through THOR Treatment Technologies (TTT). Work was performed in collaboration with PNNL, ORNL, and WRPS. SRNL thanks our colleagues, C.F. Brown, N. P. Qafoku, M.M. Valenta, and G.A. Gill, at PNNL for sharing their short term PCT data before publication.

REFERENCES

1 P.J. Certa, M.N. Wells, "River Protection Project System Plan," ORP-11242, Rev. 5, DOE Office of River Protection, Richland, WA (November 2010).

2 C.M. Jantzen, C.L. Crawford, P.R. Burket, W.G. Daniel, A.D. Cozzi, C.J. Bannochie, Radioactive Demonstrations of Fluidized Bed Steam Reforming (FBSR) as a Supplementary Treatment for Hanford's Low Activity Waste (LAW) and Secondary Wastes (SW)," WM11, Paper #11593 (2011).

3 B. Evans, A. Olson, J. B. Mason, K. Ryan, C.M. Jantzen, C.L. Crawford, "Radioactive Bench Scale Reformer Demonstration of a Monolithic Steam Reformed Mineralized Waste Form for Hanford Waste Treatment Plant Secondary Waste," WM12 Paper #12306 (February 2012).

4 J.J. Neeway, N.P. Qafoku, B.D. Williams, M.M. Valenta, E.A. Cordova, S.C. Strandquist, D.C. Dage, C.F. Brown, "Single Pass Flow-Through (SPFT) Test Results of Fluidized Bed Steam Reforming (FBSR) Waste Forms Used for LAW Immobilization," WM12 Paper #12252 (February 2012).

5 C.M. Jantzen, E.M. Pierce, C.J. Bannochie, P.R. Burket, A.D. Cozzi, C.L. Crawford, W.E. Daniel, C.C. Herman, D.H. Miller, D.M. Missimer, C.A. Nash, M.F. Williams, C.F. Brown, N. P. Qafoku, M.M. Valenta, G.A. Gill, D.J. Swanberg, R.A. Robbins, L.E. Thompson, "Fluidized Bed Steam

Reforming Waste Form Performance Testing to Support Hanford Supplemental Low Activity Waste Immobilization Technology Selection," SRNL-STI-2011-00387 (in revision).
6 C.L. Crawford, P.R. Burket, A.D. Cozzi, W.E. Daniel, C.M. Jantzen, and D.M. Missimer, "Radioactive Demonstration of Mineralized Waste Forms Made from Hanford Waste Treatment Plant Secondary Waste (WTP-SW) by Fluidized Bed Steam Reformation (FBSR)," SRNL-STI-2011-00331 (2011).
7 C.M. Jantzen, C.J. Bannochie, P.R. Burket, A.D. Cozzi, C.L. Crawford, W.E. Daniel, D.M. Missimer, C.A. Nash, "Radioactive Demonstration of Mineralized Waste Forms Made from SRS Low Activity Waste by Fluidized Bed Steam Reformation (FBSR)," SRNL-STI-2011-00383 (2012).
8 C.M. Jantzen, C.J. Bannochie, P.R. Burket, A.D. Cozzi, C.L. Crawford, W.E. Daniel, D.M. Missimer, C.A. Nash, "Radioactive Demonstration of Mineralized Waste Forms Made from Hanford Low Activity Waste by Fluidized Bed Steam Reformation (FBSR),"SRNL-STI-2011-00384 (2012).
9 C.M. Jantzen, "Mineralization of Radioactive Wastes by Fluidized Bed Steam Reforming: Comparisons to Vitreous Waste Forms and Pertinent Durability Testing," WSRC-STI-2008-00268 (2008).
10 D.M. Missimer and R.L. Rutherford, "Preparation and Initial Characterization of Fluidized Bed Steam Reforming Pure-Phase Standards," SRNL-STI-2013-00111 (2013).
11 F. Liebau, "Zeolites and Clathrasils–Two Distinct Classes of Framework Silicates,"Zeolites, 3[7], 191-92 (1983).
12 C.M. Jantzen, "Characterization and Performance of Fluidized Bed Steam Reforming (FBSR) Product as a Final Waste Form," Ceramic Transactions, Vol. 155, 319-29, J. D. Vienna et al. (Eds) (2004).
13 C.M. Jantzen, "Engineering Study of the Hanford Low Activity Waste (LAW) Steam Reforming Process," Savannah River Technology Center, Aiken, SC, WSRC-TR-2002-00317, (2002).
14 B.P. McGrail, H.T. Schaef, P.F. Martin, D.H. Bacon, E.A. Rodriquez, D.E. McReady, A.N. Primak, R.D. Orr, "Initial Evaluation of Steam-Reformed Low Activity Waste for Direct Land Disposal," Pacific Northwest National Laboratory, Hanford, WA, U.S. DOE Report PNWD-3288 (2003).
15 B.P. McGrail, E.M. Pierce, H.T. Schaef, E.A. Rodriques, J.L, Steele, A.T. Owen, D.M. Wellman, "Laboratory Testing of Bulk Vitrified and Steam-Reformed Low-Activity Forms to Support a Preliminary Assessment for an Integrated Disposal Facility," PNNL, Hanford, WA, PNNL-14414(2003).
16 F. M. Mann, R.J. Puigh, R. Khaleel, S. Finfrock, B.P. McGrail, D.H. Bacon, R.J. Serne, "Risk Assessment Supporting the Decision on the Initial Selection of Supplemental ILAW Technologies," Pacific Northwest National Laboratory, Hanford, WA, RPP-17675 (2003).
17 J.M. Pareizs, C.M. Jantzen, T.H. Lorier, "Durability Testing of Fluidized Bed Steam Reformer (FBSR) Waste Forms for High Sodium Wastes at Hanford and Idaho," SRNL, WSRC-TR-2005-00102, (2005).
18 C.M. Jantzen, J.M. Pareizs, T.H. Lorier, J.C. Marra, "Durability Testing of Fluidized Bed Steam Reforming (FBSR) Products," Ceramic Trans., V. 176, p.121-137, C. C. Herman et.al. (Eds) (2006).
19 C.M. Jantzen, T.H. Lorier, J.C. Marra, J.M. Pareizs, "Durability Testing of Fluidized Bed Steam Reforming (FBSR) Waste Forms," WM'06, Tucson, AZ, (2006).
20 D.W. Marshall, N.R. Soelberg, K.M. Shaber, "THOR® Bench-Scale Steam Reforming Demonstration," Idaho National Laboratory, Idaho Falls, ID, INEEL/EXT.03-00437, (2003).
21 N. R. Soelberg, D. W. Marshall, S. O. Bates, D.D. Taylor, "Phase 2 THOR Steam Reforming Tests for Sodium-Bearing Waste Treatment," INL, INEEL/EXT-04-01493 (January 30, 2004).
22 C.M. Jantzen, "Fluidized Bed Steam Reformer (FBSR) Product: Monolith Formation and Characterization," Savannah River National Laboratory, Aiken, SC, WSRC-STI-2006-00033, (2006).
23 C.M. Jantzen, "Fluidized Bed Steam Reformer (FBSR) Monolith Formation," WM'07 (2007).
24 A.L. Olson, N.R.Soelberg, D.W. Marshall, G.L. Anderson, "Fluidized Bed Steam Reforming of Hanford LAW Using THORsm Mineralizing Technology," INL, INEEL/EXT-04-02492, (2004).

25 T.H. Lorier, J.M. Pareizs, C.M. Jantzen, "Single Pass Flow through (SPFT) Testing of Fluidized Bed Steam Reforming (FBSR) Waste Forms," SRNL, Aiken, SC, WSRC-TR-2005-00124, (2005).
26 C.M. Jantzen, T.H. Lorier, J.M. Pareizs, J.C. Marra, J.C., "Fluidized Bed Steam Reformed (FBSR) Mineral Waste Forms: Characterization and Durability Testing," pp. 379-86 in *Scientific Basis for Nuclear Waste Management XXX.* Edited by D. S. Dunn, C. Poinssot, B. Begg. (2007).
27 A.L. Olson, N.R.Soelberg, D.W. Marshall, G.L. Anderson, "Fluidized Bed Steam Reforming of INEEL SBW Using THORsm Mineralizing Technology," INL, INEEL/EXT-04-02564, (2004).
28 THOR® Treatment Technologies, LLC, "Pilot Plant Report for Treating SBW Simulants: Mineralizing Flowsheet," Document Number 28266-RT-002 (July 2007).
29 C.L. Crawford, C.M. Jantzen, "Durability Testing of Fluidized Bed Steam Reformer (FBSR) Waste Forms for Sodium Bearing Waste (SBW) at INL," SRNL, WSRC-STI-2007-00319, (2007).
30 THOR® Treatment Technologies, LLC "Report for Treating Hanford LAW and WTP SW Simulants: Pilot Plant Mineralizing Flowsheet," Document Number RT-21-002, Rev. 1, (April 2009).
31 C.M. Jantzen and C.L. Crawford, "Mineralization of Radioactive Wastes by FBSR: Radionuclide Incorporation, Monolith Formation, and Durability Testing," WM'10, Phoenix, AZ, (2010)
32 C.L. Crawford, and C.M. Jantzen, "Evaluation of THORTM Mineralized Waste Forms (Granular and Monolith) for the DOE Advanced Remediation Technologies (ART) Phase 2 Project", SRNL-STI-2009-00505, Rev.0 (December 2011).
33 R.P. Pires, J.H. Westsik, R.J. Serene, E.C. Golovich, M.N. Valenta, K.E. Parker, "Secondary Waste Form Screening Test Results - THOR® Fluidized Bed Steam Reforming Product in a Geopolymer Matrix," PNNL- 20551 (July 2011).
34 R. Klingenberg, and J. Felsche, "Interstitial Cristobalite-type Compounds (Na2O)0.33Na[AlSiO4]," J. Solid State Chemistry, 61, 40-46 (1986).
35 W.A. Deer, R.A. Howie, and J. Zussman, "Rock-Forming Minerals, V. 4," John Wiley & Sons, Inc., NY, 435pp (1963).
36 S.V. Mattigod, B.P. McGrail, D.E. McCready, L, Wang, K.E.Parker, J.S. Young, "Synthesis and Structure of Perrhenate Sodalite," J. Microporous & Mesopourous Materials, 91 (1-3), 139-144 (2006).
37 J.Ch, Buhl, G. Englehardt, J., Felsche, "Synthesis, X-ray Diffraction, and MAS N.M.R. Characteristics of Tetrahydroxoborate Sodalite," Zeolites, 9, 40-44 (1989).
38 D.M. Tobbens, and J.C. Buhl, "Superstructure of Sodiumborate Sodalite," Berline Neutron Scattering Center (BENSC) Experimental Report E9, Helmholtz Zentrum fur Materialiene und Energie (formally the Hahn-Meitner Institute), Berlin, Germany (2000).
39 D.G. Brookins, "Geochemical Aspects of Radioactive Waste Disposal," Springer: NY, 347pp (1984).
40 E.S. Dana, "A Textbook of Mineralogy," John Wiley & Sons, Inc., New York, 851pp (1932).
41 R.M. Barrer, "Hydrothermal Chemistry of Zeolites," Academic Press, New York, 360pp (1982): see references by St. J. Thugutt, Z. Anorg. Chem, 2, 65 (1892) and E. Flint, W. Clarke, E.S. Newman, L. Shartsis, D. Bishop and L.S. Wells, J. Res. Natl. Bur. Stds, 36, 63 (1945).
42 W.A. Deer, R.A. Howie, W.S. Wise, J. Zussman, "Rock-Forming Minerals, Vol. 4B, Framework Silicates: Silica Minerals, Feldspathoids and the Zeolites," Geological Society, London, 982pp (2004).
43 S.D. Rassat, L.A. Mahoney, R.L. Russell, "Cold Dissolved Saltcake Waste Simulant Development, Preparation, and Analysis," Pacific Northwest National Laboratory, PNNL-14194 (2003).
44 Bunker, B.C., Arnold,G.W., Day, D.E. and Bray, P.J. "The Effect of Molecular Structure on Borosilicate Glass Leaching," J. Non-Cryst. Solids, 87, 226-253 (1986).
45 Bourcier, W.L. "Affinity Functions for Modeling Glass Dissolution Rates," U.S. DOE Report UCRL-JC-131186 " Glass: Scientific Research for High Performance Containment" (1998).

ADVANCES IN JHCM HLW VITRIFICATION TECHNOLOGY AT VSL THROUGH SCALED MELTER TESTING

Keith S. Matlack and Ian L. Pegg
Vitreous State Laboratory, The Catholic University of America
Washington DC, USA

ABSTRACT

Joule heated ceramic melter (JHCM) vitrification is the baseline treatment technology for US high level radioactive waste and has been adopted in Japan, Germany, and China. The Vitreous State laboratory (VSL) operates the largest array of JHCM test systems in the US, with five platforms spanning a scale-up of 60X, including the largest such test platform in the US. These systems have supported vitrification facilities at West Valley, DWPF and M-Area at Savannah River, WTP HLW and LAW at Hanford, as well Rokkasho in Japan. Melter bubbler technology invented at VSL is central to the performance of the Hanford HLW and LAW systems and was recently successfully retro-fitted into DWPF, producing a near doubling of throughput. Results from many thousands of days of melter testing have produced significant enhancements in production rates and waste loadings through optimization and integration of glass formulations, design and operating conditions, and cold cap management.

INTRODUCTION

Vitrification using Joule heating is one of several methods for converting waste into glass. Vitrification is the internationally accepted method for treating High Level Waste (HLW). Glass is an amorphous material which can incorporate a wide spectrum of elements over a wide range of compositions, is resistant to radiation damage, and can form an extremely durable product, as supported by observations on natural analogs. JHCM technology was developed in the US in the early 1970s and has been adopted as the baseline for HLW treatment in the US[1,2]; in addition to its use at several DOE sites in the US, it has also seen applications around the world including in Germany, Japan, Russia, India and China[3,4]. JHCMs are powered with alternating current supplied to electrodes positioned on the sides or bottom of the melter. Current is passed through the molten glass thereby dissipating energy via the joule heating effect, which relies on the (ionic) electrical conductivity of the melt, which in turn depends on the composition and temperature of the melt. Waste combined with glass forming additives as either glass frit or chemicals is feed onto the surface of the glass pool where it undergoes a series of reactions including evaporation, denitrification, calcination, sintering, and eventually melts into the glass pool. The rate of these reactions is influenced by several factors including glass temperature, melt pool mixing, feed composition, additive type, and glass and waste compositions. Reactions occur throughout the feed material on the glass surface (referred to as the "cold cap") and at the interface between the molten glass and the feed material and therefore the glass production rate scales as the surface area of the melt. As a result, the melt rate is typically normalized to the surface area of the melt (kg/m^2/day). Glass is periodically discharged from the melter by variety methods including bottom drains, air lifts, and vacuum discharges. Off-gas generated from the waste-to-glass conversion reactions contains moisture, various gaseous species, entrained particulate, and a substantial fraction of the overall heat loss from the system.

VSL has provided support for JHCM treatment of radioactive wastes throughout the DOE complex and internationally since the 1980s, which has included numerous completed and ongoing vitrification projects. The West Valley Demonstration Project (WVDP), the only commercial reprocessing facility in the US generated ~660,000 gallons HLW containing 24 million curies. Between 1985 and 1993, the WVDP vitrification development program was supported through glass formulation and melter testing[5-10]. The waste was successfully converted into 275 canisters of glass (~550 MT), and the facility was decommissioned; the glass formulation used at the WVDP was developed at VSL. At the Savannah River Site (SRS) M-Area site, 660,000 gallons of mixed LLW from plating operations was stored in eleven tanks. An Energy*Solutions*-VSL team won a competitive procurement to vitrify the waste on a privatized fixed-price contract. Between 1995 and 1999, the vitrification facility was designed, constructed, operated, and deactivated[5,11,12]. All the waste was converted into a stable delisted glass in the largest radioactive JHCM to have operated in the US to date. VSL performed all of the glass formulation development and testing, scaled melter testing and flow-sheet development with actual M-Area waste, and provided remote and on-site support during the production operations. The M-Area vitrification facility was also notable as the first large-scale radioactive deployment of the JHCM melter bubbling technology invented at VSL. That technology provides active mixing of the glass pool to improve heat and mass transfer, which results in enormous increases in glass production rates. This melt-rate-boost application of gas bubbling is quite different from that used for many years in commercial glass manufacturing, where its pupose is to assist in the fining process by removing air inclusions from the product glass (or indeed from gas bubblers used for level detectors, density probes, air-lifts, etc.). This unique and patented technology is also incorporated into the design of both the LAW and HLW vitrification systems at the Waste Treatment Plant (WTP) at the Hanford site and was recently successfully retrofitted into the DWPF JHCM, resulting in a near doubling of the glass production rate. VSL has provided continuous support to the design, construction, and commissioning of the WTP since 1996 and was part of the original BNFL privatization team. Similar support is currently being provided to DWPF in a variety of areas related to routine operations and improvements in facility throughput. Since 2005, continuous and ongoing vitrification technology support to the Rokkasho facility in Japan has included extensive glass formulation development and scaled melter testing work. This brief overview focuses on the scaled melter testing work performed for the WTP and DWPF.

HANFORD VITRIFICATION SUPPORT

The WTP in Hanford, Washington will be the world's largest vitrification facility; it is intended to treat 56 million gallons of liquid nuclear waste containing ~194 million curies of radioactivity, which is stored in 177 aging underground tanks. The waste will undergo a pretreatment process to separate and decontaminate the liquid LAW fraction, containing over 90% of the mass, from the HLW solids fraction containing over 90% of the radioactivity. The LAW and HLW fractions will then be vitrified in separate JHCM treatment systems.

Development work for the WTP employed a "tiered" approach to vitrification testing involving computer-based glass formulation, glass property-composition models, crucible melts, and continuous melter tests of increasingly more realistic scales. Melter systems (DM10, DM100, DM1000, DM1200) ranging from 0.02 to 1.2 m^2 installed at VSL have been used for this purpose, which, in combination with a 3.3 m^2 low activity waste (LAW) Pilot Melter at Energy*Solutions* (DM3300), span more than two orders of magnitude in melt surface area. In this way, less-costly small-scale tests can be used to define the most appropriate tests to be conducted at the larger scales in order to extract maximum benefit from the large-scale tests. For HLW vitrification development, a key component in this approach is the one-third scale DM1200, which is the HLW Pilot Melter which includes an integrated prototypical off-gas treatment system. That system replaced an earlier DM1000 system that was used for initial WTP HLW throughput testing and which was operated for over 7 years. Both melters have similar melt surface areas (1.2 m^2), but the DM1200 is prototypical of the present WTP HLW melter design whereas the DM1000 was not. In particular, the DM1200 provides for testing on a vitrification system that includes the specific train of unit operations that has been selected for both HLW and LAW WTP off-gas treatment. Melter testing was supported by extensive characterization of the melter exhaust, and the exhaust of each of the various off-gas system components, using continuous exhaust monitoring methods and various US EPA protocols, as well as by complete inorganic, organic, and radionuclide analysis of product glass, process solutions, and exhaust streams. A summary of the tiered melter approach used for the WTP is provided in Table I.

Table I. Tiered Melter Testing for Hanford WTP

DM10 > 10,000 kg Glass Produced	0.021 m^2 Surface Area, 8 kg Nominal Glass Inventory	• Wide range of HLW and LAW waste and glass compositions[13-27] • Evaluated tendency towards secondary sulfur formation[13,15-27] • ~100 tests evaluating Tc and I partitioning[24-26] • Evaluated a wide range of additives and organic reductants[14-26] • Collected regulatory data[27]
DM100 HLW ~ 32,000 kg Glass Produced	0.109 m^2 Surface Area, 180 kg Nominal Glass Inventory	• 231 days processing AZ-101, AZ-102, C-106/AY-102, C-104/AY-101, high Al, high Al and Na, high Bi, and high Cr[13,15,29-36] • Screened all composition for HLW Pilot Melter[28,29,33-37] • Determined the effect of waste and glass composition, glass temperature, bubbling rate, feed solids content, plenum heaters and feed rheology on production rate[13,15,28-36] • Collected melter exhaust data[13,15,28-36]
DM100 LAW ~ 52,000 kg Glass Produced	0.109 m^2 Surface Area, 110 kg Nominal Glass Inventory	• 231 days processing LAW Sub-Envelope Compositions (A1, A2, A3, B1, B2, C1, C3) and 15% deviations[16-23,25,37-47] • Screened all compositions for LAW Pilot Melter[37-47] • Evaluated tendency towards secondary sulfur formation[16-23,25,37-47] • Optimized reductant feed additives[16-23,25,37-47] • Evaluated Tc partitioning[25,30] • Collected melter exhaust data[16-23,25,37-47]
DM1000 ~ 17,000 kg Glass Produced	1.2 m^2 Surface Area, 1700 kg Nominal Glass Inventory	• 28 days processing AZ-101, C-106/AY-102, WVDP cold-commissioning HLW simulants to access the need for bubblers to attain required WTP HLW production rates[47]
DM1200 ~ 313,000 kg Glass Produced	1.2 m^2 Surface Area, 1700 kg Nominal Glass Inventory	• 288 days processing AZ-101, AZ-102, C-106/AY-102, C-104/AY-101, high Al, and high Bi HLW compositions[32-27,49-56] • Collected rate, regulatory, engineering and MACT permitting data[32-27,49-56] • Determined the effect of waste and glass composition, glass temperature, bubbling rate, feed solids content, and feed rheology on production rate[32-27,49-56] • Optimized the design, location and use of bubblers[32-27,49-54] • Collected melter exhaust data[32-27,49-56]
		• 48 days processing LAW Envelope A, B, C simulants[25,56-60] • Collected rate, regulatory, engineering and MACT permitting data[25,56-60] • Evaluated cause and mitigation strategies for foaming[60] • Evaluated Tc partitioning (Re as surrogate) [25]
DM3300 ~ 3,637,000 kg Glass Produced	3.3 m^2 Surface Area 6500 kg Nominal Glass Inventory	• Processed LAW Sub-Envelope Compositions (A1, A2, A3, B1, B2, C1, C3) and 15% deviations[61] • Evaluated tendency towards secondary sulfur formation[61]

Research and development in support of the WTP resulted in innovations in glass formulation and the further development and optimization of the core active melt pool mixing melter technology. LAW and HLW glass formulation included baseline glass formulations and required data packages, glass property-composition models, support of the compliance strategy, and definition of compositional operating envelopes. Small and pilot-scale melter testing demonstrated the ability to process each tank waste and the likely process variability while providing design confirmation, flow-sheet, regulatory, safety, and waste form qualification data. Specific risk areas addressed during testing included noble metal precipitation, sulfate separation, materials corrosion, feed rheology, simulant validation, feed mixing, and mixing and sampling systems.

One of the major areas of research and development involved increasing glass production rates to meet the WTP requirements for HLW streams. Melt pool agitation, glass pool temperature, the use of lid heaters, and waste and glass composition all influence melt rate and thus were evaluated for potential increases in glass production rate. Melt pool agitation through gas sparging of the melt pool using "bubblers" was developed by VSL in the early 1990s. Increases in production rates using bubblers depend on a variety of factors including the type, location, and air flow rate employed. Production rates at vitrification facilities without the use of bubblers were typically around 500 kg/m^2/day, as compared to 1500 to 2000 kg/m^2/day for the LAW and HLW Pilot Plants at nominal glass temperatures of 1150°C. Increasing glass pool temperature above the baseline temperature of 1150°C has resulted in increases in production rates of 9 to 13 kg/m^2/day per °C depending on the feed composition[13,34,36,62]; however, increases in melt pool temperature also result in greater loss of volatile species and greater materials corrosion. Lid heaters have been shown to provide some increase in production rates for systems without bubblers but testing for the WTP showed that their benefits are marginal for systems with bubblers as a result of the much improved heat transfer. While the composition of a given waste stream is essentially fixed, the overall melter feed composition, which also affects the glass production rate, can be varied through glass formulation selection and the selection of glass forming additives. Manipulation of the composition and form (i.e., frit vs. chemicals) of the glass forming additives has been demonstrated to affect cold cap behavior, melt pool viscosity, glass oxidation state, and feed rheology, all of which can affect the glass production rate. Optimization of these factors can therefore result in substantial increases in processing rates[34,36,64,65].

One of the primary objectives of the DM1200 testing was to determine the maximum production rates that would be possible for the first four HLW streams that were planned to be treated at the WTP. Over the course of this testing the design requirements changed (the required rate increased) as did the projected range of solids contents of the pretreated waste. Figure 1 illustrates results from some of the scaled melter tests that were performed to address this. Initial tests showed that melt pool bubbling was required to produce glass at rates greater than 400 kg/m^2/day, even when employing higher glass pool temperatures, higher solids content feeds, additives in the form of glass frit, none of which were in the baseline design. As a result, a design change was implemented for the WTP HLW JHCM systems to add bubblers, as was always the case for the LAW JHCM systems. Subsequently, tests with two single-outlet bubblers demonstrated that production rates of 800 kg/m^2/day (WTP target) could be obtained for HLW streams with greater than 17 wt% undissolved solids. Significant further improvements in glass production rates (up to 1500 kg/m^2/day) were achieved with feeds corresponding to 15 wt% undissolved solids from pretreatment by employing optimized bubbler configurations. These improvements are sufficient to more than make up for the production rate shortfall brought about by the reduction in the solids content in the feed from pretreatment from 20 wt% to 15 wt% undissolved solids. In view of the fact that bubblers were essentially retro-fitted into the WTP

HLW design, further optimization, including the addition of more bubbling outlets, would result in further production rate enhancements.

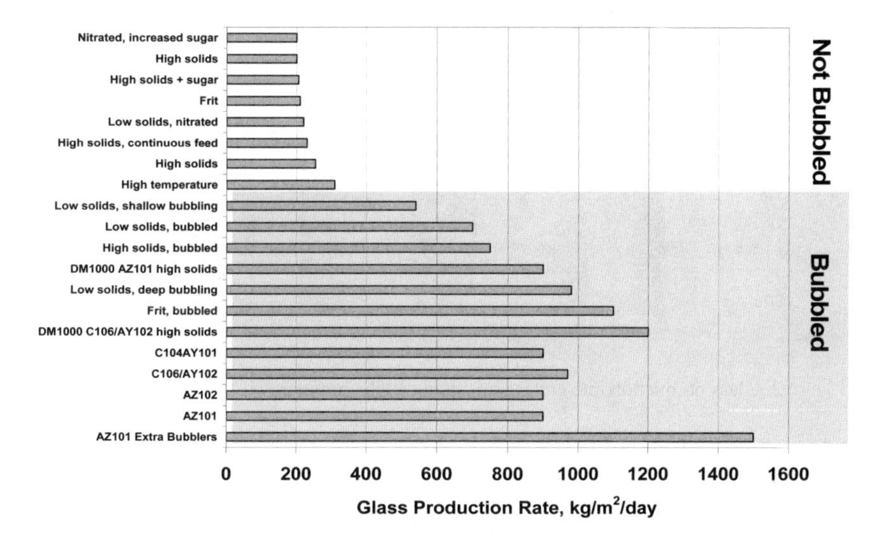

Figure 1. Glass production rates on the DM1200 melter with simulated WTP HLW streams.

Increases in production rates were also obtained through glass formulation optimization by the manipulation of the glass forming additives. This approach was applied to the high-iron waste streams to be processed initially at WTP as well as other tank wastes rich in aluminum, bismuth, chromium, and sulfur. Glass formulations were designed, prepared at crucible-scale and characterized to determine their properties relevant to processing and product quality. Glass formulations that met these requirements were screened for melt rates using small-scale tests. The small-scale melt rate screening included vertical gradient furnace (VGF) and direct feed consumption (DFC) melter tests on the DM10. Based on the results of these tests, modified glass formulations were developed and selected for larger scale melter tests on the DM100 and HLW Pilot Melter to determine their processing rates. For the high iron C-106/AY-102 waste stream, the waste oxide loading was increased from 27.75% to 42.0% while demonstrating an increase in production rate on the DM100 from 1050 to 1650 $kg/m^2/day$ at constant temperature and bubbling rate[63]. This increase in waste loading and glass processing rate provides a 2.4 fold increase in waste processing rate. The enhancement process for the high aluminum waste demonstrated on the DM100 is illustrated in Figure 2[64]. Newly developed glass formulations had waste loadings as high as 50 wt%, with corresponding Al_2O_3 concentration in the glass of 26.63 wt%. The new glass formulations showed glass production rates as high as 1900 $kg/m^2/day$ under nominal melter operating conditions. The demonstrated glass production rates are much higher than the current requirement of 800 $kg/m^2/day$ and anticipated future enhanced WTP requirement of 1000 $kg/m^2/day$.

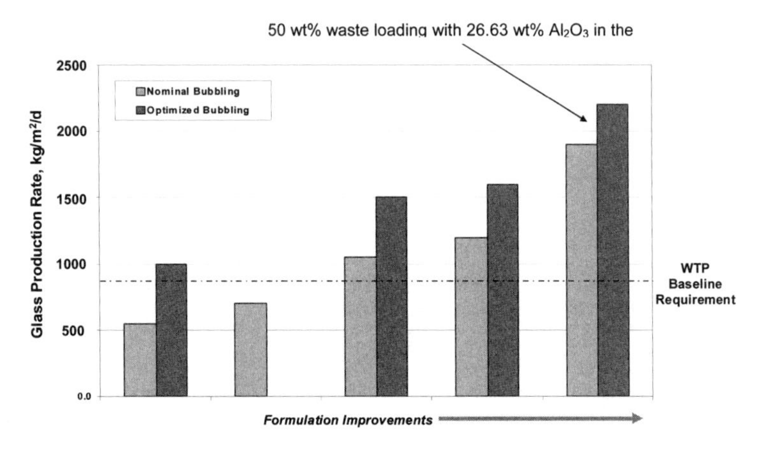

Figure 2. Glass production rate enhancements for high aluminum HLW Hanford Wastes.

DWPF MELT RATE ENHANCEMENTS

A series of tests were performed to assess the potential benefits of bubbler technology for DWPF HLW streams. The results of these tests provided the initial basis for the subsequent installation of bubblers into the DWPF melter[65]. The testing was performed on one of the two DM100 joule-heated melter systems installed at the VSL. The first test employed a simulant of DWPF Sludge Batch 3 HLW with Frit 418 and was intended to provide a calibration of the DM100 melt rate data against full-scale DWPF data obtained using actual Sludge Batch 3 with Frit 418. The DM100 specific glass production rates (i.e., on a per unit melt surface area basis) observed without bubbling were close to but slightly lower than those observed at DWPF, suggesting that the small-scale melter results are conservative. In contrast, with melt pool bubbling, the specific glass production rates with the same feed increased by nearly a factor of five. This increase is consistent with the range of melt rate improvements that have been demonstrated previously with a wide variety of other waste compositions. Subsequent testing employed a projected future DWPF HLW composition that has among the highest expected aluminum contents. Fully compliant, high-waste-loading glass formulations containing ~20 wt% Al_2O_3 were developed for that stream and a corresponding new frit composition was specified. This composition was also optimized with respect to melt rate based on small-scale melt rate tests. Without bubbling, DM100 tests with this waste and glass composition showed glass production rates that were slightly higher than those for the Sludge Batch 3 simulant without bubbling. DM100 tests with bubbling again showed a nearly five-fold improvement in glass production rates. Finally, tests were performed with the high aluminum waste in combination with a simulated SWPF stream, which resulted in 4.4 wt% TiO_2 in the glass product; similar increases in glass production rates were observed. All product glasses showed PCT releases well below the HLW requirements. The melt rate enhancements that were demonstrated in these tests are likely well beyond what the balance of the DWPF facility could support; however, the results demonstrated an approach for removing the melter as a bottleneck for overall facility throughput.

Following the decision to proceed with the retrofit of the bubbler technology into DWPF, a melt rate target corresponding to a production rate of 400 canisters per year was specified, which corresponds to an approximate doubling of the throughput; based on the previous testing with DWPF simulants and the extensive testing performed for the WTP, this was known to be well within the abilities of the technology. Further testing of bubbler conceptual designs was performed on the DM1200 melter prior to the design, fabrication, and installation of bubblers into the DWPF melter in September 2010. Four bubblers were retrofitted into available locations in the existing melter lid with several restrictions related to ease of installation/replacement and proximity to walls and other utilities. Due to limitations in the number and placement of ports in the melter lid, some of the bubblers were combined with thermowells and level detectors. The instantaneous melt rate was increased with the operation of bubblers from about 59 to 91 kg/hr (130 to 200 lb/hr)[66]; 91 kg/hr equates to about 400 canisters per year. The time required to fill each canister decreased from 28 to 30 hours without bubbling to 17 to 19 hours after the installation of bubblers. Further increases in production rate could be achieved through optimized melter lid design and bubbler configuration; however, currently the overall production rate is limited by the feed preparation process.

CONCLUSIONS

Joule heated ceramic melter technology has been successfully applied to a variety of nuclear wastes across the DOE complex and internationally. Missions employing JHCM technology have been completed at West Valley and SRS M-Area in the US and at PAMELA (Belgium-Germany) and VEK in Germany, and are ongoing at DWPF in the US and Rokkasho in Japan. The world's largest vitrification facility, the WTP, is under construction at Hanford and will employ JHCM technology for both LAW and LAW vitrification. JHCM technology is mature, robust, well understood, flexible, and been successfully deployed with a variety of tailored design features. The most significant JHCM performance advancement since its inception has been the enormous melt-rate boost provided by melter bubbler technology, which was deployed at SRS M-Area, retrofitted into DWPF, and incorporated into the Hanford WTP HLW and LAW JHCM baselines. Significant performance enhancements have been demonstrated through optimized glass formulations to further increase both melt rate and waste loading. The flexible JHCM technology is amenable to further development and additional advances are likely to be seen in the future.

ACKNOWLEDGMENTS

The authors gratefully acknowledge the invaluable contributions of all of the staff at the Vitreous State Laboratory that made this work possible. Thanks are also due to our long-time partners and collaborators at Energy*Solutions*, Inc. DOE Office of Environmental Management support for this work though various channels, including the Office of River Protection, Bechtel National, Inc., and Savannah River Remediation, LLC, is also gratefully acknowledged.

REFERENCES

[1] C.C. Chapman and J.L. McElroy, "Slurry-Fed Ceramic Melter – A Broadly Accepted System to Vitrify High-Level Waste," in "*High Level Radioactive Waste and Spent Fuel Management, Vol. II*," Eds. S.C. Slate, R. Kohout, and A. Suzuki, AMSE, 119, 1989.
[2] "Handbook of Vitrification Technologies for Treatment of Hazardous and Radioactive Waste," US Environmental Protection Agency, EPA/625/R-92/002, 1992.
[3] "Glass as a Waste Form: Summary of an International Workshop," National Academy Press, Washington, DC, 1996.
[4] "Radioactive Waste Forms for the Future," W.Lutze and R.C. Ewing, Eds., North Holland, Amsterdam, 1988.

[5] I.L. Pegg and I. Joseph, "Vitrification," Book Chapter in "*Hazardous and Radioactive Waste Treatment Technologies Handbook*," Ed. C. H. Oh, CRC Press, Boca Raton, 2001.
[6] X. Feng, I.L. Pegg, Aa. Barkatt, P.B. Macedo, S. Cucinell, and S. Lai, "Correlation between Composition Effects on Glass Durability and the Structural Role of the Constituent Oxides," *Nuclear Technology*, **85** 334 (1989).
[7] A.C. Buechele, X. Feng, H. Gu, and I.L. Pegg, "Alteration of Microstructure of West Valley Glass by Heat Treatment," in *Scientific Basis for Nuclear Waste Management XIII*, Ed. V. M. Oversby and P. W. Brown, MRS, Pittsburgh, PA p. 393 (1990).
[8] X. Feng, I.L. Pegg, E. Saad, S. Cucinell, and Aa. Barkatt, "Redox Effects on the Durability and Viscosity of Nuclear Waste Glasses," *Ceramic Transactions* Vol. 9, Nuclear Waste Management III, Ed. G.B. Mellinger, ACerS, p. 165, (1990).
[9] A.C. Buechele, X. Feng, H. Gu, I.S. Muller, W. Wagner, and I.L. Pegg, "Effects of Composition Variations on Microstructure and Chemical Durability of West Valley Reference Glass," in *Scientific Basis for Nuclear Waste Management XIV*, MRS, Pittsburgh, PA, **212**, 141 (1991).
[10] A.C. Buechele, X. Feng, H. Gu, and I.L. Pegg, "Redox State Effects on Microstructure and Leaching Properties of West Valley SF-12 Glass," *Ceramic Transactions*, **23**, Nuclear Waste Management IV, p. 85, (1991).
[11] B.W. Bowen and M.B. Brandys, "Design of a Vitrification Process for Savannah River M-Area Waste," Proc. Int. Top. Meet. Nucl. Haz. Mgmt. Spectrum '94, American Nuclear Society, Inc., LaGrange Park, IL 60525, 2240-2243.
[12] S.S. Fu, H. Gan, I.S. Muller, I.L. Pegg, and P.B. Macedo, "Optimization of Savannah River M-Area Mixed Wastes for Vitrification," *Scientific Basis for Nuclear Waste Management XX*, Eds. W. Gray and I. Triay, MRS, Pittsburgh, PA, 139 (1997).
[13] K.S. Matlack, H. Gan, W. Gong, I.L. Pegg, C.C. Chapman, and I. Joseph, "High Level Waste Vitrification System Improvements," VSL-07R1010-1, Rev. 0, Vitreous State Laboratory, The Catholic University of America, Washington, DC, 04/16/07.
[14] K.S. Matlack, W.K. Kot, H. Gan, W. Gong and I.L. Pegg, "HLW Enhancement Tests on the DuraMelter™ 10 with Hanford AZ-102 Tank Waste Simulants," VSL-06R6260-1, Rev. 0, Vitreous State Laboratory, The Catholic University of America, Washington, DC, 2/28/06.
[15] K.S. Matlack, W.K. Kot, W. Gong and I.L. Pegg, "Small Scale Melter Testing of HLW Algorithm Glasses: Matrix 2 Tests," Final Report, VSL-08R1220-1, Rev. 0, Vitreous State Laboratory, The Catholic University of America, Washington, DC, 6/27/08.
[16] K.S. Matlack, M. Chaudhuri, H. Gan, I.S. Muller, W. Gong, and I.L. Pegg, "Glass Formulation Testing to Increase Sulfate Incorporation," VSL-04R4960-1, Rev. 0, Vitreous State Laboratory, The Catholic University of America, Washington, DC, 2/28/05.
[17] K.S. Matlack, W. Gong, and I.L. Pegg, "Small Scale Melter Testing with LAW Simulants to Assess the Impact of Higher Temperature Melter Operations," VSL-04R4980-1, Rev. 0, Vitreous State Laboratory, The Catholic University of America, Washington, DC, 2/13/04.
[18] K.S. Matlack, W. Gong, and I.L. Pegg, "Glass Formulation Testing to Increase Sulfate Volatilization from Melter," VSL-04R4970-1, Rev. 0, Vitreous State Laboratory, The Catholic University of America, Washington, DC, 2/24/05.
[19] K.S. Matlack, W. Gong, I.S. Muller, I. Joseph, and I.L. Pegg, "LAW Envelope C Glass Formulation Testing to Increase Waste Loading," Final Report, VSL-05R5900-1, Rev. 0, Vitreous State Laboratory, The Catholic University of America, Washington, DC, 1/27/06.
[20] K.S. Matlack, H. Gan, I.S. Muller, I. Joseph, and I.L. Pegg, "LAW Envelope A and B Glass Formulation Testing to Increase Waste Loading," Final Report, VSL-06R6900-1, Rev. 0, Vitreous State Laboratory, The Catholic University of America, Washington, DC, 3/23/06.

[21] K.S. Matlack, S.P. Morgan, and I.L. Pegg, "Melter Tests with LAW Envelope B Simulants to Support Enhanced Sulfate Incorporation," Final Report, VSL-00R3501-1, Rev. 0, Vitreous State Laboratory, The Catholic University of America, Washington, DC, 11/27/00.
[22] K.S. Matlack, S.P. Morgan, and I.L. Pegg, "Melter Tests with LAW Envelope A and C Simulants to Support Enhanced Sulfate Incorporation," Final Report, VSL-01R3501-2, Rev. 0, Vitreous State Laboratory, The Catholic University of America, Washington, D.C., 1/26/01.
[23] K.S. Matlack, I.S. Muller, W. Gong, and I.L. Pegg, "Small Scale Melter Testing of LAW Salt Phase Separation," VSL-07R7480-1, Rev. 0, Vitreous State Laboratory, The Catholic University of America, Washington, DC, 8/20/07.
[24] K.S. Matlack, I.S. Muller, I. Joseph, and I.L. Pegg, "Improving Technetium Retention in Hanford LAW Glass – Phase 1," VSL-10R1920-1, Rev. 0, Vitreous State Laboratory, The Catholic University of America, Washington, DC, 3/19/10.
[25] K.S. Matlack, I.S. Muller, R. Callow, N. D'Angelo, T. Bardakci, I. Joseph, and I.L. Pegg, "Improving Technetium Retention in Hanford LAW Glass – Phase 2," VSL-11R2260-1, Rev. 0, Vitreous State Laboratory, The Catholic University of America, Washington, DC, 7/20/11.
[26] K.S. Matlack, H. Abramowitz, M. Brandys, I.S. Muller, D.A. Callow, N. D'Angelo, R. Cecil, I. Joseph, and I.L. Pegg, "Technetium Retention in WTP LAW Glass with Recycle Flow-Sheet: DM10 Melter Testing," VSL-12R2640-1, Rev. 0, Vitreous State Laboratory, The Catholic University of America, Washington, DC, 9/24/12.
[27] K.S. Matlack and I.L. Pegg, "Determination of the Fate of Hazardous Organics During Vitrification of TWRS LAW and HLW Simulants," VSL-99R3580-2, Vitreous State Laboratory, Washington, D.C., October 4, 1999.
[28] K.S. Matlack, W.K. Kot, and I.L. Pegg, "Melter Tests with AZ-101 HLW Simulant Using a DuraMelter 100 Vitrification System," VSL-01R10N0-1, Rev. 1, Vitreous State Laboratory, The Catholic University of America, Washington, DC, 2/25/01.
[29] K.S. Matlack, W. Gong and I.L. Pegg, "DuraMelter 100 HLW Simulant Validation Tests with C-106/AY-102 Feeds," VSL-05R5710-1, Rev. 0, Vitreous State Laboratory, The Catholic University of America, Washington, DC, 6/2/05.
[30] K.S. Matlack, W.K. Kot, and I.L. Pegg, "Technetium/Cesium Volatility in DM100 Tests Using HLW AZ-102 and LAW Sub-Envelope A1 Simulants," VSL-04R4710-1, Rev. 0, Vitreous State Laboratory, The Catholic University of America, Washington, DC, 9/28/04.
[31] K.S. Matlack, W.K. Kot, W. Gong and I.L. Pegg, "Small Scale Melter Testing of HLW Algorithm Glasses: Matrix 1 Tests," VSL-07R1220-1, Rev. 0, Vitreous State Laboratory, The Catholic University of America, Washington, DC, 11/12/07.
[32] K.S. Matlack, W. Gong, T. Bardakci, N. D'Angelo, W. Kot and I.L. Pegg, "Integrated DM1200 Melter Testing of HLW C-106/AY-102 Composition Using Bubblers," VSL-03R3800-1, Rev. 0, Vitreous State Laboratory, The Catholic University of America, Washington, DC, 9/15/03.
[33] K.S. Matlack, W. Gong, T. Bardakci, N. D'Angelo, W. Kot and I.L. Pegg, "Integrated DM1200 Melter Testing of HLW C-104/AY-101 Compositions Using Bubblers," VSL-03R3800-3, Rev. 0, Vitreous State Laboratory, The Catholic University of America, Washington, DC, 11/24/03.
[34] K.S. Matlack, H. Gan, M. Chaudhuri, W.K Kot, W. Gong, T. Bardakci, I. Joseph, and I.L. Pegg, "DM100 and DM1200 Melter Testing with High Waste Loading Glass Formulations for Hanford High-Aluminum HLW Streams," VSL-10R1690-1, Rev. 0, Vitreous State Laboratory, The Catholic University of America, Washington, DC, 8/16/10.
[35] K.S. Matlack, W. Gong, T. Bardakci, N. D'Angelo, M. Brandys, W.K. Kot, and I.L. Pegg, "Integrated DM1200 Melter Testing Using AZ-102 and C-106/AY-102 HLW Simulants: HLW

Simulant Verification," VSL-05T5800-1, Rev. 0, Vitreous State Laboratory, The Catholic University of America, Washington, DC, 6/27/05.
[36] K.S. Matlack, H. Gan, M. Chaudhuri, W. Gong, T. Bardakci, I.L. Pegg, and I. Joseph, "Melt Rate Enhancement for High Aluminum HLW Glass Formulations," VSL-08R1360-1, Rev. 0, Vitreous State Laboratory, The Catholic University of America, Washington, DC, 12/19/08.
[37] K.S. Matlack, W. Gong, and I.L. Pegg, "Compositional Variation Tests on DuraMelter 100 with LAW Sub-Envelope A1 Feed (LAWA44 Glass) in Support of the LAW Pilot Melter," VSL-02R62N0-4, Rev. 0, Vitreous State Laboratory, The Catholic University of America, Washington, D.C., 6/18/02.
[38] K. S. Matlack, W. Gong and I.L. Pegg, "Compositional Variation Tests on DuraMelter 100 with LAW Sub-Envelope A2 Feed (LAWA88) Glass in Support of the LAW Pilot Melter," Final Report, VSL-02R62N0-3, Rev. 0, Vitreous State Laboratory, The Catholic University of America, Washington, D.C., 11/1/02.
[39] K.S. Matlack, W. Gong, and I.L. Pegg, "Compositional Variation Tests on DuraMelter 100 with LAW Sub-Envelope A3 Feed in Support of the LAW Pilot Melter," VSL-01R62N0-1, Rev. 1, Vitreous State Laboratory, The Catholic University of America, Washington, D.C., 7/15/02.
[40] K.S. Matlack, W. Gong, and I.L. Pegg, "Compositional Variation Tests on DuraMelter 100 with LAW Sub-Envelope B1 Feed in Support of the LAW Pilot Melter," Final Report, VSL-02R62N0-5, Rev. 0, Vitreous State Laboratory, The Catholic University of America, Washington, D.C., 5/8/03.
[41] K.S. Matlack and I.L. Pegg, "Compositional Variation Tests on DuraMelter 100 with LAW Sub-Envelope B2 Feed in Support of the LAW Pilot Melter," VSL-03R3410-2, Rev. 0, The Catholic University of America, Vitreous State Laboratory, Washington, D.C., 10/20/03.
[42] K.S. Matlack, W. Gong, and I.L. Pegg, "Compositional Variation Tests on DuraMelter 100 with LAW Sub-Envelope C1 Feed (LAWC22 Glass) in Support of the LAW Pilot Melter," VSL-02R62N0-2, Rev. 1, Vitreous State Laboratory, The Catholic University of America, Washington, D.C., 9/23/02.
[43] K.S. Matlack, W. Gong, R.A. Callow and I.L. Pegg, "Compositional Variation Tests on DuraMelter 100 with LAW Sub-Envelope C2 Feed in Support of the LAW Pilot Melter," VSL-04R4410-1, Rev. 0, Vitreous State Laboratory, The Catholic University of America, Washington, DC, 6/17/04.
[44] K.S. Matlack, W. Gong, and I.L. Pegg, "DuraMelter 100 Sub-Envelope Changeover Testing Using LAW Sub-Envelope A1 and C1 Feeds in Support of the LAW Pilot Melter," VSL-02R62N0-6, Rev. 0, Vitreous State Laboratory, The Catholic University of America, Washington, D.C., 9/9/03.
[45] K.S. Matlack, W. Gong, and I.L. Pegg, "DuraMelter 100 Sub-Envelope Changeover Testing Using LAW Sub-Envelope A2 and B1 Feeds in Support of the LAW Pilot Melter," VSL-03R3410-1, Rev. 0, Vitreous State Laboratory, The Catholic University of America, Washington, D.C., 8/22/03.
[46] K.S. Matlack, W. Gong, and I.L. Pegg, "DuraMelter 100 Sub-Envelope Changeover Testing Using LAW Sub-Envelope A3 and C2 Feeds in Support of the LAW Pilot Melter," VSL-03R3410-3, Rev. 0, Vitreous State Laboratory, The Catholic University of America, Washington, D.C., 10/17/03.
[47] K.S. Matlack, I.S. Muller, W. Gong, and I.L. Pegg, "DuraMelter 100 Tests to Support LAW Glass Formulation Correlation Development," VSL-06R6480-1, Rev. 0, Vitreous State Laboratory, The Catholic University of America, Washington, DC, 2/27/06.
[48] K.S. Matlack, W.K. Kot, F. Perez-Cardenas, and I.L. Pegg, "Determination of Processing Rate of RPP-WTP HLW Simulants using a DuraMelter™ 1000 Vitrification System," VSL-00R2590-2, Rev. 0, 8/21/00.

[49] K.S. Matlack, W. Gong, T. Bardakci, N. D'Angelo, W.K. Kot, and I.L. Pegg, "DM1200 Tests with AZ-101 HLW Simulants," VSL-03R3800-4, Rev. 0, Vitreous State Laboratory, The Catholic University of America, Washington, DC, 2/17/04.
[50] K.S. Matlack, M. Brandys, and I.L. Pegg, "Start-Up and Commissioning Tests on the DM1200 HLW Pilot Melter System Using AZ-101 Waste Simulants," VSL-01R0100-2, Rev. 1, Vitreous State Laboratory, The Catholic University of America, Washington, DC, 10/31/01.
[51] K.S. Matlack, W.K. Kot, T. Bardakci, T.R. Schatz, W. Gong, and I.L. Pegg, "Tests on the DuraMelter 1200 HLW Pilot Melter System Using AZ-101 HLW Simulants," VSL-02R0100-2, Rev. 0, Vitreous State Laboratory, The Catholic University of America, Washington, DC, 6/11/02.
[52] K.S. Matlack, W. Gong, T. Bardakci, N. D'Angelo, W. Kot and I.L. Pegg, "Integrated DM1200 Melter Testing of HLW AZ-102 Compositions Using Bubblers," VSL-03R3800-2, Rev. 0, Vitreous State Laboratory, The Catholic University of America, Washington, DC, 9/24/03.
[53] K.S. Matlack, W. Gong, T. Bardakci, N. D'Angelo, W. Lutze, P. M. Bizot, R. A. Callow, M. Brandys, W.K. Kot, and I.L. Pegg, "Integrated DM1200 Melter Testing of Redox Effects Using HLW AZ-101 and C-106/AY-102 Simulants," VSL-04R4800-1, Rev. 0, Vitreous State Laboratory, The Catholic University of America, Washington, DC, 5/6/04.
[54] K.S. Matlack, W. Gong, T. Bardakci, N. D'Angelo, W. Lutze, R.A. Callow, M. Brandys, W.K. Kot, and I.L. Pegg, "Integrated DM1200 Melter Testing of Bubbler Configurations Using HLW AZ-101 Simulants," VSL-04R4800-4, Rev. 0, Vitreous State Laboratory, The Catholic University of America, Washington, DC, 10/5/04.
[55] K.S. Matlack, H. Gan, W.K. Kot, M. Chaudhuri, R.K. Mohr, D. A. McKeown, T. Bardakci, W. Gong, A. C. Buechele, and I.L. Pegg, "Tests with High-Bismuth HLW Glasses," VSL-10R1780-1, Rev. 0, Vitreous State Laboratory, The Catholic University of America, Washington, DC, 12/13/10.
[56] K.S. Matlack, W. Gong, T. Bardakci, N. D'Angelo, M. Brandys, W. Kot, and I.L. Pegg, "Regulatory Off-Gas Emissions Testing on the DM1200 Melter System Using HLW and LAW Simulants," VSL-05R5830-1, Rev. 0, Vitreous State Laboratory, The Catholic University of America, Washington, DC, 10/31/05.
[57] K.S. Matlack, W. Gong, T. Bardakci, N. D'Angelo, and I.L. Pegg, "Integrated Off-Gas System Tests on the DM1200 Melter with RPP-WTP LAW Sub-Envelope A1 Simulants," VSL-02R8800-2, Rev. 0, Vitreous State Laboratory, The Catholic University of America, Washington, DC, 9/03/02.
[58] K.S. Matlack, W. Gong, T. Bardakci, D'Angelo, and I.L. Pegg, "Integrated Off-Gas System Tests on the DM1200 Melter with RPP-WTP LAW Sub-Envelope C1 Simulants," VSL-02R8800-1, Rev. 0, Vitreous State Laboratory, The Catholic University of America, Washington, DC, 7/25/02.
[59] K.S. Matlack, W. Gong, T. Bardakci, N. D'Angelo, and I.L. Pegg, "Integrated Off-Gas System Tests on the DM1200 Melter with RPP-WTP LAW Sub-Envelope B1 Simulants," VSL-03R3851-1, Rev. 0, Vitreous State Laboratory, The Catholic University of America, Washington, DC, 10/17/03.
[60] K.S. Matlack, W. Gong, T. Bardakci, N. D'Angelo, P.M. Bizot, R.A. Callow, M. Brandys, and I.L. Pegg, "Bubbling Rate and Foaming Tests on the DuraMelter 1200 with LAWC22 and LAWA30 Glasses," VSL-04R4851-1, Rev. 0, Vitreous State Laboratory, The Catholic University of America, Washington, DC, 7/1/04.
[61] G. Diener "RPP-WTP Pilot Melter Envelope Testing Summary Report," REP-PLT-027, Rev. 1, Duratek, Inc., Columbia, MD, 9/16/05.

[62] E.C. Smith, G.A. Diener, I. Joseph, and B.W. Bowan II, and I.L. Pegg "Waste Vitrification Melter Throughput Enhancement through Increased Operating Temperature" WM-05 Conference, 2005, Tucson, AZ, WM-5386, 2005.
[63] K.S. Matlack, H. Gan, M. Chaudhuri, W.K Kot, I. Joseph, and I.L. Pegg, "Melter Throughput Enhancements for High-Iron HLW," VSL-12R2490-1, Rev. 0, Vitreous State Laboratory, The Catholic University of America, Washington, DC, 5/31/12.
[64] I.Joseph, B.W. Bowan II, H. Gan, W.K. Kot, K.S. Matlack, I.L. Pegg and A.A. Kruger, "High Aluminum HLW Glasses for Hanford's WTP," 10241, WM-2010 Conference, Phoenix, AZ,2010.
[65] B.W. Bowan, II, I.Joseph, K.S. Matlack, H. Gan, W.K. Kot, and I.L. Pegg, "Tests of Simultaneous Melt Rate and Waste Loading Enhancement for DWPF HLW Streams," 10254, WM-2010 Conference, Phoenix, AZ, 2010,
[66] M.E. Smith and D.C. Iverson, "Installation of Bubblers in the Savannah River Site Defense Waste Processing Facility Melter," 11136, WM-2011 Conference, Phoenix, AZ, 2010.

IMPACT OF PARTICLE AGGLOMERATION ON ACCUMULATION RATES IN THE GLASS DISCHARGE RISER OF HLW MELTER

J. Matyáš[1], D. P. Jansik[1], A. T. Owen[1], C.A. Rodriguez[1], J. B. Lang[1], A.A. Kruger[2]
1. Pacific Northwest National Laboratory
   Richland, WA, USA
2. Office of River Protection
   Richland, WA, USA

ABSTRACT

The major factor limiting waste loading in continuous high-level radioactive waste (HLW) melters is an accumulation of particles in the glass discharge riser during a frequent and periodic idling of more than 20 days. An excessive accumulation can produce robust layers a few centimeters thick, which may clog the riser, preventing molten glass from being poured into canisters. Since the accumulation rate is driven by the size of particles we investigated with X-ray microtomography, scanning electron microscopy, and image analysis the impact of spinel forming components, noble metals, and alumina on the size, concentration, and spatial distribution of particles, and on the accumulation rate. Increased concentrations of Fe and Ni in the baseline glass resulted in the formation of large agglomerates that grew over the time to an average size of ~185±155 µm, and produced >3 mm thick layer after 120 h at 850 °C. The noble metals decreased the particle size, and therefore significantly slowed down the accumulation rate. Addition of alumina resulted in the formation of a network of spinel dendrites which prevented accumulation of particles into compact layers.

INTRODUCTION

The most common and potentially problematic solids that can form and accumulate in joule heated melters vitrifying high-level radioactive waste (HLW) at the Hanford Waste Treatment Plant are spinel crystals [(Fe, Ni, Mn, Zn)(Fe, Cr)$_2$O$_4$], RuO_2, and their combination. There is a worry that these solids can growth into large particles or form large agglomerates in the glass discharge riser of the melter during a frequent and periodic idling from 20 to 100 days[1,2] when the temperature of the molten glass can drop in some areas to as low as 850 °C. This, together with the stagnant melts and small inner diameter of the riser (~76 mm), provide ideal conditions for accumulation of particles into a few centimeters thick robust layer[3], which may clog the riser and thereby prevent molten glass from being poured into canisters.

A question arises whether it would be possible to control the accumulation rate of spinel crystals through optimization of noble metals concentration in radioactive wastes. A small concentration of noble metals may be sufficient to nucleate enough spinel crystals to limit their growth to a size of 10 µm or less. However, high concentrations of noble metals are not favorable because of their tendency to form large agglomerates[4]. These rapidly settling agglomerates can also form due to interactions of spinel crystals with noble metal particles.[5] Another factor that can greatly affect the accumulation of spinel crystals in glasses containing high concentrations of spinel forming components is an addition of Al_2O_3 or Li_2O. These components tend to promote the formation of a network of large spinel dendrites which prevents building the dense settled layers.[5]

This laboratory study investigated particle agglomeration (size, concentration of agglomerates and their spatial distribution) and accumulation rates in three HLW glass compositions with X-ray microtomography (XMT), scanning electron microscopy (SEM), and image analysis. Also, an attempt was made to measure an average composition of spinel crystals from elemental dot maps that were produced with an energy dispersive spectroscopy (EDS).

EXPERIMENTAL

HLW Glass Compositions

Table I shows compositions for three glasses (Ni1.5/Al10, Ni1.5/Fe17.5, and Ni1.5/Fe17.5/Ru0.015) used in the particle agglomeration study. The varied components were Al_2O_3, Fe_2O_3, NiO, and RuO_2/Rh_2O_3. The remaining components were kept in the same proportions as in the baseline glass.[3] Glass batches were prepared from AZ-101 simulant and additives (H_3BO_3, Li_2CO_3, Na_2CO_3, and SiO_2). Additional Al, Fe, Ni, and Rh were added as Al_2O_3, Fe_2O_3, NiO, and Rh_2O_3. Ruthenium was added in the form of ruthenium nitrosyl nitrate solution drop by drop to SiO_2 that was dispersed on a Petri dish. The SiO_2 cake was dried in oven at 105 °C for 1 hour, quenched, and hand-mixed in the plastic bag with the rest of the glass batch. Then, the glass batch was milled in an agate mill for 5 min to ensure homogeneity. Glasses were produced in Pt-10%Rh crucibles following a two-step melting process: melting of homogenized glass batches at 1200 °C for 1 h and remelting of produced glasses at the same temperature and time after quenching and grinding.

Table I. Composition of glasses in mass fraction of oxides and halogens.

Component	BL	Ni1.5/Al10	Ni1.5/Fe17.5	Ni1.5/Fe17.5/Ru0.015[(a)]
Al_2O_3	0.0821	**0.1000**	0.0784	0.0783
B_2O_3	0.0799	0.0776	0.0763	0.0762
BaO	0.0009	0.0009	0.0009	0.0009
CaO	0.0057	0.0055	0.0054	0.0054
CdO	0.0065	0.0063	0.0062	0.0062
Cr_2O_3	0.0017	0.0017	0.0016	0.0016
F	0.0001	0.0001	0.0001	0.0001
Fe_2O_3	0.1451	0.1409	**0.1750**	**0.1750**
K_2O	0.0034	0.0033	0.0032	0.0032
Li_2O	0.0199	0.0193	0.0190	0.0190
MgO	0.0013	0.0013	0.0012	0.0012
MnO	0.0035	0.0034	0.0033	0.0033
Na_2O	0.1866	0.1812	0.1781	0.1780
NiO	0.0064	**0.0150**	**0.0150**	**0.0150**
P_2O_5	0.0032	0.0031	0.0031	0.0031
SiO_2	0.4031	0.3913	0.3848	0.3846
SO_3	0.0008	0.0008	0.0008	0.0008
TiO_2	0.0003	0.0003	0.0003	0.0003
ZnO	0.0002	0.0002	0.0002	0.0002
ZrO_2	0.0416	0.0404	0.0397	0.0397
Cl	0.0002	0.0002	0.0002	0.0002
Ce_2O_3	0.0020	0.0019	0.0019	0.0019
CoO	0.0001	0.0001	0.0001	0.0001
CuO	0.0004	0.0004	0.0004	0.0004
La_2O_3	0.0022	0.0021	0.0021	0.0021
Nd_2O_3	0.0018	0.0017	0.0017	0.0017
SnO_2	0.0010	0.0010	0.0010	0.0010
Total	1.0000	1.0000	1.0000	1.0000

[a] Added 0.0003 Rh_2O_3 and 1.5E-4 RuO_2

Particle Agglomeration

Figure 1A shows the testing assembly for the lab-scale study of the particle agglomeration. The assembly consisted of an alumina crucible with an outer and inner diameter of 18 and 15 mm, respectively, and height 80 mm that was positioned on the alumina plate inside a Pt-crucible with an inner diameter of 30 mm and height 100 mm. The fabricated glasses were first melted in Pt/10%Rh crucibles at 1200 °C for an hour, and then the crucibles were removed one by one from the melting furnace, and molten glass poured into each of four assemblies that were rested on the 508-mm-diameter platform inside the big Deltech furnace at 850 °C. These assemblies were covered with a lid and removed (air-quenched) at different times up to 5 days. The alumina crucibles were core-drilled from assemblies and were investigated with XMT for the size and distribution of agglomerates. Then, the thin sections were prepared from middle of the crucible for optical microscopy and SEM-EDS observations, and an image analysis.

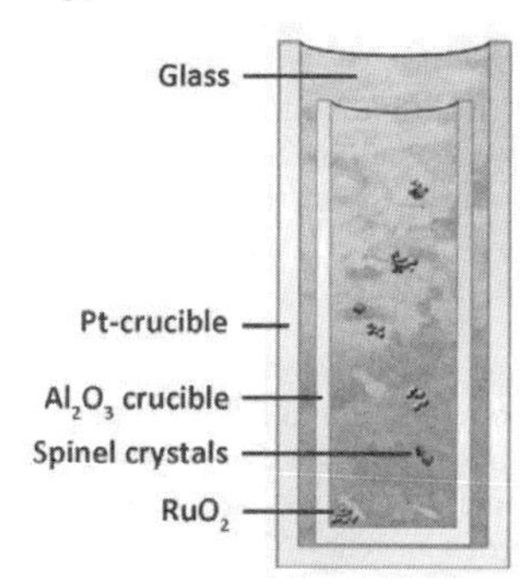

Figure 1A. Cross-section of testing assembly.

X-ray Microtomography

XMT analysis was conducted using an NSI X-View Digital X-ray Imaging and Microfocus Computed Tomography (XMCT) system (North Star Imaging Inc., Rogers, Minnesota). X-rays were generated by a microfocus X-ray source (Comet Feinfocus model 160.48 160 kV) and collected by a PaxScan® 2520V flat panel digital X-ray detector with an active imaging area of 203 × 254 mm. This system is capable of spatial resolution of 6 μm in the focal plane for an object ~6 mm in diameter. Data acquisition and image reconstruction were conducted with X-View IW and efX-ct software's (North Star Imaging Inc., Rogers, Minnesota), respectively.

The X-ray tube voltage and current were adjusted to 110 kV and 340 μA, respectively, to produce an optimal luminosity contrast with values between 3500 and 6500. An X-ray detector response was optimized with dark and light field calibrations (gain or wedge), which were performed with the crucible removed from the field of view to create a uniform background response. The core-drilled alumina crucibles were put in the middle of the rotating stage that was positioned 89.8 mm from the center axis of rotation and X-ray tube, which resulted in a resolution of 18.7 μm (the smallest particle that can be detected). Each crucible was imaged in three separate sections (bottom, middle, and top) at 1 frame per second, with two images collected per 0.25° of rotation over 360°. Subsequently, the off-set calibration of stage with a medium sized NSI calibration tool was performed on each section to correct stage movement errors, and images were reconstructed with efX-ct software.

Two-dimensional (XY) slice reconstructions of the bottom, middle, and top sections of the crucible were segmented with AVIZO Fire 7.0® image processing software (Visualization Science Group, Burlington, MA) to capture location and shape of particles. The size and count of particles was quantified with a three-dimensional volume quantification algorithm. The data were exported into Matlab, where each crucible volume was sectioned into 32x 2.5-mm tall cylinders to obtain size, count, and volume faction of particles as a function of distance from the bottom of the crucible.

Scanning Electron Microscopy and Image Analysis

The YZ thin-sections of entire crucibles were polished, carbon coated and analyzed with JEOL JSM-5900 SEM (SEMTech Solutions Inc., North Billerica, Massachusetts). The operating conditions were 15 kV acceleration voltage and working distance 34 mm. SEM images of rectangle areas (4 × 3 mm for Ni1.5Fe17.5 and Ni1.5/Al 10 glasses, and 1.3 × 1 mm for Ni1.5Fe17.5/Ru0.015 glass) were centered along z-axis and were taken every 5 mm starting from the bottom and moving up to the top of the crucible. Clemex Vision PE 6.0 image analysis software (Clemex Technologies Inc., Québec, Canada) was used to analyze the images for the size and surface area fraction of particles.

Energy-Dispersive X-ray Spectroscopy for Composition Analysis

Elemental dot maps were collected with a SEM JEOL 7001F/TTLS (JEOL Ltd., Tokyo, Japan) equipped with an Energy-Dispersive X-ray Spectroscopy (EDS) silicon drift detector (SDD, 30mm active area). The microscope was setup at working distance of 10 mm at an accelerating voltage of 15kV and probe current of 2-10 μA. The SDD peaking time was between 0.8 - 3.2 μs to maintain a count rate of 150 – 200 kcps and a dead time of 30 - 50%. Maps were collected in a drift corrected mode with a resolution of 1024 × 800 pixels, 200 μs/pixel, and copulation of 256 frames.

RESULTS AND DISCUSSION

Figure 1 shows the spatial distribution of spinel crystals, air voids, and cracks in the two-dimensional XMT cross-sections for Ni1.5/Fe17.5 glass heat-treated at 850 °C for 96 h. These cross-sections are for the box with dimensions of 30 × 7 × 7 mm which was extracted from the center of the bottom section of the crucible. The y-z plane visualizes projection of all crystals onto one plane. The high concentration of nickel and iron in the glass resulted in the formation of large spinel crystals/agglomerates with an average size > 100 μm. These sparsely-distributed particles rapidly settled and accumulated in > 3 mm thick layer.

Figure 2A and B show the size and volume distribution of particles for the entire crucible in the Ni1.5/Fe17.5 glass heat-treated at 850 °C for 24 and 96 h. The low grey-level contrast of XMT in the layer (96 h) resulted in an overestimation of the particle size (> 450 μm) and occupied volume (> 86 %). The image analysis was not able to correctly resolve the individual particles but clumped them together into big clusters. The size of particles and their volume distribution was quite uniform after 24 h up to 10 mm below the glass level, and ranged from 100 to 130 μm and 1-2 vol %, respectively. In contrast, particles of an average size ~185±155 μm were identified in the 30-mm region above the bottom after an additional 72 h of settling. This indicates that the particles grew as they settled with a rate ~1.1 μm/h.

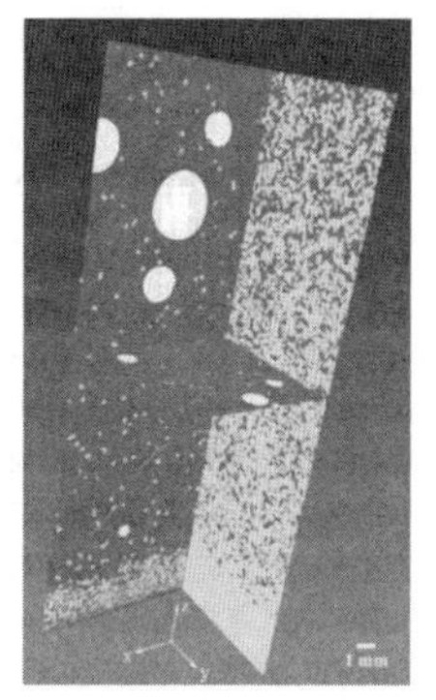

Figure 1. X-ray microtomography cross-sections for Ni1.5/Fe17.5 glass (850 °C - 96 h); glass is blue and gray, spinel crystals are yellow, cracks and air voids are white. All the crystals are projected onto y-z plane.

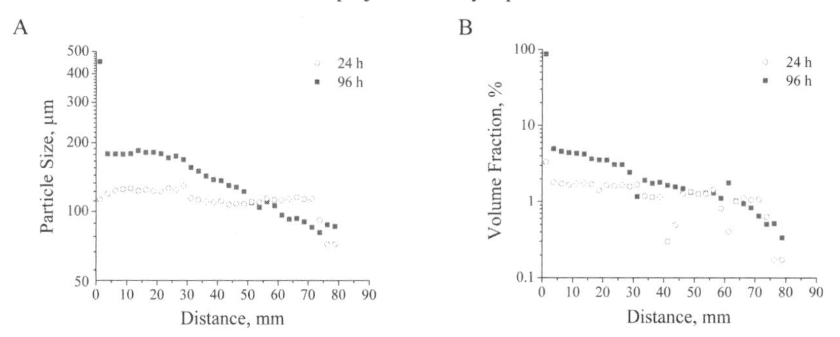

Figure 2. Size (A) and volume distribution (B) of particles vs. distance from the bottom of the crucible (0 mm – bottom, 80 mm – top) for Ni1.5/Fe17.5 glass (850 °C for 24 and 96 h).

Table II shows the distribution of particle sizes as determined with image analysis from SEM images for Ni1.5/Fe17.5, Ni1.5/Fe17.5/Ru0.015, and Ni1.5/Al10 glasses. Figure 3 shows the surface area occupied by the particles as a function of distance from the bottom of the crucible for the same glasses. Addition of Ni and Fe to baseline glass resulted in the formation of large particles that grew over the time, e.g., 90 % of all particles were smaller than 85±56 and 126±111 µm after 24 and 96 h, respectively. These large particles were initially uniformly distributed throughout the sample but got concentrated in ~25-mm region at the bottom after 49 and 96 h. The settling of particles within this region produced compact 0.5- and 3-mm-thick layers with particles covering ~34 % of area, respectively. Figure 4 visualizes the layers accumulated in Ni1.5/Fe17.5 glass after 49 and 96 h at 850 °C. In contrast, addition of noble metals to high-Ni-Fe glass resulted in the uniform distribution of particles with a size below 30 µm (D90) even after 120 h at 850 °C. Because of the smaller particles the accumulated layers were only 79 and 168 µm thick after 72 and 120 h, respectively. Figure 5 shows the layers accumulated in Ni1.5/Fe17.5/Ru0.015 glass after 72 and 120 h at 850 °C. Addition of Al to high-Ni glass produced a network of spinel dendrites with "arms" as long as 1 mm. These dendrite locked structure prevented accumulation of particles into thick layer. The dendrites slowly grew (see Table II) and dendritic network got more compacted over the time (~1 % after 24 h vs. ~3 % after 72 h). Figure 6 shows an example of the 3D network of spinel dendrites that formed in high-Ni-Al glass after 72 h at 850 °C. Figure 7 shows thin sections of dendrites for the same glass heat-treated at 850 °C for 24 and 72 h.

Table II. Distribution of particle sizes for Ni1.5/Fe17.5, Ni1.5/Fe17.5/Ru0.015, and Ni1.5/Al10 glasses heat-treated at 850 °C for different times; D10, D50, and D90 – 10, 50, 90% of crystals are smaller than these values, respectively.

Glass	Time, h	D10, μm	D50, μm	D90, μm
Ni1.5/Fe17.5	24	27±22	51±36	85±56
	96	41±48	74±64	126±111
Ni1.5/Fe17.5/Ru0.015	72	9±6	21±9	28±10
	96	12±6	20±9	27±10
	120	12±8	21±8	29±13
Ni1.5/Al10	24	21±14	36±16	55±24
	48	16±9	41±17	63±27
	72	20±14	47±28	71±38

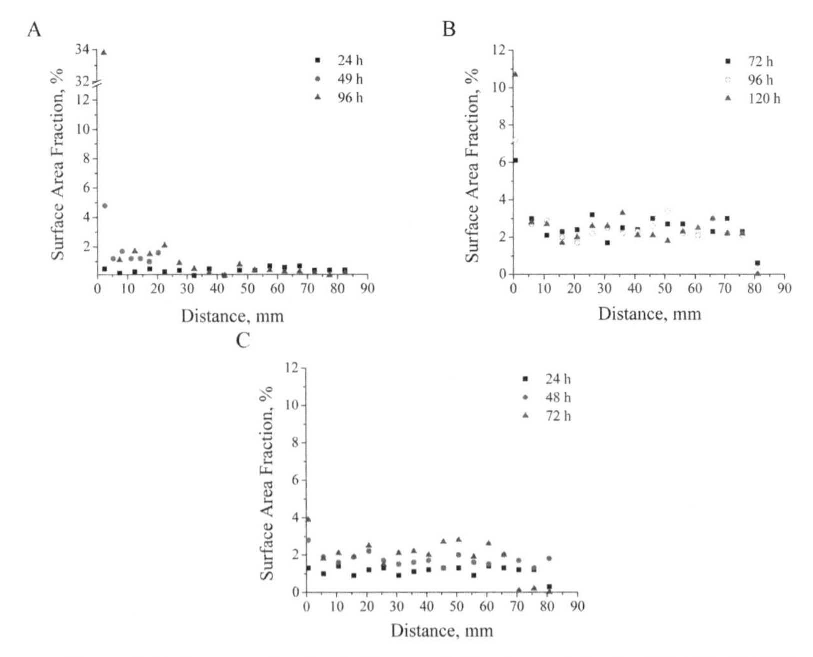

Figure 3. Surface area fraction in % occupied by the particles for Ni1.5/Fe17.5 (A), Ni1.5/Fe17.5/Ru0.015 (B), and Ni1.5/Al10 (C) glasses heat-treated at 850 °C for different times.

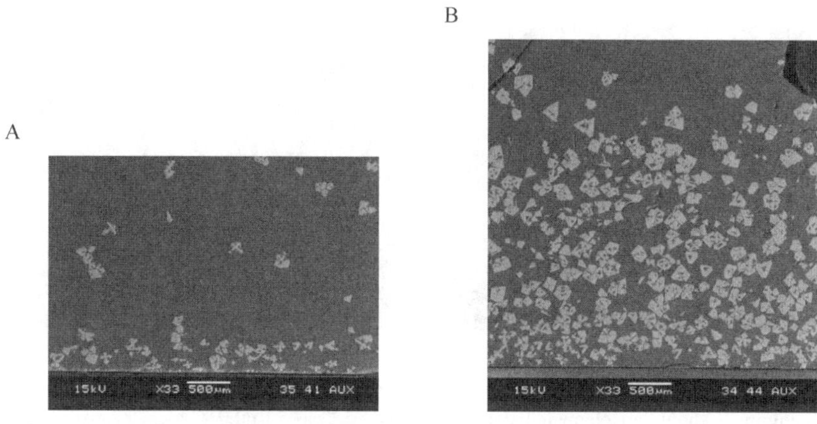

Figure 4. SEM images of accumulated layers for Ni1.5/Fe17.5 glass heat-treated at 850 °C for 49 h (A, 469 μm) and 96 h (B, 3111 μm).

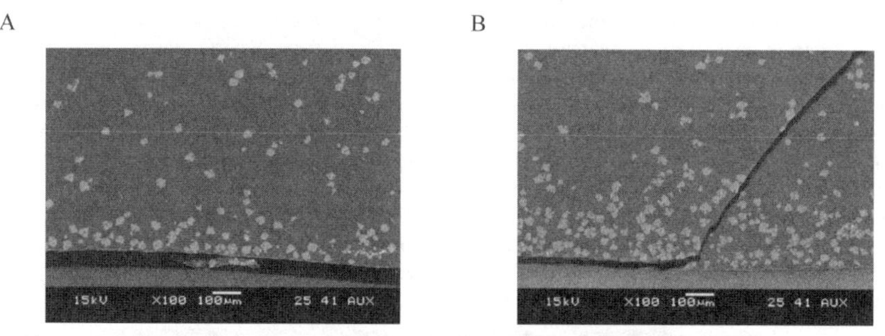

Figure 5. SEM images of accumulated layers for Ni1.5/Fe17.5/Ru0.015 glass heat-treated at 850 °C for 72 h (A, 79 μm) and 120 h (C, 168 μm).

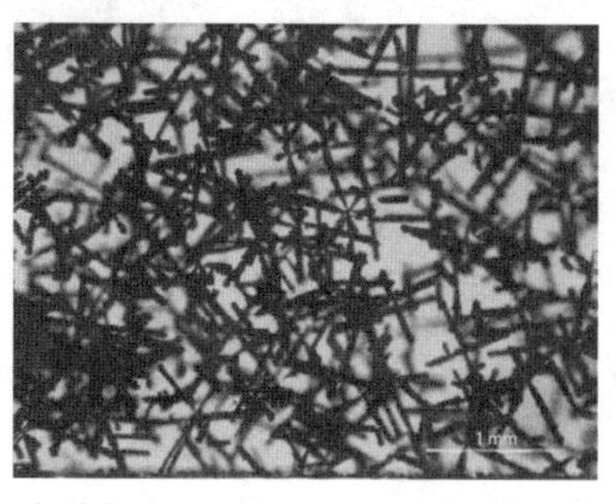

Figure 6. Optical image of spinel dendrite network at the bottom of crucible for Ni1.5/Al10 glass heat-treated at 850 °C for 72 h.

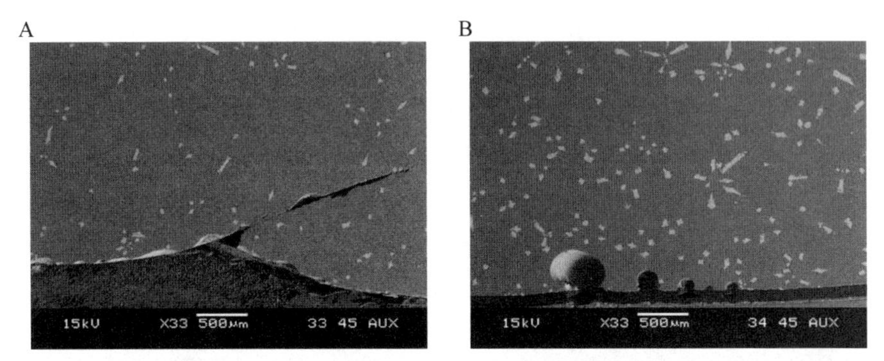

Figure 7. SEM images of dendrites in a 2D plane for Ni1.5/Al10 glass heat-treated at 850 °C for 24 h (A) and 72 h (B). The surface area fraction of dendrites increased with time.

Figure 8 shows an elemental dot map for Ni1.5/Fe17.5/Ru0.015 glass heat-treated at 850 °C for 96 h. Glasses containing high concentration of Ni and Fe, and some noble metals precipitated spinel crystals with a layer enriched with Cr and Na. Chromium is partitioned in the glass as Cr^{3+} and Cr^{6+} in an approximately 50:50 ratio.[6] Since some Cr^{3+} is used for spinel crystals there is an extra Cr^{6+} that can wet the crystals and form $NaCrO_4$ layer. To test this hypothesis, a several higher magnification line scans were made to determine the changes in composition across the crystals. Figure 9 shows a location of line scan through a selected crystal containing a RhO_2 particle. Figures 10A and B visualize changes in the concentration of Na, Cr, Rh, Fe, Ni, and Mn along this line scan. The line scans revealed that higher concentration of Na in the skin of crystals is an artifact from charging but Cr enrichment in ~2.5-μm layer was real. This was confirmed by higher average concentration of Cr in the crystals that formed/grew in Ni1.5/Fe17.5/Ru0.015 glass. Table III shows an approximate elemental composition of spinel crystals (as determined with EDS) in the Ni1.5/Fe17.5, Ni1.5/Fe17.5/Ru0.015, and Ni1.5/Al10 glasses heat-treated at 850 °C at different times.

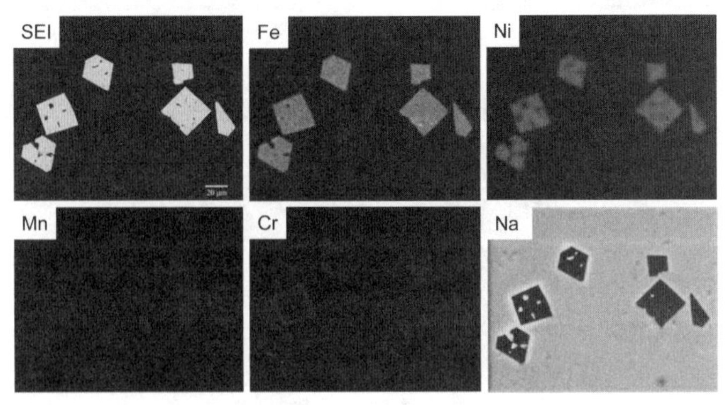

Figure 8. Elemental dot map for Ni1.5/Fe17.5/Ru0.015 glass heat-treated at 850 °C for 96 h.

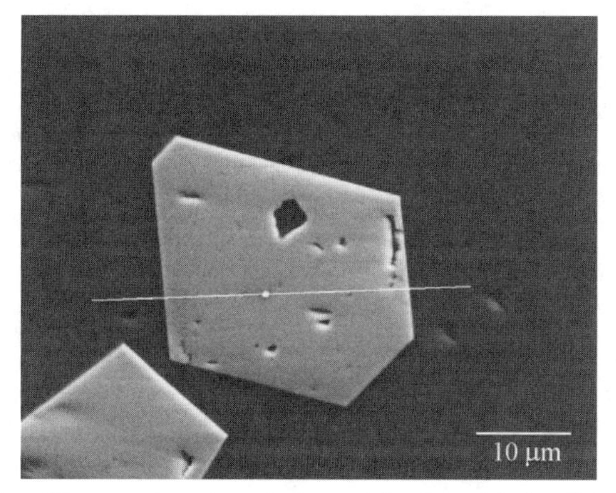

Figure 9. Location of line scan through a spinel crystal that formed in the Ni1.5/Fe17.5/Ru0.015 glass after 72 h at 850 °C.

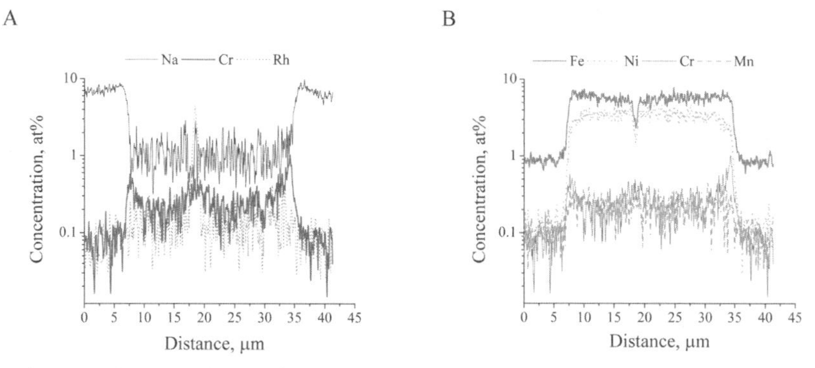

Figure 10. Concentration profiles of Na, Cr, and Rh (A), and Fe, Ni, Cr, and Mn (B) along the line scan in Figure 9.

Table III. Composition of spinel crystals in atomic % for Ni1.5/Fe17.5, Ni1.5/Fe17.5/Ru0.015, and Ni1.5/Al10 glasses heat-treated at 850 °C at different times.

Glass	O	Cr	Mn	Fe	Ni
Ni1.5/Fe17.5-850°C-49h	81.37	0.17	0.11	11.32	7.02
Ni1.5/Fe17.5/Ru0.015-850°C-96 h	81.22	0.42	0.15	11.45	6.75
Ni1.5/Al10-850°C-48 h	82.26	0.35	0.11	10.79	6.49

CONCLUSION

The experimental study of spinel accumulation indicated that high concentrations of spinel-forming constituents in the glass can produce settling layers of a few mm thick in a few days. Adding ~ 0.9 wt% of NiO and ~ 3.4 wt% of Fe_2O_3 to baseline glass resulted in the formation of large particles/agglomerates that got bigger over the time to an average size of ~185±155 μm, and produced >3 mm thick layer after 120 h at 850 °C. The accumulation of these layers can be significantly suppressed by noble metals (RuO_2, RhO_2). These components slowed down or stopped the spinel accumulation because of their effect on decreasing the average crystal size. Adding ~ 0.015 wt% of RuO_2 to high-Ni-Fe glass decreased the crystal size and prevented agglomeration of particles. This resulted in the layer less than 0.2 mm thick after 120 h at 850 °C. Adding of ~ 1.8 wt% Al_2O_3 to high-Ni glass resulted in the formation of a network of spinel dendrites that prevented accumulation of particles into a compact spinel layer. There is a reasonable chance that the spinel crystals locked in this non-compacted layer can be removed with glass during the pouring into canisters. X-ray tomography proved to be a useful non-destructive technique to evaluate an agglomeration of large particles and to provide a spatial distribution of particles in the relatively large glass volumes.

ACKNOWLEDGEMENT

The authors are grateful to Jarrod Crum for help with collection of elemental dot maps and the U.S. Department of Energy's Hanford Tank Waste Treatment and Immobilization Plant Federal Project Office, Engineering Division, for financial support. Pacific Northwest National Laboratory is operated by Battelle for the U.S. Department of Energy under Contract DE-AC05-76RL01830.

REFERENCES

[1]M.E. Smith, D.F. Bickford, The behavior and effects of the noble metals in the DWPF melter system, WSRC-TR-97-00370, Aiken, SC, March 1998.
[2]N.D. Hutson, D.C. Witt, D.F. Bickford, S.K. Sundaram, On the issue of noble metals in the DWPF melter, WSRC-TR-2001-00337, Aiken, AC, August 2001.
[3]J. Matyáš, J.D. Vienna, A. Kimura, M. Schaible, and R.M.Tate, Development of Crystal-Tolerant Waste Glasses, *Ceramic Transactions* 222, 41-51 (2010).

[4]K. M. Fox, D. K. Peeler, T. B. Edwards, D. R. Best, I. A. Reamer, R. J. Workman, J. C. Marra, B. J. Riley, J. D. Vienna, J. V. Crum, J. Matyáš, A. B. Edmondson, J. B. Lang, N. M. Ibarra, A. Fluegel, A. Aloy, A. V. Trofimenko, and R. Soshnikov, International Study of Aluminum Impacts on Crystallization in U.S. High Level Waste Glass, , Savannah River National Laboratory, Aiken, South Carolina, SRNS-STI-2008-00057 (2008).

[5]J. Matyáš, A.R. Huckleberry, C.P. Rodriguez, J.B. Lang, A.T. Owen, and A.A. Kruger, HLW Glass Studies: Development of Crystal-Tolerant HLW Glasses, PNNL-21308 (2012).
[6]H. D. Schreiber, B.K. Kochanowski, C.W. Schreiber, A.B. Morgan, M.T. Coolbaugh, and T.G. Dunlap, Compositional dependence of redox equilibria in sodium silicate glasses, *J. Non-Cryst. Solids* 177, 340-346 (1994).

SYSTEMATIC DEVELOPMENT OF ALKALINE-EARTH BOROSILICATE GLASSES FOR CAESIUM LOADED ION EXCHANGE RESIN VITRIFICATION

O. J. McGann, P. A. Bingham†, N. C. Hyatt
Immobilisation Science Laboratory, Department of Materials Science and Engineering, University of Sheffield, Sheffield, S1 3JD, United Kingdom
† Current institution: Materials and Engineering Research Institute, Sheffield Hallam University, Sheffield, S1 1WB, United Kingdom

ABSTRACT

Caesium loaded ion exchange resin wastes are problematic for vitrification due to their organic nature; the presence of problematic anionic species which lead to phase separation in glasses; and the volatility of caesium at melting temperatures. The presence of a large inventory of radiologically-contaminated ion exchange resins from past, current and future civil nuclear power generation means that the development of a suitable route of vitrification is essential. This paper explores the development of a glass-forming system intended for the purpose of ion exchange resin vitrification, covering systematic studies of compositional variation and the structural and physical effects of these changes, resulting in the development of novel glass compositions optimised for ion exchange resin vitrification.

INTRODUCTION

Radioisotope-loaded ion exchange resins are a problematic form of waste originating from the treatment of aqueous wastes streams from nuclear applications. Due to their organic, acidic and anion-rich nature they are incompatible with most forms of waste immobilisation hosts currently in use for nuclear wastes. Vitrification offers an attractive route for the disposal of these resin wastes since it reduces the waste volume by a significant fraction and produces a durable waste-form. However, their anionic and organic content, as well the volatility of the primary radioisotope Cs-137, mean that current vitrification techniques and materials are not effective for resin disposal [1-2].

The aim of the present work was to explore the effects of compositional variation on one glass selected from a range of candidates: an alkali alkaline-earth borosilicate system. This glass was subjected to a series of compositional modification experiments with the aim of identifying the physical and structural alterations caused, and to build up information from which an optimised glass composition could be developed. The composition of the initial system can be found in Table I. This system was selected due to its capacity to incorporate anionic species, capacity for resisting the reducing effects of organic wastes, superior durability, high density (allowing high waste loading per unit volume) and low liquidus temperature (T_{liq}) allowing waste processing at reasonable temperatures [3-4]. However, it was noted that T_{liq}, although low, is still higher than might be considered ideal for the vitrification of some volatile caesium-containing wastes. In addition, the iron in the system may not all be present as Fe^{3+}, which may compromise the ability of the glass to accommodate reducing wastes. The following experimental series are described and discussed in this paper:

1. The variation of the systems component alkaline-earth species (BaO & CaO). This series was intended to explore the effects of varying the ratio of BaO / CaO and thereby to explore the effects of varying charge density of the alkaline-earth component.
2. The replacement of CaO by other divalent oxides. This series was intended to explore the effects when 5 mol% CaO was replaced by MgO, SrO, MnO or ZnO.

Table I. Nominal glass compositions (mol%)

Component	AE Base
SiO_2	51.72
B_2O_3	2.23
Al_2O_3	0.38
Fe_2O_3	2.92
CaO	11.08
BaO	18.23
Li_2O	7.80
Na_2O	5.64

3.

EXPERIMENTAL PROCEDURES

Sample Preparation

Samples were melted from oxide and carbonate raw materials at 1200 °C for 3 hours in ZrO_2-stabilised Pt crucibles. The samples were stirred for the final 2 hours of the 3-hour melting period, then poured into preheated block moulds and annealed at 500 °C. All samples were homogenous and free of inclusions and bubbles.

Sample Characterisation

Densities were determined using the Archimedes method. The glass transition temperature, T_g, was determined using DTA (Differential Thermal Analysis). DTA experiments were all performed in air using Pt crucibles and measurements were taken automatically by a computer operated system (Perkin Elmer DTA 7).

The liquidus temperature, T_{liq}, was determined using a tube furnace in which a sample was heated across a known temperature gradient. By determining the temperature across the length of the sample (inserted in Al_2O_3 boat of ~10 cm length), T_{liq} was established by optically locating the point where crystal growth occurs and matching that to the temperature at that displacement, producing results accurate to within ± 10 °C.

FT-IR (Fourier Transform Infra-Red) spectroscopy and Raman spectroscopy were both applied to investigate the glass forming network. A Perkin Elmer Spectrum 2000 FT-IR spectrometer was used to collect spectra at 500 – 1500 nm and a Renishaw InVia Raman microscope and 514 nm laser was used to collect spectra over the range 0 – 1800 cm^{-1}. FT-IR results were collected in reflectance and were corrected utilising a software Kramers-Kronig transform. Samples for both techniques were monoliths polished to 1 μm. FT-IR spectra were fitted with Gaussian line-shapes assigned based on the work of MacDonald et al. [5].

^{57}Fe Mössbauer spectroscopy was utilised to determine the oxidation state and coordination of Fe within the samples. A WissEl ^{57}Fe Mössbauer spectrometer was utilised operating in a constant acceleration mode with a ^{57}Co source (activity ~9000 μCi). The results produced were fitted using the software package Recoil 1.03. The recoil free fraction was assumed to be equal in the Fe^{2+} and Fe^{3+} environments.

Solid state NMR of the ^{27}Al nuclei was applied to study the coordination and environment of network forming and network modifying components of the system. Spectra were recorded at 104.20 MHz using a Varian VNMRS 400 spectrometer and a 4mm (rotor o.d.) MAS probe. They were obtained using a cross-polarisation with a 0.5 or 1 s recycle delay and a 1 ms contact time. Between 1000 and 10000 repetitions were accumulated. Spectral referencing was with respect to an external sample of 1M aqueous $AlCl_3$.

RESULTS

Alkaline-Earth Ratio Modification (BaO & CaO)

Glass density was found to vary linearly with increasing BaO replacement of CaO, as may be expected due to the higher molar mass of BaO relative to CaO [6].

Both T_g and the T_{liq} were found to vary in a non-linear manner in proportion to the ratio of BaO and CaO, both achieving a minimum point at a BaO / (BaO + CaO) ratio of between 0.6 and 0.75. The variation in T_g across the full range of compositions was approximately 40 °C. The variation of T_{liq} across the range of samples varied by 260 °C between its maximum point in the series, at full replacement of BaO with CaO, and the minimum point. Both the variation in T_{liq} and T_g are shown in Figures 1 and 2. This observed variation in T_{liq} is in agreement with the ternary SiO_2-BaO-CaO phase diagram published by Toropov et al. [7] which shows a non-linear variation in T_{liq} across the range of compositions explored, although across a smaller range of temperatures than observed here.

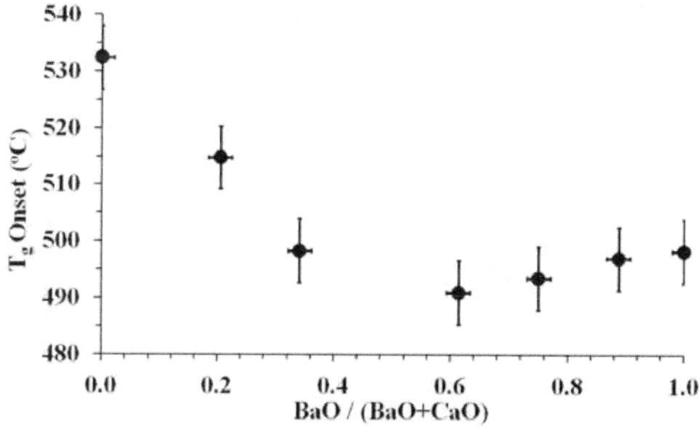

Figure 1. Variation in T_g onset across the BaO:CaO variation series. Error determined from experimental error associated with technique and fitting.

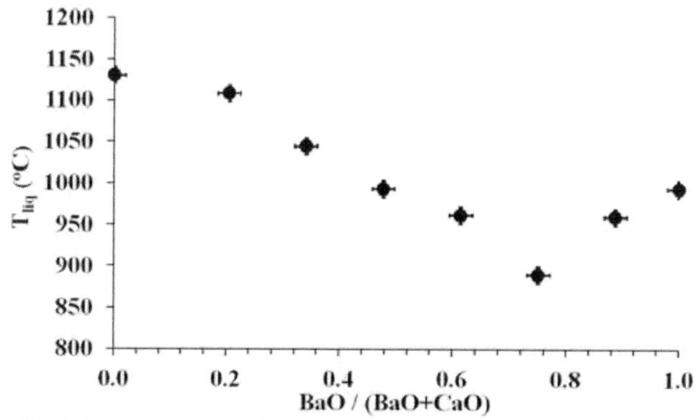

Figure 2. Variation in T_{liq} across the BaO:CaO variation series. Error determined from experimental error associated with technique and fitting.

Mössbauer spectroscopy showed that throughout the series the Fe oxidation state was invariant, $Fe^{3+}/\Sigma Fe$ being 96 ± 2 % Fe^{3+} in all but one sample, showing no variation within the limits of error. The full CaO sample showed a higher extent of reduction at 87.7 ± 2 % Fe^{3+}. This result is consistent with previous results from Mysen et al.[8] and Bingham et al.[9] which link iron redox ratios to optical basicity.

FT-IR spectroscopy showed that the Q-speciation of all the samples varied between two distributions indicative of Q^2 and Q^3 Si tetrahedral units. This can be explained by the relatively large content of alkaline earth elements present in these glasses (BaO and CaO), which has in previous studies been linked to the formation of a strong Q^2 band, associated with combinations of alkali and alkali-earth modifiers coordinated to Si^{4+} tetrahedral units [3,10-11]. In this series it can be seen that this Q^2 band varies in relative intensity and in energy position relative to the Q^3 band which remains unchanged. As the relative content of BaO is increased the Q^2 band increases in intensity and decreases in energy. This is shown in Figure 3 with the spectra themselves and in Figure 4 where the fitted area of the bands is shown, and although the fitted variations are small, they display a general trend consistent with the effects of composition. These results in FT-IR spectroscopy are also consistent with the observations of Stebbins et al.[12] where increases in the proportion if Ba^{2+} cations is linked to increased proportion of NBOs. Negligible variation was apparent, within statistical limits, in FT-IR band below 700cm^{-1}, as can be seen in the Si-O-Si bending band in Figure 4.

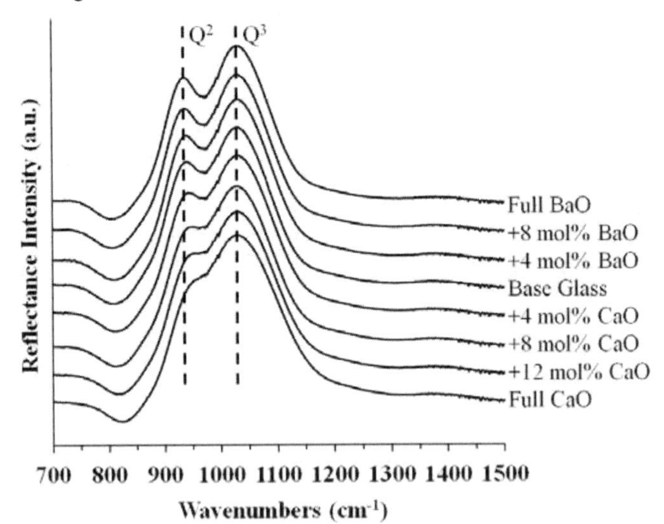

Figure 3. FT-IR spectra of samples from BaO:CaO variation series. Lines serve as guide for eyes, marking Q^2 and Q^3 band positions.

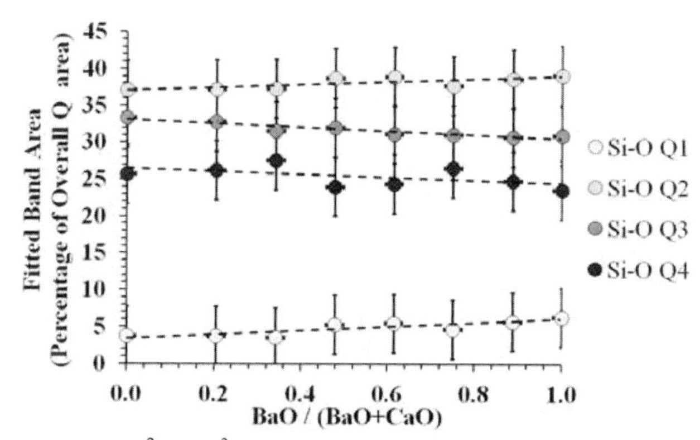

Figure 4. Fitted area of Q^2 and Q^3 bands and the Si-O-Si Bending Mode. Other bands excluded for visualisation. Lines are guide for the eyes only. Error derived from deviation of fitted line-shape.

Replacement of CaO by Divalent Oxides

T_g onset was determined to vary within a narrow range of 17 ± 5 °C depending upon the replacement oxide. The maximum T_g onset is apparent at 506 ± 5 °C in the MgO sample and the minimum T_g onset in the series occurs for the MnO sample at 493 ± 5 °C. This is shown in Figure 5.

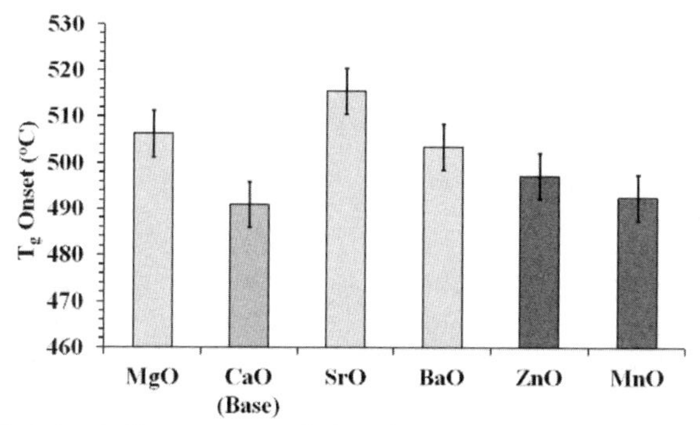

Figure 5. Variation in T_g onset across CaO replacement series. Error determined from experimental error associated with technique and fitting.

The variation in T_{liq} as a result of CaO replacement covers a temperature range of approximately 132 ± 5 °C. All samples, with the exception of the SrO replacement sample, show reductions in T_{liq} as a result of the replacement of CaO. These samples exhibit a reduction in T_{liq} of up to 110 ± 5 °C from the T_{liq} of the base glass composition. This effect is strongest in the ZnO sample, which exhibits a lower T_{liq} even than previously described within the BaO:CaO variation series. This is shown in Figure 6.

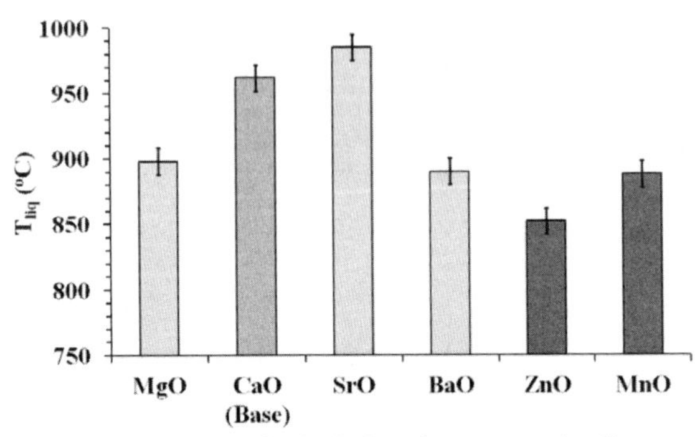

Figure 6. Variation in T_{liq} across samples in CaO replacement series. Errors are derived from experimental errors associated with measurement.

The variation in density is plotted in Figure 7. Marked variations in samples density were observed with a discernible trend being apparent in the ZnO (3.50 ± 0.01 gcm^{-3}) and MnO (3.49 ± 0.01 gcm^{-3}) samples between increased density and higher molecular mass relative to Ca. This connection between density and molecular mass is consistent with the work of Huggins et al [6]. This does not apply in the alkaline-earth samples as the MgO sample shows higher density (3.47 ± 0.01 gcm^{-3}) than the SrO sample (3.41 ± 0.01 gcm^{-3}) despite a lower relative atomic mass and the SrO sample does not show the increased density that might be expected from its higher relative molecular mass.

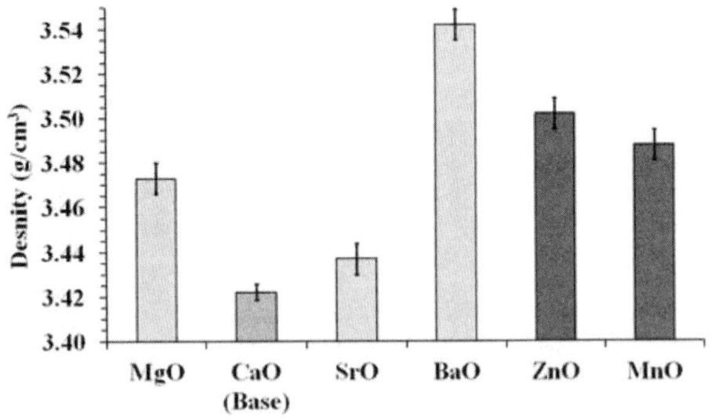

Figure 7. Variation in density across samples in Ca replacement series. Errors derived from standard deviation of three measurements.

Using Mössbauer spectroscopy it can be seen that both of the alkali-earth replacement samples (MgO & SrO) show evidence of partial reduction of Fe^{3+} to Fe^{2+} (5 ± 2 %). In both the MnO and ZnO samples the Fe content is determined to be fully oxidised to Fe^{3+}. This is consistent with the effects of decreases in the optical basicity of the system [8-9]. Figure 8 shows the variation in overall Fe^{2+} content in the samples. Analysis of the coordination of Fe in the glasses, based upon the fitted Mössbauer spectra, indicates that there are two distinct groupings amongst the samples. The MgO, MnO and ZnO replacement glasses all display quadrupole splitting at higher values then observed in either the base glass or the SrO replacement sample.

This implies that Fe^{3+} occupies more distorted environments in the MgO, MnO and ZnO samples.

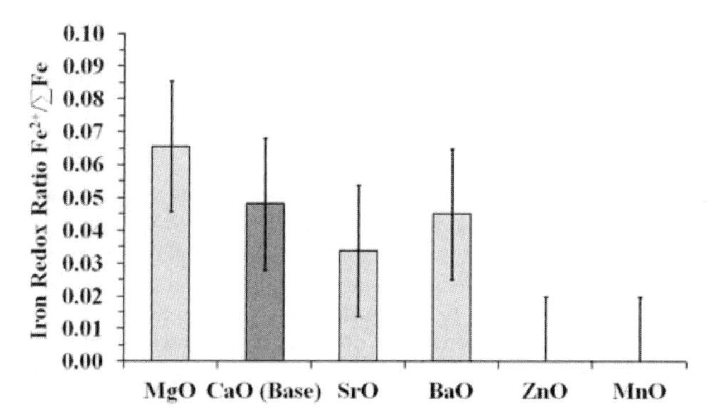

Figure 8. Comparison of Fe oxidation state in across CaO replacement series. Derived from Lorentzian fitting of Mössbauer spectra. Error bar represents three standard deviations, derived from three independent measurements.

FT-IR spectroscopy of the CaO replacement samples shows the presence of the distinct Q^2 and Q^3 bands. Significant variation in the relative intensity of the Q^2 band relative to the Q^3 band is apparent. The Q^2 band can be seen to be most intense in the MgO sample, and least evident in the SrO sample, with the ZnO and MnO replacement sample having a similar Q^2 / Q^3 distribution to the base glass. These results are anomalous and are the opposite to what might have been expected based on the previous series, in which lower period alkaline-earth cations are seen to increase the area of the Q^2 group . The FT-IR spectra are shown in Figure 9.

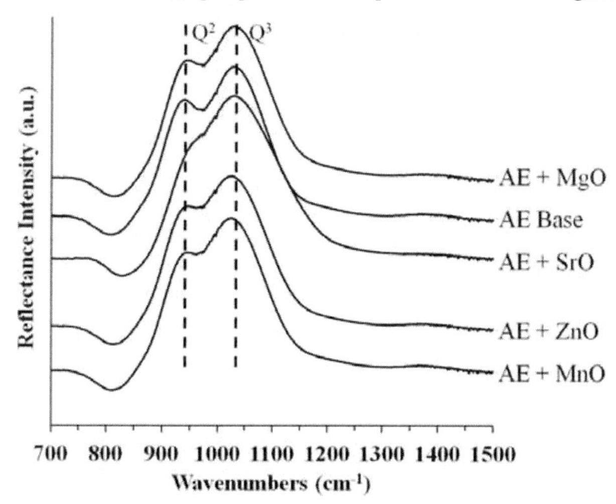

Figure 9. FT-IR spectra of samples from the CaO replacement series.

MAS NMR Analysis

In order to obtain solid state NMR results for the series being studied it was necessary to produce samples which were identical in composition except free from Fe. The glasses were produced using the same techniques as the samples in the original series. Characterisation experiments indicated that all the same trend in physical properties and FT-IR spectra were present in these samples and as such they determined to be representative for the original samples.

Variation in Ratio of Alkaline-Earth Component Species

^{27}Al NMR indicated the presence of Al^{3+} in three coordination states across the series (4, 5 and 6 coordinated Al^{3+}). A non-linear trend in the proportion of $^{[6]}Al^{3+}$ was observed across the series, achieving a maximum point at a ratio of BaO / (CaO + BaO) = 0.75 ± 0.07. This point closely matches the point at which the minimum T_{liq} was observed. In addition a general trend was observed in a small reduction in the proportion of $^{[5]}Al^{3+}$ and a slight increase in $^{[4]}Al^{3+}$ on increasing proportion of BaO. Variation in Al^{3+} coordination is shown in Figure 10. These results are atypical of alkali rich borosilicate compositions [13].

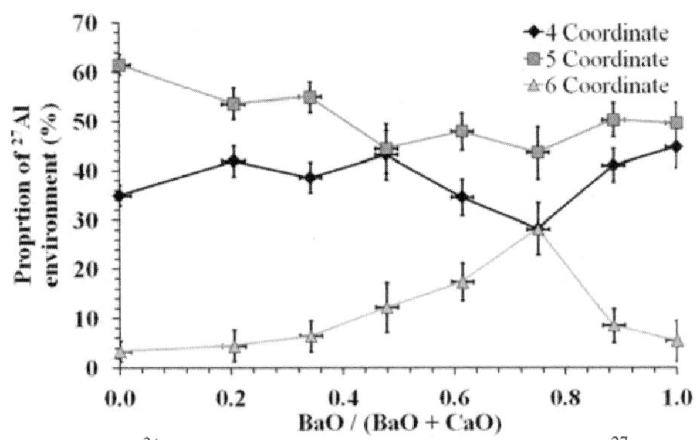

Figure 10. Variation in Al^{3+} coordination as determined from fitting of ^{27}Al MAS NMR with Gaussian line-shapes. Errors derived from systematic error associated with fitting.

Replacement of CaO by Divalent Oxides

^{27}Al NMR indicated a range of Al^{3+} environments were present across the range of the samples. Figure 12 shows the variation in Al^{3+} environments in the series. ZnO and MgO samples both show significant increases in the proportion of $^{[4]}Al^{3+}$ over the initial system. Results of the ZnO sample notably indicate a significantly lower proportion of $^{[6]}Al^{3+}$ than present in other samples within the series, and an increased proportion of $^{[5]}Al^{3+}$. Again an unusually high proportion of $^{[5]}Al^{3+}$ and $^{[6]}Al^{3+}$ is present [13]. Figure 11 shows the variation in Al^{3+} coordination.

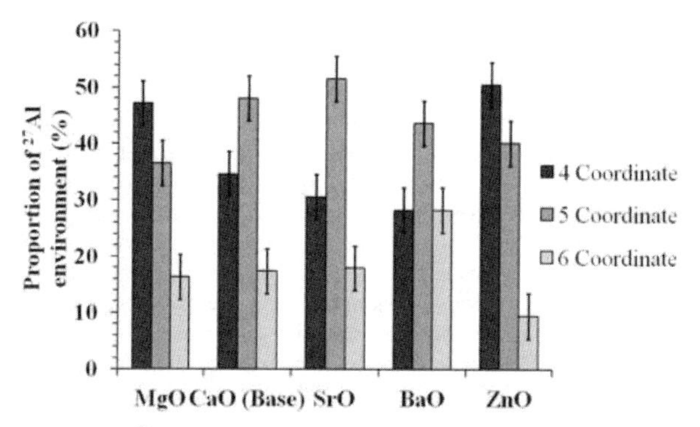

Figure 11. Variation in Al^{3+} coordination as determined from fitting of ^{27}Al MAS NMR with Gaussian line-shapes. Error derived from systematic error associated with fitting.

DISCUSSION

Considering the results from both developmental series separately it can be seen that a number of the sample possess properties that may be advantageous for the vitrification of ion exchange resin wastes.

Considering the first series, the modification of the ratio of alkaline-earth species, it can be seen that the replacement of 4 mol% CaO by BaO in the base composition results in a reduction of T_{liq} by ~80°C relative to the base glass, the minimum value recorded across the series. These changes represent an improvement in the context of ion exchange resin vitrification since the reduced T_{liq} allows for a reduction in the melting temperature which would improve the retention of volatile radionuclides such as Cs-137. The work of Toropov et al. [7] supports this result in that they determined a similar intermediate minimum T_{liq} in a three-phase SiO_2-CaO-BaO system. Our results differ in that the variation is of a greater magnitude, which may be attributable to the additional components present in our glass system. FT-IR spectroscopy and ^{27}Al MAS NMR provide some insights into the origin of this minimum.

Considering Figure 2 it can be seen that on increasing the proportion of BaO in the system, from full CaO content, the T_{liq} decreases in a roughly linear manner. FT-IR spectroscopy provides an explanation for this observation in that fitted FT-IR spectra show an increase in the area of the Q^2 band in proportion to increased BaO content. This increase in Q^2 area at the expense of Q^3 and Q^4 bands indicates an increase in the fraction of non-bridging oxygens (NBOs) in the system. This observation is supported by the work of Lee et al. [15] and Rajyasree et al. [16] who report increased proportion of NBOs in response to increased proportions of Ba^{2+} cations in the system, which are known to be more modifying in silicate glasses than Ca^{2+} cations. Fitted FT-IR spectra also show a decrease in the position of the Q^2 band in proportion to increasing BaO content, indicating lower energy excitations and longer Si-NBO bonds. This observation can also be attributed to the modifying effect of Ba^{2+} cations[15-16]. These results indicate a general decrease in polymerisation and an increase in distortion in the silicate network. This conclusion that increasing BaO content leads to decreased polymerisation, increased network distortion is supported by Eremyashev et al.[17].

^{27}Al NMR experiments showed that the proportion of $^{[6]}Al^{3+}$ achieved a maximum at BaO / (CaO + BaO) ≈ 0.75, closely matching the minimum T_{liq} point. The presence of large proportions of $^{[6]}Al^{3+}$ is atypical of alkali borosilicate systems containing Al^{3+} in which $^{[4]}Al^{3+}$ is more common due to the charge compensation of Al network formers by alkali cations. Higher coordination environments have been reported in glasses where compensation species are not available to Al^{3+}, forcing Al^{3+} to assume distorted $^{[5]}Al^{3+}$ and $^{[6]}Al^{3+}$ coordination states[13,18-19]. The increase in $^{[6]}Al^{3+}$ seen in this series can therefore be correlated to an increase in network distortion. Consequentially it can be seen that the network distortion increases in response to the addition of BaO to the system, which is consistent with the modifying character of Ba^{2+} cations interacting with Ca^{2+} cations, and achieves a maximum point at BaO / (CaO + BaO) $\approx$ 0.75, before sharply decreasing. These variations in network disorder shown by FT-IR and ^{27}Al NMR results may provide some qualitative support for the observed variation in the T_{liq}. The decrease in T_{liq} observed between BaO / (CaO + BaO) $\approx$ 0 and 0.75 can be linked to increasing network disorder due to an increasing number of dissimilar components in glass (Ba^{2+} and Ca^{2+}). It is known that increased disorder frustrates crystallisation mechanisms in some systems, and as such the T_{liq} is decreased. Similarly the reduced disorder observed increasingly from BaO / (CaO + BaO) $\approx$ 0.75 to 1, which can be linked to the gradual removal of one component (Ca^{2+}) from the system, qualitatively supports a decrease in systematic disorder and an increase in T_{liq}.

Considering the second sample series it is apparent that the samples in which 5 mol% CaO was replaced by ZnO and MnO possess significantly lower values of T_{liq} (lower by >100 ^{o}C) then the base glass and furthermore possess a greater extent of oxidised Fe^{3+}. These changes represent significant improvements on the base system in the context of ion exchange resin vitrification since it would allow greater retention of volatiles and an increased capacity for accommodating reduction by the combustion of organic species during melting.

The reduced T_{liq} seen in both ZnO and MnO replacement samples may be attributable to the addition of additional transition metals species into the system, potentially increasing the thermodynamic barrier to crystallisation and reducing the T_{liq}. FT-IR and ^{27}Al NMR results in that FT-IR indicates that the proportion of Si Q^2 units does not change significantly, and ^{27}Al NMR shows a low proportion of $^{[6]}Al^{3+}$ indicating a lower degree of network distortion. The increase in Fe^{3+} content, relative to the base glass and alkaline-earth replacement samples, can perhaps be attributed to charge compensation of the Fe by the transition metal ions; to mutual redox reactions (in the case of MnO); to the increase the optical basicity of the system as a result of the reduction in alkaline-earth content; or to a combination of these factors[8-9]. The increased density seen in both ZnO and MnO samples can be attributed to the increased average molecular mass of the system, using the analysis of density presented by Huggins et al.[6]

The SrO sample presented a significantly increased T_{liq} relative to the base glass, which is unfavourable for ion-exchange resin vitrification. This result is accompanied by FT-IR spectra for the sample which show a strong increase in the Q^3 and Q^4 bands, indicating a more polymerised silicate network. This high degree of polymerisation is not fully explained by the current results, and is not fully consistent with the results expected on the basis of the observed effects of the other alkaline earth oxides MgO, CaO and BaO. The MgO sample presents a decreased T_{liq} and a significantly increased density. ^{27}Al NMR shows that the MgO possess the least distortion of any of the alkaline-earth modified glasses. FT-IR also indicates an increased incidence of Si-O in the Q^3 environment. These observations indicate the MgO system is more polymerised, which supports its increased density. The decreased T_{liq} can be attributed to the effect of the additional alkaline-earth cation increasing the thermodynamic barrier to crystallisation, as with the ZnO and MnO samples.

CONCLUSIONS

This work has demonstrated that through a range of compositional modifications it is possible to produce a glass composition which is close to optimal, in terms of melting and vitrification properties, for the vitrification of ion exchange resin wastes. Specifically it has been shown that increasing the proportion of BaO in the system (4 mol% increase) results in the reduction of the T_{liq} by approximately 120 °C, with a concomitant increase in the density of the glass. Similarly the replacement of 4 mol% CaO with a proportional amount of ZnO result in a reduction in T_{liq} similar to that which was previously described, as well as an increase in density and an increase in the proportion of reduced Fe content in the system. Both of these represent significant improvements to the properties of the initial alkali alkaline-earth borosilicate glass system studied, through relatively minor changes in composition. Through FT-IR, Raman and NMR spectroscopy these changes in the physical properties of the system were studied and linked to changes to the glass network. Further work is required to fully explore the properties of these developed systems, specifically their melt viscosity, crystallisation behaviour, anionic capacity and durability which have not been explored in this work. Beyond the intended application of IEX resin vitrification, the systems explored in the paper may prove beneficial to the vitrification of other problematic waste streams.

ACKNOWLEDGEMENTS

Solid-state NMR spectra were obtained at the EPSRC UK National Solid-state NMR Service at Durham University, UK.
Prof Neil Hyatt is grateful to the NDA and the Royal Academy of Engineering for funding.
This work was supported in part by a CASE award from the EPSRC and Magnox Ltd.

REFERENCES

[1]C. Cicero-Herman, P. Workman K. Poole, D. Erich, J. Harden, Commercial Ion exchange Resin Vitrification Studies, Proc. Int. Symp. Waste Man. Tech. Ceram. Nucl. Ind. WSRC-MS-98-00392.

[2]N. Hutson, C. Crawford, Immobilisation of the Radionuclides from Spent Ion-Exchange Resins Using Vitrification, WM'02 Conference, 1-9 (2002).

[3]C. Kaushik, R. Mishra, P. Sengupta, A. Kumar, D. Das, G. Kale, K. Raj, Barium Borosilicate Glass – A Potential Matrix for Immobilization of Sulfate Bearing High-Level Radioactive Liquid Waste, J. Nucl. Mater., **358**, 129-38 (2006).

[4]P. Bingham, N. Hyatt, R. Hand, Vitrification of UK intermediate Level Radioactive Wastes Arising from Site Decommissioning: Property Modelling and Selection of Candidate Host Glass Composition, Eur. J. Glass Sci. Technol. A: Glass Technol. **18,** 83-100 (2012).

[5]S. MacDonald, C. Schardt, D. Masiello, J. Simmons, Dispersion Analysis of FTIR Reflection Measurements in Silicate Glasses, J. Non-Cryst. Solids, **275**, 72-82 (2000).

[6]M. Huggins, K. Sun, Calculations of Density and Optical Constant of a Glass from its Compositions in Weight Percentage, J. Am. Ceram. Soc., **26,** 4-11 (1943).

[7]N. Toropov, F. Galakhov, I. Bondar, The Diagram of the State of the Ternary System BaO-Al_2O_3-SiO_2, Izvest. Akad. Nauk SSSR Doklady, **5,** 647-55 (1956).

[8]B. Mysen, D. Virgo, F. Seifert, Redox Equilibria of Iron in Alkaline Earth Silicate Melts: Relationships Between Melt Structure, Oxygen Fugacity, Temperature and Properties of Iron-Bearing Silicate Liquids, Am. Mineral., **69**, 834-47 (1984).

[9]P. Bingham, J. Parker, T. Searle, J Williams, I. Smith, Novel Structural Behaviour of Iron in Alkali-Alkaline-Earth-Silica Glasses, C. R. Chimie., **5**,787-96 (2002).

[10]J. Ramkumar, S. Chandramouleeswaran, V. Sudarsan, R. Mishra, Barium Borosilicate Glass as a Matrix for Uptake of Dyes, J. Hazard. Mater., **172,** 457-64 (2009).

[11]R. Mishra, V. Sudarsan, C. Kaushik, K. Raj, S. Kulshreshtha, A. Tyagi, Effect of Barium on Diffusion of Sodium in Borosilicate Glass, J. Non-Cryst. Solids, **156,** 129-34 (2008).

[12]J. Stebbins, J. Oglesby, Z. Xu, Disorder Among Network-Modifier Cations in Silicate Glasses: New Constraints from Triple-Quantum ^{17}O NMR, Am. Mineral., **82**, 1116-24 (1997).

[13]M. Toplis. B. Dingwell, L. Tommaso, Peraluminous Viscosity Maxima in Na_2O-Al_2O_3-SiO_2 Liquids: The Role of Triclusters in Tectosilicate Melts, Geochim. Cosmochim. Acta, **61**, 2605-12 (1997).
[14]F. Angeli, J. Delaye, T. Charpentier, J. Petit, D. Ghaleb, P. Faucon, Influence of Glass Chemical Compositions on the Na-O Bond Distance: A ^{23}Na 3Q MAS NMR and Molecular Dynamics Study, J. Non-Cryst. Sol., **276**, 132-44 (2000).
[15]S. Lee, B. Mysen, G. Cody, The Effect of Na/Si on the Structure of Sodium Silicate and Aluminosilicate Glasses Quenched from Melts at High Pressure: A Multi-Nuclear (Al-27, Na-23, O-17) 1D and 2D Solid-State NMR Study, Phys. Rev. B, **229,** 162-72 (2006).
[16]C.Rajyasree, K. Rao, Spectroscopic Investigations on Alkali Earth Bismuth Borate Glasses Doped with CuO, J. Non-Cryst. Sol., **357** 836-41 (2011).
[17]V. Eremyashev, A. Osipov, L. Osipova, Borosilicate Glass Structure with Rare-Earth-Metal Cations Substituted for Sodium Cations, Glass Ceram., **68**, 3-6 (2011).
[18]P. Zhao, S. Kroeker, J. Stebbins, Non-Bridging Oxygen Sites in Barium Borosilicate Glasses: Results from 11B and 17O NMR, J. Non-Cryst. Solids **276**, 122-31 (2000)
[19]B. Mysen, D. Virgo, I. Kushiro, The Structural role of Aluminium in Silicate Melts – a Raman Spectroscopic Study at 1 atm, Am. Mineral., **66,** 678-701 (1981).
[20]J. Shelby, Effect of Morphology on the Propertries of Alkaline Earth Silicate Glasses, J. Appl. Phys., **50**, 8010-15, (1979).

EFFECT OF TEMPERATURE ON THE CREVICE CORROSION SUSCEPTIBILITY OF PASSIVATING NICKEL BASED ALLOYS

Edgar C. Hornus[a], C. Mabel Giordano[a], Martín A. Rodríguez[a,b], Ricardo M. Carranza[a], Raul B. Rebak[c]

[a] Gerencia Materiales, Comisión Nacional de Energía Atómica, Instituto Sabato, UNSAM/CNEA
[b] Consejo Nacional de Investigaciones Científicas y Técnicas, Buenos Aires, Argentina
[c] GE Global Research, Schenectady, NY 12309, USA

ABSTRACT

Nickel based alloys 625 (Ni-21Cr-9Mo-3.7(Nb+Ta)), C-22 (Ni-22Cr-13Mo-3W), C-22HS (Ni-21Cr-17Mo) and HYBRID-BC1 (Ni-22Mo-13Cr) were tested for localized corrosion resistance in chloride solutions at temperatures ranging from 20°C to 100°C. For each alloy, the repassivation potential linearly decreased with the temperature. The crevice corrosion resistance of the alloys correlated well with the PREN (Pitting Resistance Equivalent Number). Crevice corrosion was observed at temperatures lower than the literature reported critical crevice temperature for each alloy. The finding of a critical potential below which crevice corrosion does not occur at any temperature and chloride concentration is discussed. The application of the results for industrial and long term applications is also discussed.

INTRODUCTION

Passivating nickel based alloys are considered among the candidate materials for engineered barriers of nuclear waste repositories. The projected lifetime of the main engineered barrier, which is the waste container, spans over 10,000 years. Materials for the construction of the containers should have good corrosion resistance. [1, 2] Nickel alloys bearing chromium and molybdenum have excellent resistance to general and localized corrosion. [3] Localized corrosion, in the forms of pitting and crevice corrosion, is among the degradation processes that will limit the lifetime of engineered barriers. [1, 2] The growth of a corrosion pit generates a morphology that is essentially a crevice, so active pits and crevices are basically identical in behavior. However, crevice corrosion occurs in occluded regions, and it may be stabilized at lower potentials than pitting corrosion. [4, 5] Crevices may occur due to the presence of solid deposits, corrosion products, etc. The crevice corrosion susceptibility of an alloy in a given environment is measured by the value of the repassivation potential ($E_{R,CREV}$). The lower the $E_{R,CREV}$ for a given alloy means that the environment is more aggressive. [4]

The PREN (Pitting Resistance Equivalent Number) is frequently used as an indicative measure of Cr-passivating alloys resistance to localized corrosion. [4] For nickel alloys, PREN is defined in Equation 1 as a function of weight percentages of the alloying elements Cr, Mo and W. A recent work shows that PREN is a good measure of the crevice corrosion resistance of nickel passivating alloys. [6]

$$PREN = \%Cr + 3.3(\%Mo + 0.5\%W) \quad (1)$$

The study of the effect of temperature on the corrosion resistance of the alloy is important since the containers will pass through a temperature gradient caused by the heat released from radioactive decay. [1] The temperature has a strong effect on the crevice corrosion resistance of nickel based alloys and stainless steels. [4-9] It has been reported that there is approximately a linear relationship between the repassivation potential and the temperature. [7] The repassivation potential decreases as the temperature increases. The critical temperature above which localized corrosion may occur varies between 30°C and 60°C for Ni-Cr-Mo alloys. [4] This parameter may differ substantially depending on the experimental method employed for measuring crevice corrosion. Limited systematic research on the influence of the temperature on the crevice corrosion of passivating alloys is available. Some authors indicate that once the crevice corrosion initiates at a given temperature, it may propagate at lower temperatures. They suggest that temperature has an effect similar to potential on the crevice corrosion susceptibility. [8]

The objective of this work was to assess the effect of the temperature on the crevice corrosion resistance of passivating nickel based alloys.

EXPERIMENTAL

The chemical compositions of the alloys in weight percent are listed in Table I. The alloys were used in the mill annealed condition. Electrochemical tests were conducted in a one-liter, three-electrode vessel. The temperature of the solution was controlled by immersing the cell in a water bath, which was kept at a constant temperature. Nitrogen was purged through the solution 1 hour prior to testing and it was continued throughout the entire test. The reference electrode was a saturated calomel electrode (SCE), which has a potential of 0.242 V more positive than the Standard Hydrogen Electrode (SHE) at room temperature. The counter electrode consisted of a platinum foil spot-welded to a platinum wire (with a total area of 50 cm^2 approximately). All the potentials are reported in the SCE scale.

Table I. Chemical composition of the tested alloys in weight percent.

Alloy	Ni	Cr	Mo	W	Fe	Co	Si	Mn	C	V	Al	B	Nb+ Ta
625	62	21	9	0	5	1	0.5	0.5	0.1	0	0.4	0	3.7
C-22	56	22	13	3	3	2.5	0.08	0.5	0.01	0.35	0	0	0
C-22HS	61	21	17	1	2	1	0.08	0.8	0.01	0	0.5	0.006	0
HYBRID-BC1	62	13	22	0	2	0	0.08	0.25	0.01	0	0.5	0	0

Prism crevice assemblies (PCA) specimens were used in the crevice corrosion tests. They were fabricated based on ASTM G 48, [10] and contained 24 artificially creviced spots formed by a ceramic washer (crevice former) wrapped with a PTFE tape (ASTM G192-08). [10] The total surface area of the PCA specimen immersed in the electrolyte was 14 cm^2. The applied torque to the crevicing washers was 5 N-m. The specimens had a finished grinding of SiC abrasive paper number 600 and they were degreased in acetone and washed in distilled water just before the tests. The tests were performed by duplicate or triplicate. The experiments were conducted at temperatures ranging from 20°C to 100°C in a 5 mol/L calcium chloride ($CaCl_2$) solution. The tests were performed at ambient pressure.

The crevice corrosion repassivation potential was determined by the Potentiodynamic-Galvanostatic-Potentiodynamic (PD-GS-PD) method. [9, 11] It consists of three stages: (1) a potentiodynamic polarization (at a scan rate of 0.167 mV/s) in the anodic direction until reaching a total anodic current of 30-300 μA, (2) the application of a constant anodic current of I_{GS} = 30-300 μA (approximately i_{GS} = 2-20 μA/cm^2) for 2 hours, and (3) a potentiodynamic polarization (at 0.167 mV/s) in the cathodic direction, from the previous potential until reaching alloy repassivation.

The general corrosion of the tested alloys was studied in hydrochloric acid (HCl) solutions simulating the conditions within active crevices. Non creviced prismatic specimens were used to measure general corrosion rates. The total surface area of the prismatic specimen immersed in the electrolyte was 10 cm^2. All the specimens had a finished grinding of SiC abrasive paper number 600 and they were degreased in acetone and washed in distilled water. Polishing was performed 1 hour prior to testing. The corrosion potential (E_{CORR}) was monitored for 2 hours of each alloy in 1 and 3 mol/L HCl for eight different temperatures (30°C-100°C).

RESULTS

Crevice corrosion in 5 mol/L $CaCl_2$

Figure 1 shows the determination of the crevice corrosion repassivation potential of alloy HYBRID-BC1 in 5 mol/L $CaCl_2$ at 80°C. $E_{R,CREV}$ was determined at the intersection of the forward (stage 1) and reverse (stage 3) scans (Fig. 1). The drop in potential during the galvanostatic step (stage 2) indicated crevice corrosion propagation. The current density of the galvanostatic step was 2 $\mu A/cm^2$ for the tests at temperatures of 70°C to 100°C, and it was 20 $\mu A/cm^2$ for tests at temperatures of 20°C to 60°C. This variation of the galvanostatic current applied does not significantly affect the value of $E_{R,CREV}$.[11]

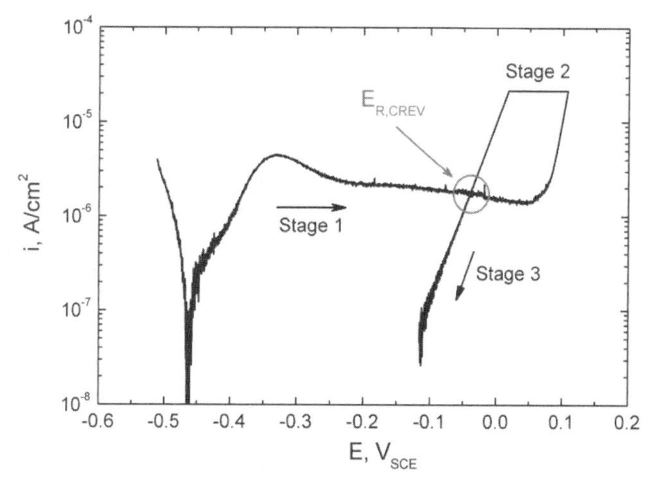

Figure 1. PD-GS-PD test for alloy HYBRID-BC1 in 5 mol/L $CaCl_2$ at 80°C.

Figure 2 shows $E_{R,CREV}$ obtained by the PD-GS-PD method as a function of temperature (T) for the four tested alloys. At each temperature, the crevice corrosion resistance of the alloys increased in the following order: 625 < C-22 < C-22HS < HYBRID-BC1, which is in agreement with their PREN (Table II). Alloy HYBRID-BC1 suffered crevice corrosion only at $T \geq 60$°C. Alloys C-22 and C-22HS suffered crevice corrosion only at $T \geq 40$°C. Alloy 625 suffered crevice corrosion in the entire studied temperature range, from 20 to 100°C. No tests were performed at temperatures below 20°C, thus a lower limit of temperature for crevice corrosion of Alloy 625 could not be obtained. Table II shows the Critical Crevice Temperature (CCT) values obtained from the literature [12] for the tested alloys and those inferred from the present work. CCT is usually determined by 72 hours of exposure in acidified 6% $FeCl_3$ (ASTM G 48, Methods D and C).[10] The literature reported CCT values are considerable higher than those obtained in the current work. However, the trend of CCT with PREN is the same in both cases. The differences among CCTs may be due to the different testing methods and/or crevicing mechanisms used.

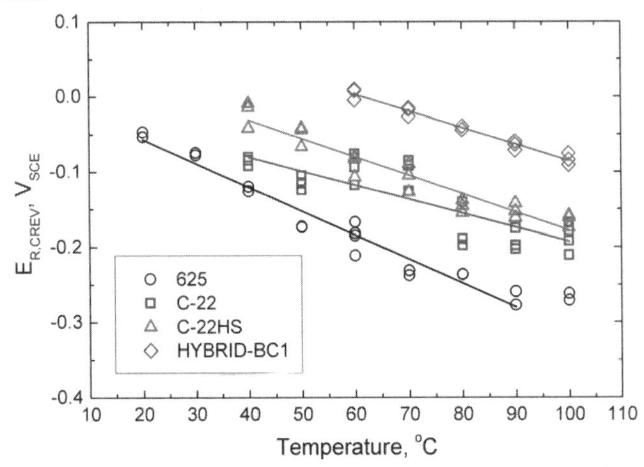

Figure 2. $E_{R,CREV}$ as a function of the temperature for alloys 625, C-22, C-22HS and HYBRID-BC1 in $CaCl_2$ 5 mol/L. Symbols: data; lines: linear fit.

Table II. Critical crevice temperature (CCT) from literature and from the present work.

Alloy	PREN	CCT literature[12]	CCT this work
625	50.7	40°C	CCT < 20°C
C-22	69.9	80°C	30°C ≤ CCT < 40°C
C-22HS	78.8	CCT > 85°C	30°C ≤ CCT < 40°C
HYBRID-BC1	87.6	CCT > 85°C	50°C ≤ CCT < 60°C

$E_{R,CREV}$ decreased linearly with temperature for the tested alloys, as reported in the literature. [7] The fit parameters obtained by linear least squares are listed in Table III. Good correlation coefficients (R^2) were obtained in most of the cases. However, alloy C-22 showed a low value of R^2 which was attributed to the use of specimens from different heats.

Table III. Fit parameters for $E_{R,CREV}$ as a function of T.

$E_{R,CREV} = A\,T + B$	A (V/°C)	B(V_{SCE})	R^2
Alloy 625	-0.0029	-0.007	0.937
Alloy C-22	-0.0019	-0.005	0.694
Alloy C-22HS	-0.0025	0.068	0.911
Alloy HYBRID-BC1	-0.0022	0.135	0.969

Figures 3 and 4 show alloys 625 and HYBRID-BC1 after testing in $CaCl_2$ 5 mol/L at different temperatures. The attack started at the crevice former / metal interface and progressed outward towards the non-occluded surface. As long as the temperature increased the attack spread more to the non-occluded surface. The spread of the localized attack was higher for alloy 625 when compared to alloy HYBRID-BC1 (Figs. 3 and 4). In general, the lower the temperature and the higher the PREN the narrower and deeper the attack was. Figure 5 shows scanning electron microscope images of crevice corroded alloys C-22HS (5a) and HYBRID-BC1 (5b) tested in $CaCl_2$ 5 mol/L at 70°C. The type of attack was shiny crystalline. [13] The morphology shown in Figure 5 is representative of the attack suffered by the four studied alloys at the different temperatures.

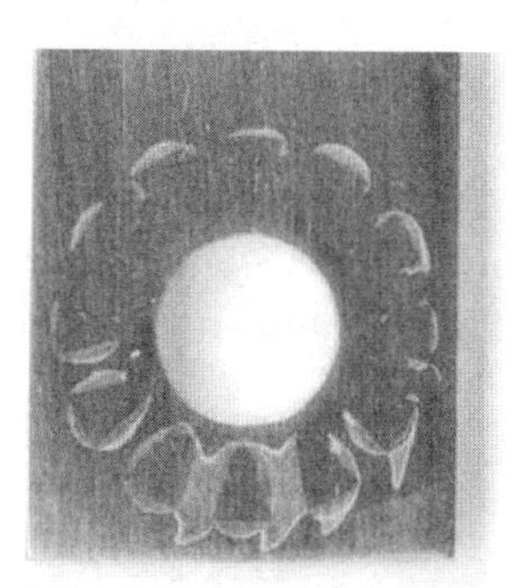

(a) (b)

Figure 3. Alloy 625 tested in $CaCl_2$ 5 mol/L at (a) 40°C and (b) 70°C. For size reference the diameter of the hole is 7 mm (see also ASTM G 192). Mag ~3X

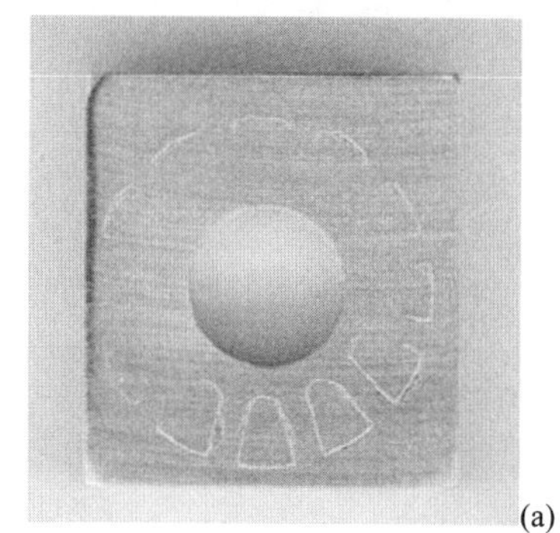

(a) (b)

Figure 4. Alloy HYBRID-BC1 tested in $CaCl_2$ 5 mol/L at (a) 60°C and (b) 100°C. For size reference the diameter of the hole is 7 mm (see also ASTM G 192). Mag ~3X

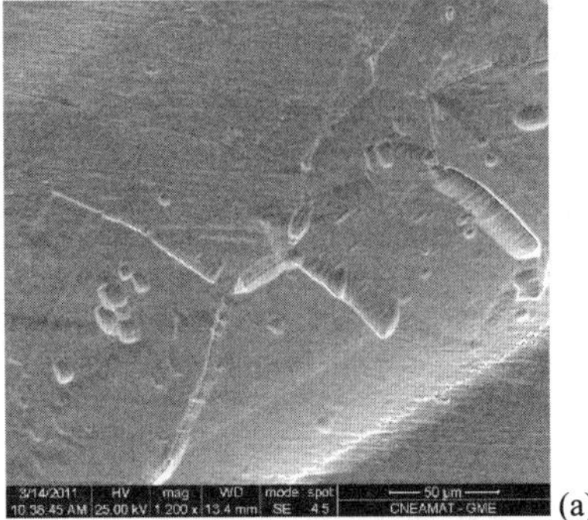
(a)

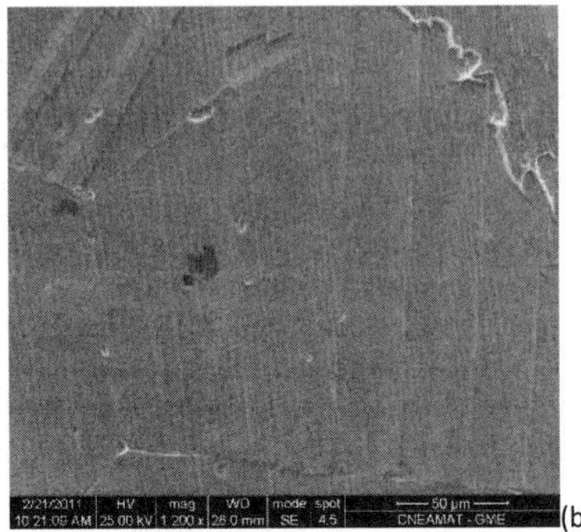
(b)

Figure 5. Crevice corrosion attack on alloys (a) C-22HS and (b) HYBRID-BC1 tested in $CaCl_2$ 5 mol/L at 70°C.

General corrosion in HCl

The corrosion potential (E_{CORR}) of the tested alloys was measured for 2 hours at different temperatures in HCl solutions. These measurements were performed in 1 mol/L HCl for alloy 625 and in 3 mol/L HCl for alloys C-22, C-22HS and HYBRID-BC1. The selected acidic solutions simulate the conditions within active crevices. Figure 6 shows E_{CORR} after 2 hours of immersion as a function of temperature. E_{CORR} of the alloys slowly decreased as the temperature increased, until it reached a relatively stable value. Alloy HYBRID-BC1 showed the highest E_{CORR} values, followed by alloy C-22HS and C-22. Alloy 625 showed the lowest E_{CORR} values.

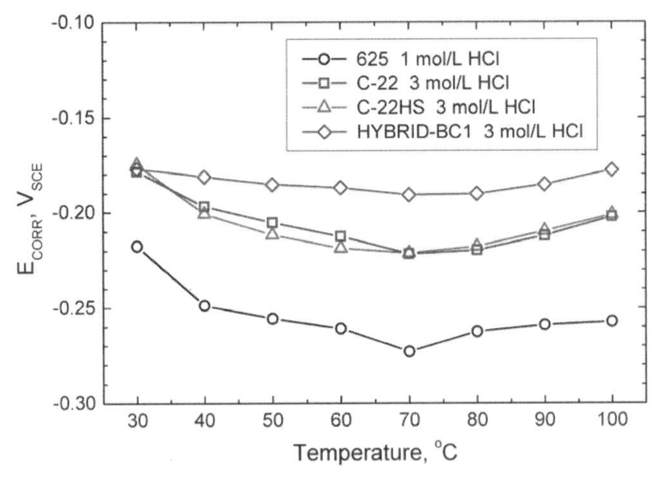

Figure 6. Corrosion potential of the tested alloys after 2 hours of immersion in HCl, as a function of the temperature.

Localized acidification model

According to the localized acidification model, the repassivation potential is the sum of three contributions: the corrosion potential of the alloy in the locally acidic solution (E_{CORR}*), a polarization necessary to locally sustain the critical acidity (η) and the ohmic potential drop (ΔΦ). [5] This is stated in Equation 2. The terms η + ΔΦ are negligible for sufficiently high chloride concentrations and temperatures. In these conditions Equation 2 becomes Equation 3. This is, from a theoretical viewpoint, the lowest potential at which the localized corrosion is able to proceed ($E_{R,CREV}^{MIN}$).

$$E_{R,CREV} = E_{CORR}^{*} + \eta + \Delta\Phi \quad (2)$$
$$E_{R,CREV}^{MIN} = E_{CORR}^{*} \quad (3)$$

The critical crevice solution is more aggressive for more corrosion resistant alloys. [4] In the present case, E_{CORR}* was estimated from the average E_{CORR} for T > 50°C of alloys HYBRID-BC1, C-22HS and C-22 in 3 mol/L HCl; and it was estimated from the average E_{CORR} for T > 50°C of alloy 625 in 1 mol/L HCl (Fig. 6). As long as the temperature increases, η + ΔΦ from Equation 2 decreases until it becomes nil and Equation 3 is valid. In order to estimate the conditions at which Equation 3 is valid, E_{CORR}* and $E_{R,CREV}$ were extrapolated to higher temperatures. The fitting equations obtained for $E_{R,CREV}$ as a function of T were used for the extrapolation (Fig. 2 and Table III). The extrapolations of $E_{R,CREV}$ and E_{CORR}* are given in Figure 7, and the potential ($E_{R,CREV}^{MIN}$) and temperature (T^{MIN}) corresponding to the intersections for each alloy are listed in Table IV. $E_{R,CREV}^{MIN}$ and T^{MIN} increased with PREN. T^{MIN} for alloy 625 is within the studied range, while T^{MIN} for alloys C-22 and C-22HS were slightly above the studied temperature range. T^{MIN} for alloy HYBRID-BC1 was far from the studied range. $E_{R,CREV}^{MIN}$ and T^{MIN} listed in Table IV for alloys HYBRID-BC1, C-22HS and C-22 are approximate values since they derive from extrapolations.

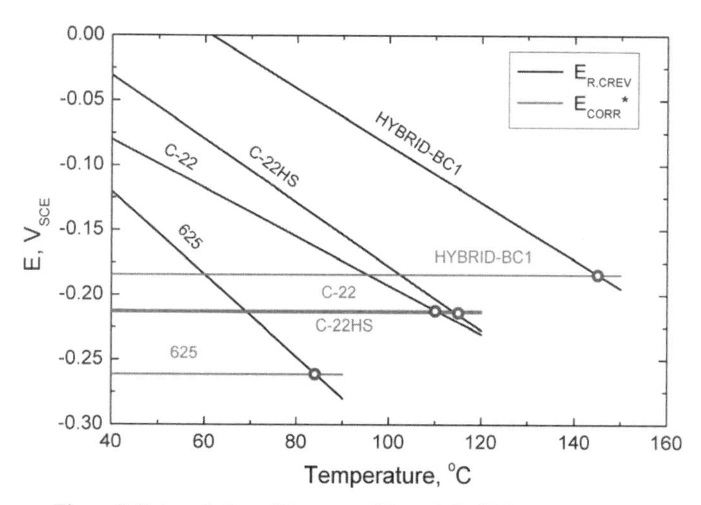

Figure 7. Extrapolation of $E_{R,CREV}$ and E_{CORR}* for higher temperatures.

Table IV. $E_{R,CREV}^{MIN}$ and T^{MIN} corresponding to the intersections of $E_{R,CREV}$ and E_{CORR}* in Figure 7.

Alloy	$E_{R,CREV}^{MIN}$ (V_{SCE})	T^{MIN} (°C)
625	-0.261	85
C-22	-0.212	110
C-22HS	-0.214	115
HYBRID-BC1	-0.184	145

Applicability of the results

Localized corrosion may occur only if E_{CORR} is higher than the $E_{R,CREV}$ for the alloy in the considered conditions. $E_{R,CREV}$ is a function of many environmental and metallurgical variables. E_{CORR} is also a function of many variables but it is mainly affected by the oxidizing power of the solution. [13]

$E_{R,CREV}^{MIN}$ is insensitive to environmental variables, such as temperature and chloride concentration and it is also independent from geometrical variables, such as the crevicing mechanism, crevice former material and type, etc. Localized corrosion is not expected to occur below $E_{R,CREV}^{MIN}$ in any environmental condition. [13] Consequently, $E_{R,CREV}^{MIN}$ is a strong parameter for assessing the localized corrosion susceptibility of nickel alloys and other Cr-passivating alloys in a long term timescale, such as engineered barriers of nuclear repositories. However, the value of $E_{R,CREV}^{MIN}$ is extremely conservative and its use for industrial application would limit the use of most of nickel alloys, even in mildly oxidizing environments. The $E_{R,CREV}^{MIN}$ criterion may be useful for long term applications of nickel passivating alloys in non-oxidizing environments.

The high aggressive conditions needed in the development of the $E_{R,CREV}^{MIN}$ criterion might not be established in the large timescales of a nuclear repository. There are many conditions to fulfill for enabling a crevice-corrosion propagation large enough to perforate the wall a waste container. To fully dissolve a patch of a through-wall container, a creviced region with a high hydrochloric acid concentration and a large diffusion path should form on the alloy surface and be maintained during long times. It is not known if such a severe crevice may be formed under debris or deposits. In-service crevices have not been characterized. In general, the geometry of actual crevices has not been studied in detail. [11] A cathodic reaction occurring at a sufficiently high rate is necessary to sustain the localized corrosion in the creviced area. Passive films may inhibit certain cathodic reactions, this is the case of oxygen reduction on alloy C-22, thus precluding the nucleation and propagation of the localized corrosion. [14, 15] Crevice corrosion arrest has been observed on alloy C-22 after some amount of propagation (less than 100 h propagation), which limits the penetration depth. [16] Non concluding studies exist to determine if a corroding crevice may be able to re-initiate in the same spot after it previously died. Future research should include the cathodic kinetics of the most important reduction reactions on passive films, the characterization of in-service crevices, the arrest of crevice corrosion propagation after some amount of propagation and the possibility of crevice corrosion re-initiation after it had initially stalled.

Now that the Yucca Mountain Project (YMP) is on hold, it is difficult to predict if a nickel alloy would be a candidate material for another repository program. It will depend on the conditions of the geologic site selected for the repository. A passivating nickel alloy was initially considered for the YMP because of its unique emplacement characteristics such as above the water table with unrestricted access of air. The effect of irradiation on the redox potential of the system was taken into account for the YMP total system performance assessment. Short term laboratory tests to measure localized corrosion repassivation potential are relevant for long term applications because values of E_{crit} are a property of the alloy/environment pair. The value of E_{crit} will not change as a function of time. Considerations of the cost of nickel vs. carbon steel are outside the scope of this work.

SUMMARY AND CONCLUSIONS

Alloy HYBRID-BC1 was the most resistant to chloride-induced crevice corrosion, followed by alloys C-22HS, C-22 and 625. The localized corrosion resistance of the tested alloys increased with PREN. The crevice corrosion repassivation potential showed a linear or quasi-linear decrease with temperature. The slopes obtained by linear fits were -2.2 mV/°C for alloy HYBRID-BC1, -2.5 mV/°C for alloy C-22HS, -1.9 mV/°C for alloy C-22, and -2.9 mV/°C for alloy 625. According to the localized acidification model, $E_{R,CREV}$ decreases with increasing temperatures until reaching the corrosion potential in the acidified solution which is the lowest potential at which the localized corrosion is able to propagate. The potential and temperatures at which these conditions apply were determined by extrapolation. They were -0.184 V_{SCE} and 145°C for alloy HYBRID-BC1, -0.214 V_{SCE} and 115°C for alloy C-22HS, -0.212 V_{SCE} and 110°C for alloy C-22, and -0.261 V_{SCE} and 85°C for alloy 625. These extrapolated potentials are strong parameters for assessing the localized corrosion susceptibility of the materials in a long timescale, since they are independent of temperature, chloride concentration, crevice mechanism, crevice former material and type, etc. However, they are extremely conservative and their use for industrial application would limit the use of most of the nickel passivating alloys, even in mildly oxidizing environments. Future research should include the effect of cathodic kinetics on passive films and the likelihood of formation of such a severe crevice as those used for crevice corrosion testing.

ACKNOWLEDGMENTS

Financial support from the Agencia Nacional de Promoción Científica y Tecnológica of the Ministerio de Educación, Ciencia y Tecnología (Argentina), and from the Universidad Nacional de San Martín (Argentina) is acknowledged. The authors are grateful to N. S. Meck from Haynes International, who kindly supplied the tested alloys.

REFERENCES

[1] P. A. Whiterspoon, and G. S. Bodvarsson, Geological Challenges in Radioactive Waste Isolation, Third Worldwide Review, University of California, Berkeley, CA, USA (2001).

[2] J. M. Gras, *C. R. Physique*, **3**, 891 (2002).

[3] R. B. Rebak, "Corrosion of Non-Ferrous Alloys. I. Nickel-, Cobalt-, Copper-, Zirconium-and Titanium-Based Alloys," in: R. W. Cahn, P. Haasen, and E. J. Kramer (Ed.), Corrosion and Environmental Degradation, vol. II, Wiley-VCH, Weinheim, Germany, Ch. 2 (2000).

[4] Z. Szklarska-Smialowska, Pitting and Crevice Corrosion of Metals, NACE Intl. Houston TX (2005).

[5] J. R. Galvele, "Transport Processes and the Mechanism of Pitting of Metals," *J. Electrochem. Soc.*, **123**, 464 (1976).

[6] N. S. Zadorozne, C. M. Giordano, M. A. Rodríguez, R. M. Carranza, and R. B. Rebak, "Crevice corrosion kinetics of nickel alloys bearing chromium and molybdenum," *Electrochim. Acta*, Volume 76, issue (August 1, 2012), p. 94-101, ISSN: 0013-4686 DOI: 10.1016/j.electacta.2012.04.157 (2012)

[7] K. J. Evans, A. Yilmaz, S. D. Day, L. L. Wong, J. C. Estill, and R. B. Rebak, "Using electrochemical methods to determine alloy 22's crevice corrosion repassivation potential," *Journal of Metals*, **57**, pp. 56-61 (2005).

[8] S. Valen, and P. O. Gartland, "Crevice corrosion repassivation temperatures of highly alloyed stainless steels," *Corrosion*, **51**, pp. 750-756 (1995).

[9] A. K. Mishra, and G. S. Frankel, "Crevice Corrosion Repassivation of Alloy 22 in Aggressive Environments," *Corrosion*, **64**, pp. 836-844 (2008).

[10] Annual Book of ASTM Standards, vol. 03.02, ASTM Intl., West Conshohocken, PA (2005).

[11] M. Rincón Ortíz, M. A. Rodríguez, R. M. Carranza, and R. B. Rebak, "," *Corrosion*, **66**, 105002 (2010).

[12] HASTELLOY® C-22HS® alloy preliminary data, brochure H-2122, Haynes International, Inc; HASTELLOY® HYBRID-BC1® alloy preliminary data, Haynes International, Inc (2012) extracted from www.haynesintl.com on 08-June-2012.

[13] R. B. Rebak, "Factors Affecting the Crevice Corrosion Susceptibility of Alloy 22," Paper N°05610, Corrosion/05, NACE Intl., Houston, TX (2005).

[14] M. A. Rodriguez, R. M. Carranza, and R. B. Rebak, "Crevice Corrosion of Alloy 22 at Open Circuit Potential in Chloride Solutions at 90C," Paper N°09424, Corrosion/09, NACE Intl., Houston, TX (2005).

[15] P. Jakupi, J. J. Noel, and D. W. Shoesmith, "Crevice corrosion initiation and propagation on Alloy 22 under galvanically-coupled and galvanostatic conditions," *Corros. Sci.*, **53**, pp. 3122-3130 (2011).

[16] K. G. Mon, G. M. Gordon, and R. B. Rebak, "Stifling of Crevice Corrosion in Alloy 22," in Proceedings of the 12th International Conference on Environmental Degradation of Materials in Nuclear Power Systems – Water Reactors, pp. 1431-1438, Salt Lake City, UT 14-18 August 2005 (TMS, 2005: Warrendale, PA)

DETERMINING THE THERMAL CONDUCTIVITY OF A MELTER FEED

Jarrett Rice, [1] Richard Pokorny, [2] Michael Schweiger, [1] Pavel Hrma [1]
[1] Pacific Northwest National Laboratory, Richland WA 99352 USA
[2] Department of Chemical Engineering, Institute of Chemical Technology in Prague. Technicka 5, 166 28 Prague 6, Czech Republic

ABSTRACT

To stabilize radioactive waste in the form of glass, a melter is charged with a mixture of waste and glass-forming additives in the form of slurry. On the surface of the molten glass, the melter feed forms the cold cap, a crust of reacting material. To convert the feed to molten glass, heat must be transferred through the cold cap. This process is a key factor in determining the rate of melting. To measure the heat conductivity of the reacting feed as a function of temperature, we have designed a setup consisting of a large cylindrical crucible with an assembly of thermocouples that record the temperature field within the feed as a function of time. The thermal diffusivity is obtained by the heat transfer equation. To calculate the heat conductivity, we used the feed density and reaction enthalpy data previously determined with thermoanalytical methods.

INTRODUCTION

The Hanford Waste Treatment and Immobilization Plant in southeastern Washington has been tasked with stabilizing some 200,000 m^3 of radioactive and chemical waste in the form of glass. The mixture of waste and glass-forming additives is charged to a melter where it forms a cold cap—a layer of reacting feed components floating on the pool of molten glass. The heat required to dry the slurry feed and melt the glass is transferred through the cold cap. The rate of melting, a key factor in determining the cleanup life cycle, is affected by the heat conductivity of the reacting feed. Thus, the heat conductivity is an important parameter in the enthalpy balance of the cold cap.[1-5] Its value has a large impact on the cold cap characteristics, such as its thickness or, via average heating rate of the melter feed, on the foaming. Therefore, it is necessary to have the best possible estimate of the heat conductivity.

Our study is concerned with the estimation of heat conductivity of a melter feed in which various reactions take place during heating. Thus, feed density and effective heat capacity vary with the temperature. The change of density was obtained by combining the measurement of feed volume changes with thermogravimetric analysis (TGA). The rate of heat generation/consumption by batch reactions was measured with differential scanning calorimetry (DSC). To measure feed conductivity, we have designed a setup consisting of a large cylindrical crucible with an assembly of thermocouples (TCs). The crucible was filled with a simulated high-level-waste feed and the evolution of the temperature field was monitored while the crucible was heated at a constant rate. We used an analytical approximation of the measured temperature field to obtain the thermal diffusivity as a function of temperature.

EXPERIMENTAL

The simulated high level waste glass feed (A0) was prepared using the batch sheet shown in Table I. Components were added to 1000 mL deionized water slurry based on solubility. Using a Corning hot plate (Model PC-351), the slurry was kept between 60 and 80°C during addition of the chemicals and constantly stirred using a Fisher Scientific StedFast Stirrer (Model 1200). After all chemicals had been added, the heat was turned off and the slurry was stirred for 24 hours. The hot plate and a heat lamp were used to dry the slurry, after which the partially dried batch was placed in an oven at 105°C for 24 hours. The coarse dried batch was milled to powder in a tungsten carbide mill. A loss-on-ignition test was run using an estimated 20% dried feed by mass plus 80% silica to determine the correct fraction of silica to add to the batch.

Table I. Batch composition for A0-feed: Batch sheet for 500 g of glass

Chemical	Mass (g)	Chemical	Mass (g)
$NaNO_2$	1.69	$Bi(OH)_3$	6.40
$Na_2C_2O_4$	0.63	$Mg(OH)_2$	0.85
Na_2SO_4	1.78	$NiCO_3$	3.18
$Zn(NO_3)_2 \cdot 4H_2O$	1.33	NaF	7.39
$Zr(OH)_4 \cdot 0.65H_2O$	2.74	$Fe(H_2PO_2)_3$	6.21
Na_2CrO_4	5.57	H_3BO_3	134.91
KNO_3	1.52	Li_2CO_3	44.15
$Pb(NO_3)_2$	3.04	$Al(OH)_3$	183.74
NaOH	49.71	$Fe(OH)_3$	36.91
CaO	30.39	SiO_2	152.53

The fused silica crucible (~98.5% SiO_2) (Denver Fire Clay Ceramics Inc.) used for the experiment was 190 mm tall and ~130 mm in diameter. Eight type-K TCs (Omega Engineering Inc.) were placed within the crucible and two outside the crucible wall (Fig. 1). The TCs were sleeved in an alumina rod and held in place by a custom-built TC bridge (Fig. 2). The TC bridge allowed the depth for all 10 TCs to be changed, but the radial dimension was fixed. The ideal setup and spacing is labeled in Fig. 1; however, the eight TCs on the inside of the crucible were positioned as shown in Table II. The spacing between thermocouples was 12 to 13 mm except B–C spacing, which was smaller because the B TCs were slightly bent (Fig. 3).

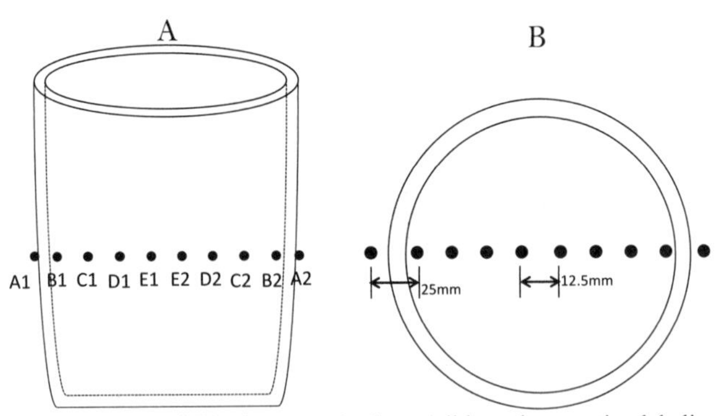

Figure 1. Scheme of TC placement in the crucible and respective labeling; profile (A) and top view (B).

Table II. Thermocouple positioning labeled as radius (r) from the center of the crucible.

TC label	B1	C1	D1	E1	E2	D2	C2	B2
r, mm	-38.5	-31.5	-19.5	-7.5	5.5	17.5	30.5	38.5

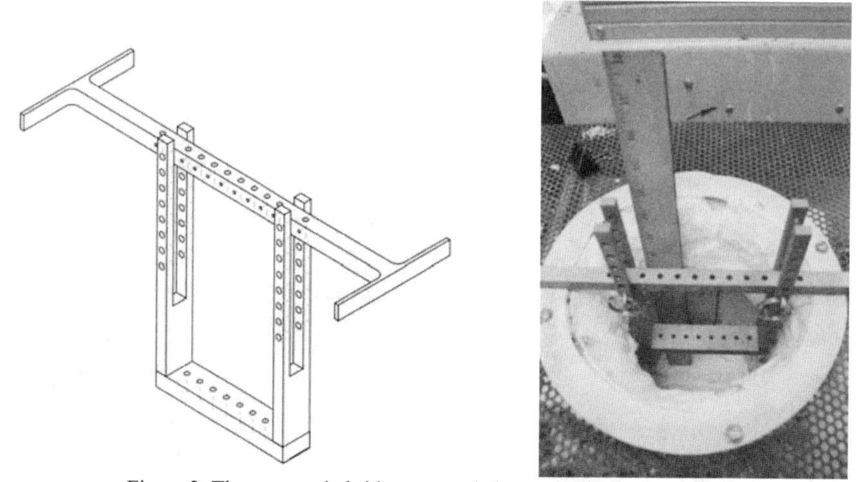

Figure 2. Thermocouple bridge suspended over the furnace opening.

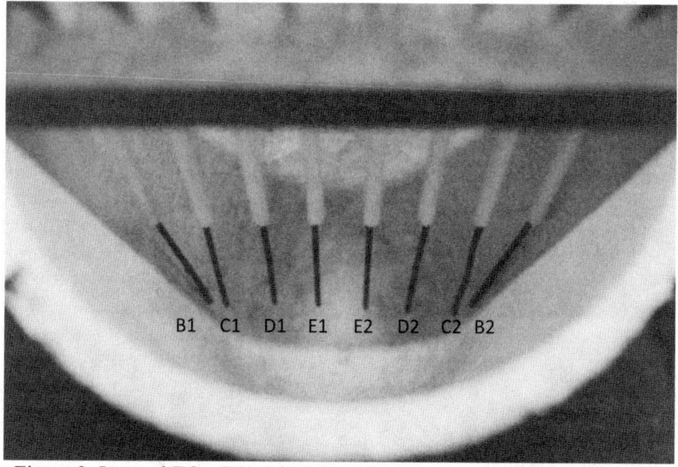

Figure 3. Internal TCs; B1 and B2 bent from contacting the crucible wall.

The crucible was filled with the feed to ~60% of volume (~460 g feed) and placed in the modified Deltech furnace through the top opening (Fig. 2). The TCs were fixed with the TC bridge to allow the tips to rest 4.8 cm below the surface of the feed. With the crucible in place within the furnace, the TC assembly was lowered into the furnace; where the bottom of the TC bridge rested on the top of the crucible and the I-bar kept the bridge suspended (Fig. 2).. The crucible was jostled to consolidate the feed and insulation was packed on the top of the crucible all the way to the top of the furnace opening (Fig. 4). The furnace was heated at 10 K min^{-1} for the initial test with silica sand and at 5 K min^{-1} for tests with the feed. A Fluke data acquisition unit and Hydra Series II data logger (Fluke 2620A) collected and stored temperature data from the TCs every 60 seconds.

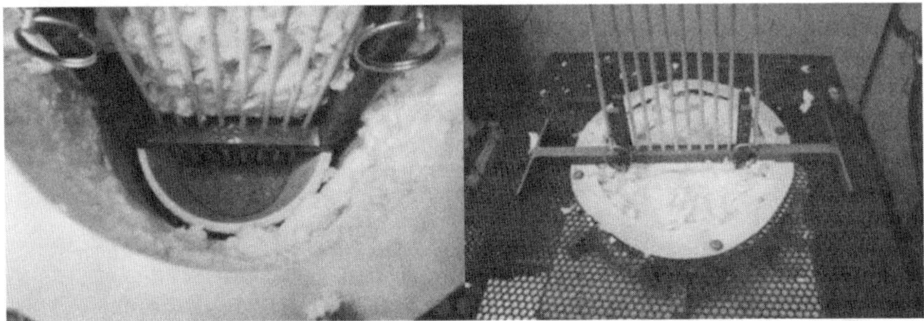

Figure 4. Alumina insulation was packed above the crucible to fill in the furnace opening.

Using silica sand, a set of initial tests were run with the crucible and thermocouple assembly. The tests revealed an asymmetry in the heat flow, reflecting the layout of the heating elements. The TCs closer to the right side of the furnace, TC A2 (Fig. 1), consistently measured higher temperatures than those closer to the left wall, TC A1 (Fig 1). Inserting an alumina cylinder (15 cm OD, 4 mm thick) between the crucible and the furnace heating elements (Fig. 5), reduced the temperature difference on the opposite sides of the crucible to ≤10°C.

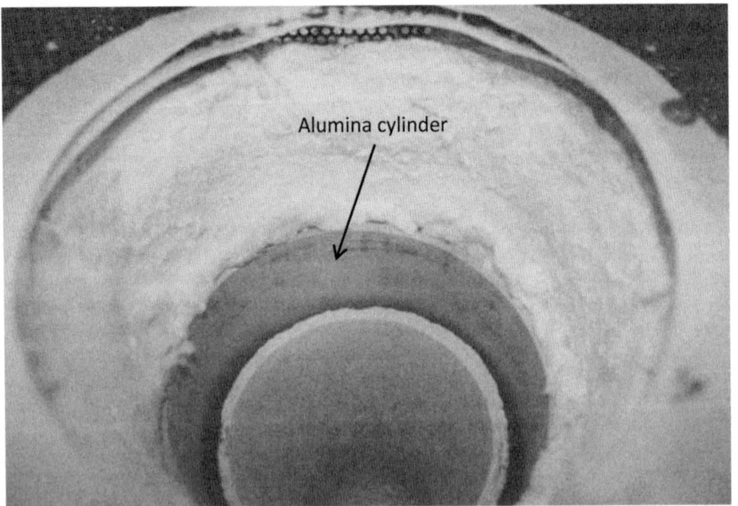

Figure 5. Alumina cylinder was inserted between furnace elements and crucible as a heat shield.

With a more uniform and symmetric heat flow established, data from one of the last tests (Test 0) with silica sand (Sigma Aldrich 50-70 mesh, 210-297 μm), was used to measure the heat conductivity of the sand.

Two experiments were performed with melter feed. In the first experiment (Test 1), the furnace was turned off once the internal TCs reached 800°C to avoid violent foaming at temperatures >850°C. The feed was then allowed to cool to room temperature, during which it was monitored for the subsequent 24-hour period. To check an anomaly described in the following section, in the second experiment, heating was held until all internal TCs equilibrated at 600°C (Test 2), a temperature by which all major reactions were complete. The feed was then allowed to cool for 24 hours to room temperature.

RESULTS

Fig. 6 displays the temperature evolution during Test 1. From previous work, it was understood that A0 feed exhibits a maximum volume increase at ≥850°C[6] As the first test with A0 feed, the Surprisingly, when the innermost TCs reached ~130°C they did not record a change in temperature between 70 and 100 min of heating (see the inset in Fig. 6). The anomalous development of the temperature field is well illustrated on the temperature profiles seen in Fig. 7. To address this anomaly and ensure data was accurate and repeatable, a second test with A0 feed was run. Fig. 8 displays results. The flat interval occurred again at the same temperature as in Test 1 (Fig. 9).

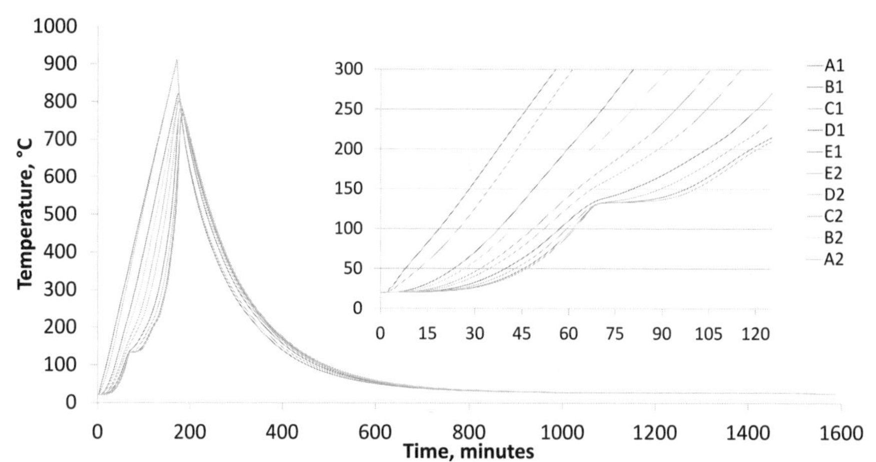

Figure 6. Temperature history during Test 1 with A0 feed.
Inset: magnified plot between 0 and 125 min.

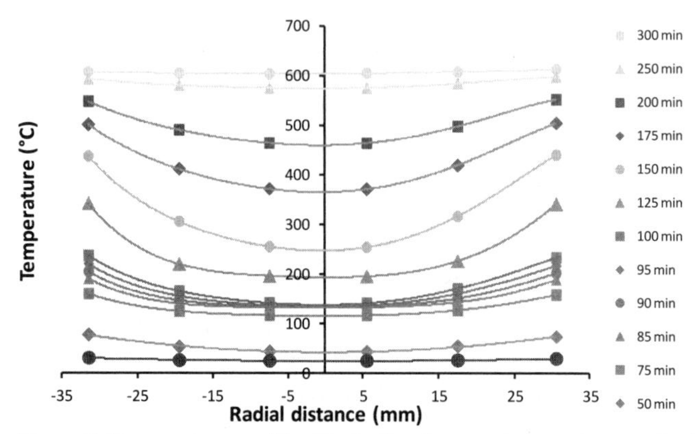

Figure 7. Temperature profiles illustrating the zero temperature gradient between 75 and 100 min.

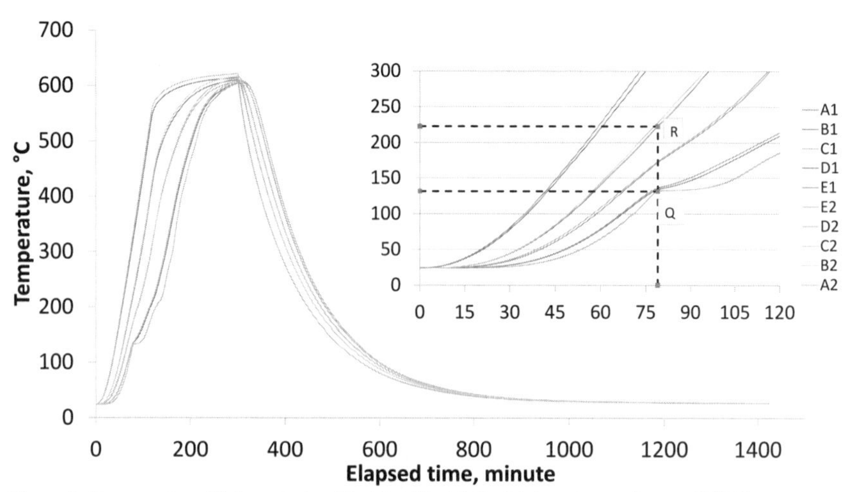

Figure 8. Temperature history during Test 2 with A0 feed taken to maximum 600°C as measured by internal TCs. Inset: magnified plot between 0 and 125 min.

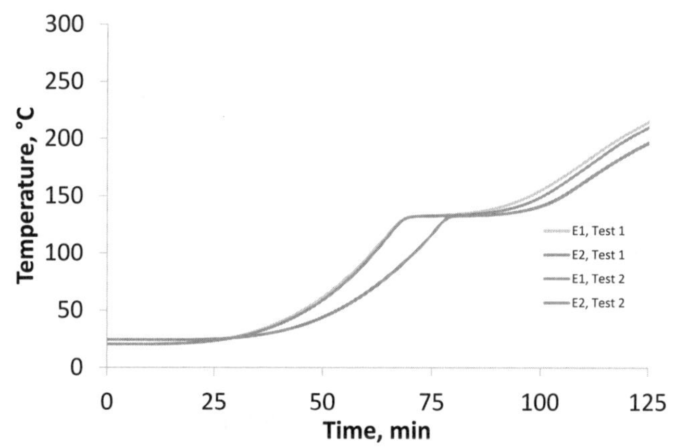

Figure 9. Comparison of time-temperature histories recorded by the innermost TCs in Tests 1 and 2.

The observed changes in the temperature profile at ~130°C (Figs. 6–9) suggest that hot gases (predominantly steam) participated in the heat transfer within the feed. Gases that evolved from the hotter feed located close to the crucible wall flowed toward the center, heating up the cooler material. As a result, the temperature profile around the crucible axis became flat and the temperature increase temporarily stopped. Fig. 10 shows the rate of gas evolution from the feed obtained by thermogravimetry[7]. The peaks between 225 and 275°C (~500–550 K) represent dehydration reactions. These reactions produced copious amounts of steam, a gas with a high heat capacity. When the temperature of the feed at the crucible wall reached ~225°C (Fig. 8 insert, point R), i.e., the temperature at which the major dehydration reaction occurs (Fig. 10), the temperature in the internal portion of the feed stopped increasing (Fig. 8 insert, point Q). The effect of convection and heat exchange between the gas and condensed phases became less pronounced, when the outer portions of the feed no longer released large quantities of water vapor. Also, the decreased specific surface area of the condensed phase, a result of melt generation, reduced the heat exchange between the condensed phase and the gas (the heat exchange rate is proportional to the temperature difference and specific surface area).

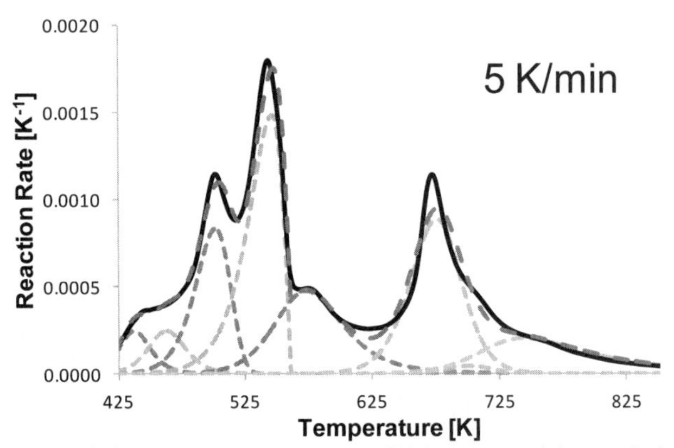

Figure 10. Gas evolution rate versus temperature; TGA curves and deconvolution peaks.

COMPUTATION

Assuming uniform heating from the sides and neglecting the slight tapering of the crucible, the heat equation in the one-dimensional cylinder can be written in the form

$$\rho c_p \frac{\partial T}{\partial t} = \frac{1}{r}\frac{\partial}{\partial r}\left(\lambda r \frac{\partial T}{\partial r}\right) + \rho \Delta H \frac{\mathrm{d}\alpha}{\mathrm{d}t} + S \tag{1}$$

where ρ is the temperature-dependent feed density, c_p is the heat capacity, T is the temperature, t is the time, r is the radial coordinate, λ is the heat conductivity, ΔH is the reaction heat, α is the overall degree of conversion, and S is the rate of heat exchange between the condensed and gas phases. Defining the effective heat capacity as $c_p^{Eff} = c_p + \Delta H(\mathrm{d}\alpha/\mathrm{dT})$, and neglecting S, Eq. (1) reduces to

$$\rho c_p^{Eff} \frac{\partial T}{\partial t} = \frac{1}{r}\frac{\partial}{\partial r}\left(\lambda r \frac{\partial T}{\partial r}\right) \tag{2}$$

Finally, disregarding the change of λ with r and because $a = \lambda/(\rho c_p^{Eff})$, where a is the effective thermal diffusivity, Eq. (2) becomes

$$\frac{\partial T}{\partial t} = a \frac{1}{r}\frac{\partial}{\partial r}\left(r \frac{\partial T}{\partial r}\right) \tag{3}$$

Approximating the measured temperature field at different times t by a fourth-order polynomial, i.e., $T = b_0 + b_2 r^2 + b_4 r^4$, where the terms with odd exponents are 0 because of symmetry, we have

$$F = r^{-1}\partial_r (r \partial_r T) = 4b_2 + 16 b_4 r^2 \tag{4}$$

For each TC, we approximate $T = f(t)$ by a fourth-order polynomial, obtaining

$$G = \partial_t T = c_1 + 2c_2 t + 3c_3 t^2 + 3c_4 t^3 \quad (5)$$

In this way, we are able to calculate the effective thermal diffusivity and heat conductivity for different TC positions (D1, E1,...) at different times as

$$a = G/F \quad (6)$$

The result of calculation of a for different TC positions at different times is displayed in Fig. 11. The results for the silica sand show a slow monotonic increase of a with increasing temperature, as expected in a nonreacting mixture. The small difference in a for different TCs was caused by the asymmetry in the crucible heating. The dependence of the thermal diffusivity of the feed on the temperature exhibits a minimum, which is caused by endothermic reactions occurring at ~300°C.

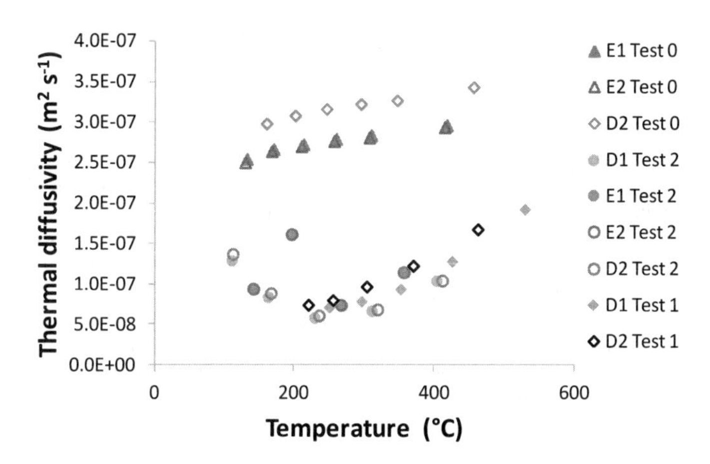

Figure 11. The temperature dependence of thermal diffusivity a, based on $a = G/F$, for silica sand (Test 0) and melter feed (Test 2 and Test 1).

For silica sand, the heat conductivity, $\lambda = a\rho c_p^{Eff}$, was calculated using the values $\rho_b = 1.54 \times 10^3$ kg m^{-3} (Ref.[8]) and $c_p = 0.83$ kJ kg^{-1} K^{-1}. By the linear regression, shown in Fig. 12,

$$\lambda = \lambda_0 + \lambda_1 T \quad (7)$$

where $\lambda_0 = 0.091$ W m^{-1} K^{-1} and $\lambda_1 = 6.80 \times 10^{-4}$ W m^{-1} K^{-2} (T in K). The extrapolated room temperature value is 0.29 W m^{-1} K^{-1}. This value is somewhat higher than the tabulated values for river sand, 0.15–0.25 W m^{-1} K^{-1}.

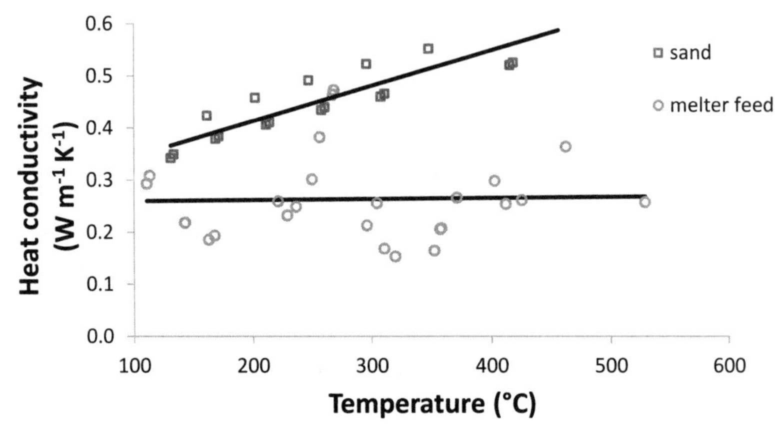

Figure 12. Calculated values of heat conductivity λ as a function of temperature T (Test 2.1).

The effective heat capacity of the melter feed was measured with DSC[9]. The density of the loose batch is $\rho_0 = 970$ kg m^{-3}. The feed volume does not change up to 700°C,and consequently, the spatial density of the feed decreases during heating as a result of the escaping gas. The degree of conversion with respect to the gas phase was measured using TGA[7]. For the feed, we used measured values of c_p^{Eff} reported by Chun et al.[9] Resulting values are displayed in Fig. 12. By fitting Eq. (7) to data, omitting the outliers at 200°C, we obtained $\lambda_0 = 0.252$ W m^{-1} K^{-1} and $\lambda_1 = 1.97 \times 10^{-5}$ W m^{-1} K^{-2} (T in K). Although the linear dependency of λ on T reasonably fits the general trend, the effective heat conductivity calculated by direct method fluctuates with temperature. This fluctuation is similar, though not identical, for calculations based on data from different TCs. We attribute these fluctuations to the convective heat transfer between the gas and condensed phase, which was unaccounted for in the heat balance and, hence, is mimicked as heat conduction.

CONCLUSIONS

We have used a cylindrical body of melter feed heated at a constant rate to monitor the temperature field evolution within the feed and to calculate the thermal diffusivity of the reacting feed as a function of temperature. Then, using DSC and TGA data, we obtained the heat conductivity of the melter feed in the form $\lambda = \lambda_0 + \lambda_1 T$, where $\lambda_0 = 0.252$ W m^{-1} K^{-1} and $\lambda_1 = 1.97 \times 10^{-5}$ W m^{-1} K^{-2} (383 K < T < 801 K). The experimental data indicated that the heat exchange between the evolving gas phase and the condensed phase can affect the heat transfer within the cold cap.

ACKNOWLEDGEMENTS

This research has been conducted as specified by Pacific Northwest National Laboratory Purchase Order Number 159580. Pacific Northwest National Laboratory is operated for the U.S. Department of Energy by Battelle under Contract DE-AC05-76RL01830. This research was supported by the U.S. Department of Energy Federal Project Office Engineering Division for the Hanford Tank Waste Treatment and Immobilization Plant and by the WCU (World Class University) program through the National Research Foundation of Korea funded by the Ministry of Education, Science and Technology (R31 - 30005). This work is based on DSC data by David Pierce and computations by Richard Pokorny. The authors are grateful to Albert Kruger for his assistance and guidance and to Jaehun Chun and Dong-Sang Kim for insightful discussions.

REFERENCES

[1]P. Hrma, A.A. Kruger, R. Pokorny, Nuclear waste vitrification efficiency: Cold cap reactions, *J. Non-Cryst. Solids* (2012), doi:10.1016/j.jnoncrysol.2012.01.051.
[2]P. Schill, Modeling the behavior of noble metals during HLW vitrification, in: W. Lutze, Modeling the behavior of noble metals during HLW vitrification in the DM1200 melter, VSL 05R5740-1, Vitreous State Laboratory, Washington DC (2005).
[3]P. Hrma, M.J. Schweiger, C.J. Humrickhouse, J.A. Moody, R.M. Tate, T.T. Rainsdon, N.E. TeGrotenhuis, B.M. Arrigoni, J. Marcial, C.P. Rodriguez, B.H. Tincher, Effect of glass-batch makeup on the melting process, *Ceram.-Silik.*, **54** (3), 193-211 (2010).
[4]O.S. Verheijen, Thermal and chemical behavior of glass forming batches, PhD thesis, TU Eindhoven, Netherlands, ISBN 90-386-2555-3, (2003).
[5]D. Kim, M.J. Schweiger, W.C. Buchmiller, J. Matyas, Laboratory-Scale Melter for Determination of Melting Rate of Waste Glass Feeds, PNNL-21005; EMSP-RPT-012, Pacific Northwest National Laboratory, Richland, WA, (2012).
[6]S.H. Henager, P. Hrma, K.J. Swearingen, M.J. Schweiger, J. Marcial, N.E. TeGrotenhuis, Conversion of batch to molten glass, I: Volume expansion, *J. Non-Cryst. Solids* (2010), doi:/10.1016/j.jnoncrysol.2010.11.102
[7]R. Pokorny, D.A Pierce, P. Hrma, Melting of glass batch: Model for multiple overlapping gas-evolving reactions, *Thermochim. Acta*, **541,** 8-14 (2012).
[8]J.F. Shackelford, W. Alexander, CRC Materials Science and Engineering Handbook, 3rd edition, CRC Press, Boca Raton, FL (2001).
[9]J. Chun, D.A. Pierce, R.Pokorny, P.Hrma, Cold-cap reactions in vitrification of nuclear waste glass: experiments and modeling, PNNL-SA-91044, Pacific Northwest National Laboratory, Richland, WA (2012).

ELECTROCHEMICAL PROPERTIES OF LANTHANUM CHLORIDE-CONTAINING MOLTEN LICL-KCL FOR NUCLEAR WASTE SEPARATION STUDIES

S.O. Martin, J.C. Sager, K. Sridharan*, M. Mohammadian, T.R. Allen
Department of Engineering Physics, University of Wisconsin-Madison
Madison, Wisconsin, United States

ABSTRACT

Pyroprocessing is currently being researched as a method of recycling used nuclear fuel. A key component in the process is the electrorefiner where used nuclear fuel is anodically dissolved in molten LiCl-KCl eutectic salt at 500°C and then uranium is electrochemically extracted on a steel cathode. As the electrorefining process progresses, various fission products and lanthanides will accumulate in the electrorefiner salt. We report here, the investigation of electrochemical properties of the LiCl-KCl salt as functions of added lanthanum concentration.

A three-electrode electrochemical cell was constructed to study the electrochemical behavior of lanthanum (III) chloride in LiCl-KCl at 500°C. Lanthanum chloride concentrations from 0.5 – 9 wt% were investigated using cyclic voltammetry (CV), chronopotentiometry (CP), and anodic stripping voltammetry (ASV). CV was used to confirm the redox reaction mechanisms of the lanthanum in the salt. The lanthanum redox reaction was determined to be a reversible one-step process involving the transfer of three electrons. The apparent standard potential of the lanthanum redox reaction at 500°C was found to be $E^{0'} = -1.959$ V vs. $Ag^+|Ag$ reference (-3.106 V vs. $Cl^-|Cl_2$). ASV and CV were used to construct a peak current density versus lanthanum concentration plots which serve as the scientific basis for the development of an *in situ* electrochemical probe which could be used to determine lanthanum concentration in the electrorefiner salt at any given time during the electrorefining operation.

INTRODUCTION

Pyroprocessing is a high-temperature electrochemical process that is being successfully used at Idaho National Laboratory (INL) to reprocess spent metallic fast reactor fuel.[1] The process starts when the spent fuel is chopped into segments and anodically dissolved in molten 500°C LiCl-KCl eutectic salt, leaving only noble metal fission products, zirconium, and cladding hulls undissolved. A reductive potential is applied between a solid steel cathode and the anodic fuel basket to cause a desired element, in this case uranium, to electrochemically plate onto the surface of the cathode. The deposited uranium metal can then be re-enriched and/or made into new fuel. Using a liquid cadmium cathode, Pu and U can be electrochemically plated simultaneously, thus reducing proliferation concerns compared to other reprocessing methods.[2] Strontium and cesium may be separately extracted to reduce the overall long-term decay heat and radioactivity of the spent fuel.[3] Pyroprocessing can also be adapted to reprocess spent oxide fuel with the use of a front-end lithium reduction step.[4,5]

The fuel components that remain in the electrorefiner salt consist mainly of fission products and lanthanides. Lanthanides are a particularly important group because they behave chemically and electrochemically similarly to uranium, plutonium, and other actinides so they may get co-deposited with the previously mentioned elements and act as an undesirable neutron poison in the reprocessed fuel. As more fuel is reprocessed in the same LiCl-KCl, the concentration of these lanthanides and fission products increases and causes the thermodynamic properties of the salt to change.[1]

The research conducted attempts to answer fundamental questions about how the thermodynamic and electrochemical properties of the electrolyte change with the addition of rare earth chlorides between the ranges of 0.5 wt% to 9 wt%. In this paper we discuss the experimental system for conducting electrochemical studies followed by the results of the electrochemical behavior of molten salt with additions of rare earth element La. In particular, the investigation was aimed at providing a scientific basis for the development of an *in situ* electrochemical probe capable of determining the concentration of La in the electrorefiner system.

ELECTROCHEMICAL SETUP

Chemicals

The LiCl-KCl used for this research was obtained from INL and had a purity greater than 99%. The eutectic composition of the salt, 44.3 mol% KCl- 55.7 mol% LiCl, with a melting point of approximately 353°C was used. For comparison, the melting points of KCl and LiCl separately are 770°C and 610°C, respectively.

The rare-earth element used in this research, La, is present in spent nuclear fuel as a fission product and accumulates in the electrorefiner salt as more fuel is processed. La may also serve as a surrogate for U and Pu since they have similar ironic radii and coordination numbers[6] yet lack the radiation hazards. The rare-earth chloride was obtained in its anhydrous form with purity at or above 99.9%.

Electrochemical Cell

The electrochemical cell used for this research is shown in Figures 1 and 2.[7] The salt is contained within three nested crucibles at the bottom of a cylindrical heater. The innermost crucible is a 99.95% glassy carbon crucible which was chosen for its high temperature resistance, low electrical resistivity, high corrosion resistance, and closed porosity. In addition, glassy carbon is not wetted by the molten LiCl-KCl, making removal of the salt ingot from the crucible very easy. The glassy carbon crucible was placed inside an alumina and a stainless steel crucible to prevent the molten salt from contacting the heater in the event that the glassy carbon crucible fractures. The stainless steel crucible was grounded to the potentiostat and therefore served as a Faraday cage to reduce electrical noise. All three crucibles are shown in Figure 2.

The quartz structure, indicated in Figure 1, rests on top of the heater and holds the electrodes and thermocouple in position in the molten salt. The electrochemical cell utilized three electrodes: a tungsten working electrode (WE), a glassy carbon counter electrode (CE), and a silver/silver chloride reference electrode (RE). The WE used was a 2 mm diameter tungsten rod with a purity of 99.95%. Tungsten was chosen because it does not form any intermetallics with the elements of interest in this study.

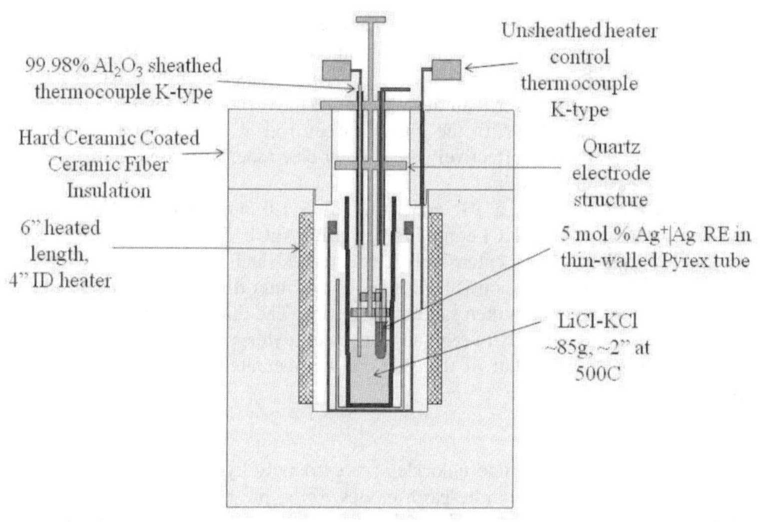

Figure 1. Design of cell used for electrochemical experiments showing thermocouple and reference electrode.[7]

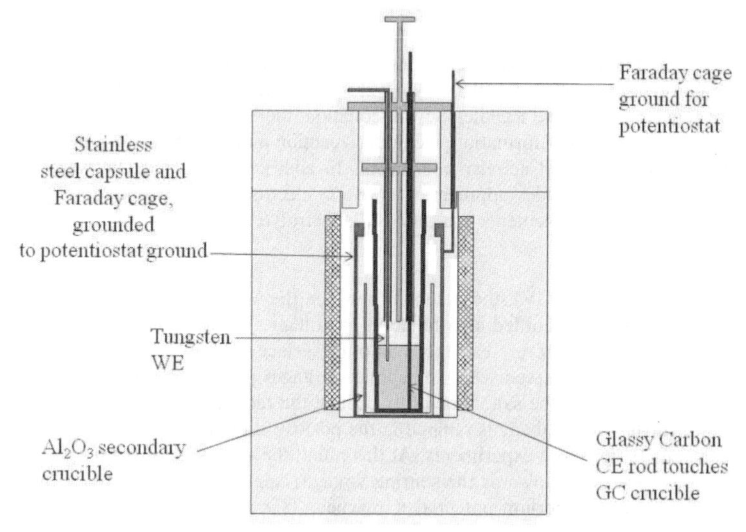

Figure 2. Design of cell used for electrochemical experiments showing counter electrode and working electrode.[7]

The CE was a 3 mm diameter 99.95% glassy carbon rod which was in contact with the glassy carbon crucible. Since the main purpose of the CE is only to complete the circuit of the electrochemical cell and allow current to flow, it is necessary to ensure that the CE area in contact with the salt is much larger than the WE so that the reaction of interest on the WE is not limited by the area of the CE. With the glassy carbon rod in contact with the glassy carbon crucible, the entire crucible was effectively the counter electrode, ensuring the surface area of the CE was much greater than the WE.

The third electrode was a RE consisting of a 1.0 mm diameter 99.9% silver wire immersed in a solution of LiCl-KCl containing 5 mol% AgCl. The use of an $Ag^{+}|Ag$ reference electrode in LiCl-KCl has been extensively characterized and is very stable with 1-10mol% AgCl.[8] The LiCl-KCl/AgCl solution is held in a 10 mm outer diameter Pyrex test tube which is partially submerged in the bulk molten salt during testing. The thin walls (<1mm) of the test tube allow for electrical conduction between the two electrolytes while keeping them chemically separated. Therefore, the potential of the WE can be accurately measured against the stable reference solution.

Glovebox

LiCl-KCl salt and lanthanide chlorides are extremely hygroscopic. In addition, the salt will readily form alkali-oxides or rare-earth oxides in the presence of oxygen.[7] Therefore, all electrochemistry work was performed in a glovebox under an inert argon atmosphere. A Dri-Train gas purification system was used to maintain O_2 and H_2O levels below 1 ppm and 2 ppm, respectively.

MOLTEN LiCl-KCl SALT ELECTOCHEMISTY

Three different electrochemical techniques were used in this research: cyclic voltammetry (CV), anodic stripping voltammetry (ASV), and chronopotentiometry (CP). Data from these techniques was analyzed with the Randles-Sevcik equation, the Nernst equation, and the Sand's equation to extract a variety of information, including: reaction mechanism, diffusion coefficient, apparent standard potential, and activity coefficient. In addition, ASV and CV were used to create calibration curves for the development of an *in situ* electrochemical probe for determining the concentration of rare-earth elements in a LiCl-KCl electrolyte.

Cyclic Voltammetry

In cyclic voltammetry (CV), the potential between the WE and RE is changed linearly with time and the current is recorded as a function of voltage. The voltage begins sufficiently negative to force the reduction of rare-earth ions onto the surface WE as a metal. As the potential increases, a current peak is observed when the applied voltages passes the equilibrium potential of the electroactive species in the salt. When this occurs, the reduced metal oxidizes off of the WE into the electrolyte. The voltage is ramped in the positive direction until the anodic limit is reached, which was 0V for these experiments. At this point, the voltage scan is reversed and the potential decreases back down towards the starting voltage, causing the rare-earth to plate back on to the WE when the equilibrium potential is reached. This cycle is then repeated multiple times to achieve more accurate results.

Analysis of the CV can be performed with the Randles-Sevcik equation (Eq. 1) if the system is mostly reversible[9].

$$\frac{I_p}{\sqrt{\upsilon}} = 0.446(nF)^{\left(\frac{3}{2}\right)}(RT)^{\left(-\frac{1}{2}\right)}C_{M^{n+}}\, D^{\left(\frac{1}{2}\right)}S \tag{1}$$

In the Randles-Sevcik equation, I_p is the peak current in A, v is the voltage sweep rate in V/s, n is the number of electrons transferred, F is the Faraday constant in s-A/mol, R is the universal gas constant in J/mol-K, T is the temperature in K, C_{Mn+} is the ionic concentration of the analyte in the bulk solution in mol/m^3, D is the diffusion coefficient in m^2/s, and S is the surface area of the working electrode in m^2. The system can be said to be reversible if it obeys the Nernst equation (Eq. 2), shown below, which describes the equilibrium potential of a half-cell as a function of temperature and analyte concentration[10].

$$E = E^{0\prime} - \frac{RT}{nF}\ln\left(\frac{a_{Red}}{a_{Ox}}\right) \qquad (2)$$

In the Nernst equation, E is the potential at the temperature of interest in V, $E^{0\prime}$ is the standard potential from the EMF series in V, R is the universal gas constant in J/mol-K, T is the temperature in K, n is the number of electrons transferred per reduction/oxidation, F is the Faraday constant in s-A/mol, a_{Red} is the activity of the reductant, and a_{Ox} is the activity of the oxidant. For most liquid systems, the activities can be changed to concentrations of the reductant and oxidant. One criteria for determining the reversibility of a system is that the anodic peak current, i_{pa}, and the catholic peak current, i_{pc}, should be proportional to the square root of the scan rate, $v^{1/2}$. A second criteria is that the mid-peak potential should not change significantly with scan rate. This mid peak potential can be calculated by equation 3:

$$E_{1/2} = \frac{E_{p,a}+E_{p,c}}{2} \qquad (3)$$

where $E_{p,a}$ and $E_{p,c}$ are the anodic and cathodic peak potentials.

With the reversibility of the system verified, the Randles-Sevcik equation can be used to determine the number of electrons transferred in each reaction given a value of D either from literature or another electrochemical technique such as CP. The number of electrons transferred can also be determined from a Nernst plot, which is a plot of the equilibrium potential, $E_{1/2}$, versus the natural logarithm of the molar fraction of analyte. The slope of the plot should be equal to RT/nF, and since R, T, and F are known, n can be readily calculated. In addition, the y-intercept of a Nernst plot is equal to the apparent standard potential (Eq. 2).[7]

Anodic Stripping Voltammetry

Anodic stripping voltammetry (ASV) has been used extensively in industry for analysis of liquids such as water, blood, urine, molten steel, and oil.[11] Anodic stripping voltammetry is similar to cyclic voltammetry in that it is a potentiodynamic technique where the voltage of the working electrode is controlled and the current response of the system is observed. However, ASV uses a plating step at the beginning where a constant cathodic voltage is applied for a period of time to reduce a significant amount of the rare-earth metal onto the working electrode. This extra plating step makes ASV better for analyzing smaller concentrations of the elements of interest. Following the plating step, the voltage is swept linearly in the positive direction and the current response is recorded. ASV can be used to determine the concentration of a given element in the electrolyte since the height of the stripping peak is proportional to the concentration of the analyte assuming a constant plating time.

Electrochemical Results for Lanthanum

The behavior of $LaCl_3$ in molten LiCl-KCl was studied for $LaCl_3$ concentrations ranging from 0.1 wt% up to 9 wt% $LaCl_3$. At each concentration, the reversibility of the system was verified using the techniques mentioned above. Figure 3 shows the peak cathodic and anodic CV currents as functions of the square root of the scan rate ($v^{1/2}$) with 0.99 wt% $LaCl_3$. The linear relationship of both peaks is indicative of a reversible system. With the reversibility confirmed, the Randles-Sevcik equation was used to determine the reaction mechanism to be $La^{3+}_{(aq)} + 3e^- \leftrightarrow La_{(s)}$, which is consistent with literature. The apparent standard potential of La at 500°C was calculated by creation of a Nernst plot (Figure 4) as described above and found to be $E^{0'} = -1.9593V$ vs. $Ag^+|Ag$ reference scale (-3.11V vs. $Cl^-|Cl_2$).

Three approaches for the determination of La^{3+} concentration were used that employ three current measurements as a function of concentration: maximum anodic CV peak height, maximum cathodic CV peak height, and maximum ASV peak height all at a sweep rate of 50 mV/s.

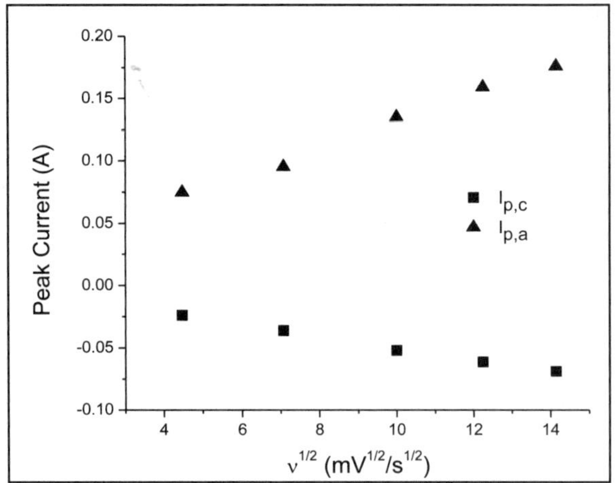

Figure 3: Peak cathodic and anodic currents as a function of $v^{1/2}$ for 0.99 wt% $LaCl_3$ in LiCl-KCl

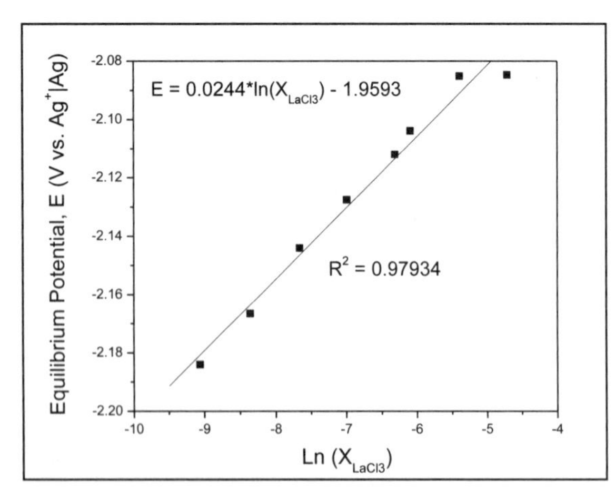

Figure 4: Nernst plot of $LaCl_3$ in LiCl-KCl eutectic. The slope should be equal to RT/nF and the y-intercept is the standard potential versus a 5 mol% $Ag^+|Ag$ reference electrode

Figure 5 shows the calibration curve for La with a plating time of 60 seconds. The ASV calibration curve begins with a somewhat exponential increase at low concentrations, continues into a linear regime up to approximately 10,000 ppm $LaCl_3$, and then enters a quasi-logarithmic response at higher concentrations. A possible explanation of this is that at low concentrations there is not enough analyte in solution to fully utilize the surface area of the electrode, at medium concentrations the analyte is at sufficient concentration to coat the working electrode in single, ordered layers during the plating step, and at high concentrations the concentration of analyte in solution is so great that surface saturation effects are noticed.

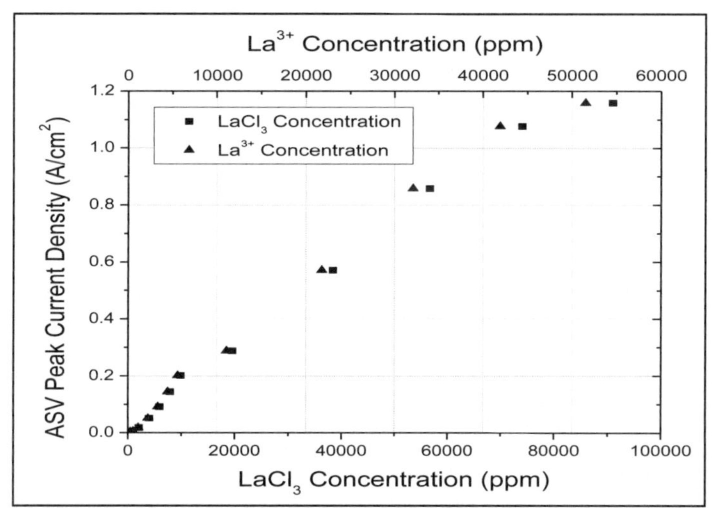

Figure 5. ASV calibration curve relating peak current density from an ASV to $LaCl_3$ and La^{3+} concentration in LiCl-KCl.

Figure 6 depicts the calibration curve for the CV anodic peak. In comparing the CV anodic curve and the ASV curve, they exhibit similar characteristics; however, the CV anodic curve does not have such severe surface saturation effects since it is essentially an anodic stripping curve without a significant pre-concentration as is used in ASV.

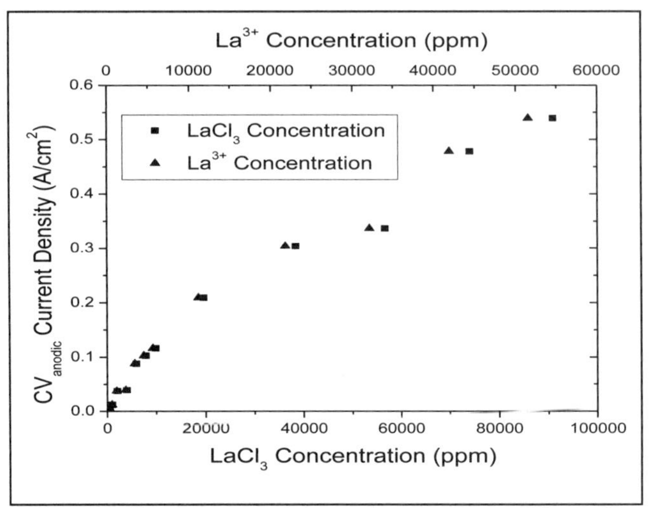

Figure 6. CV anodic peak calibration curve relating peak current from a CV to $LaCl_3$ and La^{3+} concentration in LiCl-KCl.

CONCLUSION

A three-electrode electrochemical cell was constructed consisting of a tungsten working electrode, a glassy carbon counter electrode, and a silver-silver chloride reference electrode. The cell was used to conduct electrochemical measurements on molten LiCl-KCl eutectic salt at 500°C with concentrations of $LaCl_3$ varying from 0.5 – 9 wt% for the development of pyroprocessing. The three electrochemical techniques used were cyclic voltammetry, anodic stripping voltammetry, and chronopotentiometry. Using these techniques, the reaction mechanism of $LaCl_3$ was found to be $La^{3+}_{(aq)} + 3e^- \leftrightarrow La_{(s)}$. The apparent standard potential of La was found to be -1.9593V vs. $Ag^+|Ag$ reference scale (-3.11V vs. $Cl^-|Cl_2$). Calibration curves for the determining unknown concentrations of La^{3+} in LiCl-KCl were constructed using cyclic voltammetry and anodic stripping voltammetry. Similar work has been performed by the additions of chlorides of other rare earth elements Ce, Nd, and Dy, in this program and will be reported elsewhere.

ACKNOWLEDGEMENTS
The work is supported by U.S. Department of Energy's Nuclear Energy University (NEUP) program under DOE prime contract DE-AC07-05ID14517.

FOOTNOTES
*Corresponding author

REFERENCES
[1] R. Benedict, C. Solbrig, B. Westphal, T. Johnson, S. Li, K. Marsden and K. Goff, Pyroprocessing progress at Idaho National Laboratory, Idaho National Laboratory, Tech. Rep. INL/CON-07-12983 (2007).

[2] S. Li, S. Herrmann, K. Goff, M. Simpson and R. Benedict, Actinide Recovery Experiments with Bench-scale Liquid Cadmium Cathode in Real Fission Product-laden Molten Salt, *Nuclear Technology,* **165** (2009).
[3] M. Simpson, T. Yoo, R. Benedict, S. Phongikaroon, S. Frank, P. Sachdev and K. Hartman, Strategic Minimization of High Level Waste from Pyroprocessing of Spent Nuclear Fuel, Idaho National Laboratory, Tech. Rep. INL/CON-07-12123 (2007).
[4]M. Simpson and S. Herrmann, Modeling the Pyrochemical Reduction of Spent UO_2 Fuel in a Pilot-Scale Reactor, *Nuclear Technology,* **162**, pp. 179-183 (2008).
[5]S. Herrmann, K. Durstine, M. Simpson and D. Wahlquist, Pilot-scale Equipment Development for Pyrochemical Treatment of Spent Oxide Fuel, *Global '99 International Conference on Future Nuclear Systems,* Jackson Hole, WY (1999).
[6] P. Bingham, R. Hand, M. Stennett, N. Hyatt, M. Harrison, The Use of Surrogates in Waste Immobilization Studies: A Case Study of Plutonium, *2008 MRS Spring Meeting,***1107** (2008).
[7] M. Mohammadian, Master's Thesis: Electrochemical and Thermodynamic Behavior of Rare Earth Chlorides in Molten LiCl-KCl Salt for Used Fuel Reprocessing (2011).
[8] O. Shirai, T. Nagai, A. Uehara and H. Yamana, Electrochemical Properties of the Ag^+|Ag and other Reference Electrodes in the LiCl–KCl Eutectic Melts, *J. Alloys Compounds,* **456**, pp. 498-502 (2008).
[9] P. Masset, R. Konings, R. Malmbeck, J. Serp and J. Glatz, Thermochemical Properties of Lanthanides (Ln = La, Nd) and Actinides (An = U, Np, Pu, Am) in the Molten LiCl–KCl Eutectic, *J. Nucl. Mater.,* **344**, pp. 173-179 (2005).
[10] J. Wang, *Analytical Electrochemistry.* Hoboken, NJ: John Wiley & Sons, Inc. (2006).
[11] J. Wang, *Stripping Analysis: Principles, Instrumentation, and Applications.* Deerfield Beach: VCH Publishers, Inc. (1985).

RADIONUCLIDE BEHAVIOUR AND GEOCHEMISTRY IN BOOM CLAY WITHIN THE FRAMEWORK OF GEOLOGICAL DISPOSAL OF HIGH-LEVEL WASTE

S. Salah, C. Bruggeman, N. Maes, D., Liu P., L. Wang, and P. Van Iseghem
Belgian Nuclear Research Center (SCK•CEN), Mol, Belgium

ABSTRACT

The recommended option for the long-term management of high-level and long-lived radioactive waste (HLW-LL)[1] in Belgium is geological disposal in poorly indurated clay formations. Through its Waste Plan, ONDRAF/NIRAS intends to satisfy its legal obligations while providing the Government with all the elements needed to make a "decision in principle", in other words a general policy decision, about the long-term management of HLW-LL [1]. Within this framework, Boom Clay (BC) is investigated as one of the potential host formations. The Belgian concept (i.e. *Supercontainer concept*) for geological disposal of HLW-LL considers the installation of a multi-barrier repository system within poorly indurated clays (i.e. Ypresian Clays, Boom Clay). A multi-barrier system typically comprises the natural barrier, provided by the host rock and its surroundings (aquifers, biosphere), as well as the engineered barrier system (EBS). In order to accurately assess the performance of the proposed disposal system, the retention/migration behaviour of safety relevant radionuclides (RN's) over long time scales must be adequately known and quantified. A sound understanding of the underlying geochemical processes and mechanisms is prerequisite to derive consistent sets of retention/migration parameters required for performance assessment calculations and safety evaluations. An overview of the main research activities and recent achievements at SCK•CEN (Belgian Nuclear Research Center) related to radionuclide retention/migration behaviour, involving experimental studies, geochemical as well as transport modeling will be provided.

INTRODUCTION

In Belgium, several players are active in the field of radioactive waste management. The Federal Agency for Nuclear Control (FANC) is an independent agency of the Belgian Government and plays the role of regulator for radioactive waste and oversees nuclear energy matters. The main mission of the Belgian Agency for Radioactive Waste and Enriched Fissile Materials, ONDRAF/NIRAS (O/N) is the short- and long-term management of all types of radioactive waste through the development and implementation of solutions that respect the environment and society. Besides this, O/N has to constantly keep the quantitative and qualitative inventory of radioactive waste and nuclear liabilities up-to-date. Belgoprocess, a subsidiary of O/N is in charge of the "operational" waste management comprising waste processing and conditioning operations, as well as interim storage. With respect to low- and medium level short-lived waste (LILW-SL), corresponding to category A (CatA) waste, in 2006 the Belgian Government decided to dispose of this type of waste in a near-surface disposal facility. License application is foreseen for 2013 and the operational phase is supposed to start in 2016. The long-term solution

---

[1] ***Category A (LILW-SL)****: low- and intermediate-level short-lived waste. This category comprises* radio-elements with half-lives of less than 30 years, emitting *generally alpha radiation.*
Category B (LILW-LL)*: low- and intermediate-level long-lived waste. This category comprises* radio-elements with half-lives of more than 30 years, emitting *generally alpha radiation.*
Category C (HLW-SL and HLW-LL)*:high-level short lived and long-lived waste.* This category comprises large amounts of beta and alpha emitting radio-elements having short or long half-lives. They are highly heat-generating. This waste arises from the reprocessing of irradiated nuclear fuel. Spent fuel that is not reprocessed also belongs to this category.

recommended by O/N for category B&C waste, corresponding to long-lived LILW waste, vitrified high-level radioactive waste (HLW) and spent fuel (SF) is deep geological disposal in poorly indurated clay formations [1]. The Belgian Nuclear Research Center (SCK•CEN) can call upon decades of expertise in research related to geological disposal and waste management solutions for Belgian Radwaste types. Already in 1980, SCK•CEN started construction of the HADES (**H**igh-**A**ctivity **D**isposal **E**xperimental **S**ite) underground facility, situated at a depth of about 225 meters within the Boom Clay (BC) formation, enabling to study plastic clays as potential host formations for B&C wastes. Since 2000, HADES is fully managed by EIG EURIDICE (**Eu**ropean **U**nderground **R**esearch **I**nfrastructure for **D**isposal of nuclear waste **I**n **C**lay **E**nvironment) representing an **E**conomic **I**nterest **G**rouping (EIG) involving SCK•CEN and O/N. By investigating the feasibility of constructing, operating and sealing a waste repository in a deep clay layer, EIG EURIDICE contributes to the national disposal programme run by O/N.

BOOM CLAY – A POTENTIAL HOST FORMATION

In Figure 1a, a map of the NE' part of Belgium is illustrated showing the so-called Campine basin, where BC was deposited ~35-30 million years ago. At Mol (Province of Antwerp), BC is ~100 m thick and located at a depth of 190-290 m. The typical banded structure of BC, which is due to intercalated organic-rich matter (OM), carbonate-rich and more silty layers, can be recognized in Figure 1b. Although macroscopically BC looks heterogeneous, with respect to its composition, it is quite homogeneous, comprising ~30-60% clay minerals (mainly illite and smectite), 20-60% quartz, 5-10% feldspars, and 1-5% carbonates, pyrite and OM, respectively.

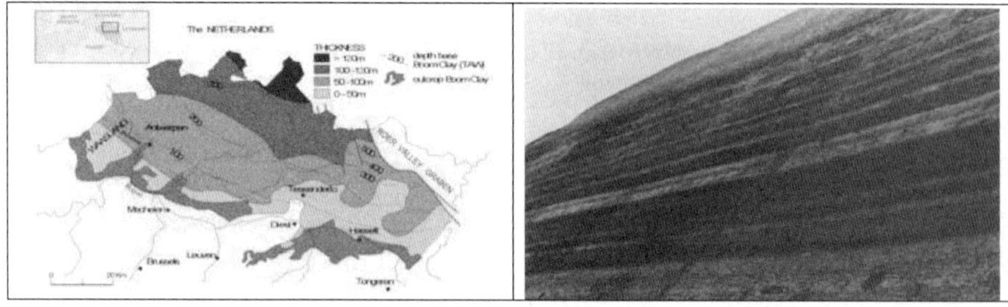

Figure 1: a) Depth of the base and thickness of BC [2], b) Outcrop BC at Terhagen-Rumst open pit quarry

Due to this composition, BC has a high sorption capacity and represents a reducing environment (Eh = -281 mV), with a pH around 8.5 buffered by the carbonates. The pore water corresponds to a 0.015 mol/L sodium-bicarbonate ($NaHCO_3$) water (Mol region). Furthermore, Boom Clay has a low hydraulic conductivity and permeability. The measured vertical hydraulic conductivity (K_v) value is 8.5×10^{-12} m/s (harmonic mean) and the horizontal hydraulic conductivity (K_h) is 4.7×10^{-10} m/s (arithmetic mean) [3]. The cation exchange capacity (CEC) of BC was determined to range between 13-27 meq/100 g and shown to correlate well with the smectite contents of the analysed samples (Figure 2). This clearly shows that smectite is the most important contributor to the CEC of BC. If the linear correlation trend is extrapolated to 0% smectite content, a CEC value of ~3.7 meq/100 g is found. This value is attributed to the illite and natural organic matter (NOM) components in BC. From the slope of the trend, Boom Clay smectite would have a CEC value of ~88 meq/100 g. Besides cation exchange, also surface complexation reactions are contributing to the uptake of radionuclides onto smectite and illite.

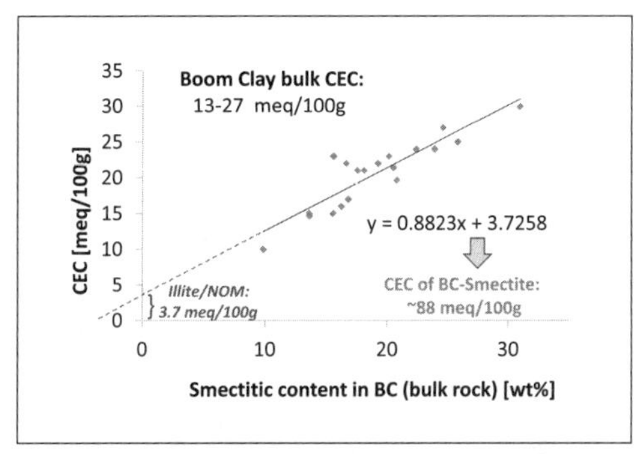

Figure 2: CEC values as measured by Cu-trien technique on the BC bulk rock plotted against smectite content [4]

ROLE OF NATURAL ORGANIC MATTER

Boom Clay is relatively rich (1-5 wt%) in natural organic matter (NOM), which may interact with the radionuclides and thus influence their transport behaviour. The NOM can be subdivided into two fractions, i.e. an immobile fraction and a mobile fraction. The immobile fraction is dominated by a kerogen-type OM, which may act as a sorption sink and as such may retard the radionuclide migration. The mobile fraction is characterized by a dissolved organic carbon content (DOC) of 50-250 mg/L and is composed of ~70% humic acids (HA) and ~30% fulvic acids (FA). By using different techniques like, for example, size exclusion chromatography (SEC) and ultrafiltration (UF), it was shown that the organic matter in Boom Clay pore water is of "colloidal nature" with ~45% being in size < 10 kDa and 35% belonging to the 30-100 kDa size fraction. The fraction > 300 kDa is considered to be immobile [5].

THE DISPOSAL CONCEPT

The disposal concept was developed by O/N and according to this concept the envisaged underground disposal facility will be subdivided into separate sections based on the waste type (B and C type waste sections) with a central access gallery to which the disposal galleries will be linked [6]. Together they will form a two-dimensional framework constructed within the host-rock formation at a depth of ~230-240 m. The current approach is to develop a multi-barrier system to contain and isolate the radioactive waste from the biosphere as long as required. The immobilized waste (most likely within cement, glass or bitumen) will be inserted into a canister, which will be surrounded by different engineered barriers, like the overpack (carbon steel), the buffer (= OPC based concrete) and the waste container. The latter is foreseen to be made out of concrete for B waste (Figure 3a) and of stainless steel for C waste (Figure 3b). The B-waste containers are named "monoliths", while the C-waste configuration is the so-called "Supercontainer". Due to the plasticity of the clayey host formation, the disposal galleries will be stabilized by a concrete lining (i.e. concrete wedge blocks) and the space between this lining and the waste containers backfilled by cementitious material.

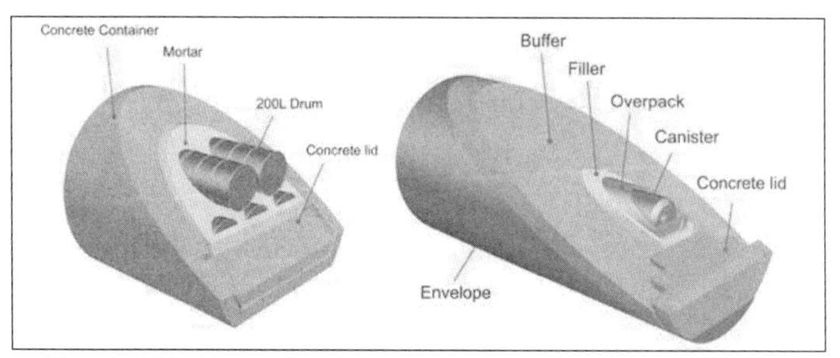

Figure 3: a) 3D representation of a concrete container (i.e. monolith) containing 10 primary waste packages of bituminized waste (left) and b) a supercontainer containing 2 canisters of vitrified HLW (right).

THE SAFETY AND FEASIBILITY CASE I (SFC-1)

The current work of O/N is focused on the so-called Safety and Feasibility Case 1 (SFC-1), which will represent an important milestone to come to an operational disposal site [7]. A SFC is an integration of all scientific and technological arguments that describe, substantiate and quantify the safety and feasibility of geological disposal. Within the safety concept for disposal, the "*performance*" of the disposal system (= fulfilling passive, long-term safety) can be described by means of so-called safety functions [8]. A safety function can be either a property or a process. The combined action of the safety functions ensures compliance of the disposal system with all safety requirements, both during the operational and the post-closure phase. Three main safety functions (and some subfunctions) have been defined in the Belgian programme for radioactive waste management, i.e. the *"isolation"* (I) function, the *"engineered containment"* (C) function and the *"delay and attenuation of the releases"* (R) safety function (see Figure 4).

The Engineered Barrier System (EBS) has to fulfill the *containment* function (C) during the operational and thermal phase. The main function of the waste form is to *limit the contaminant releases* (R1- subfunction). Due to the low hydraulic conductivity and sorption/retention capacity, the host formation is also contributing to the *delay and attenuation of the releases* (R-function) by *limiting the water flow through the system* (R2-subfunction) and by *retarding the contaminant migration* (R3-subfunction).

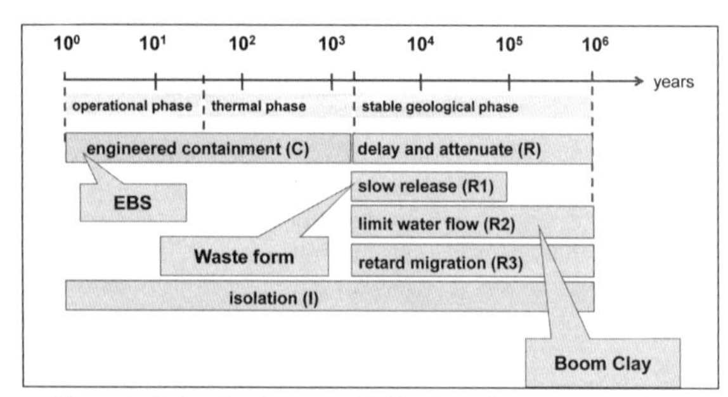

Figure 4: Safety functions provided by the main components of the disposal system (figure adapted according to [6])

THE RESEARCH STRATEGY

The research strategy developed by the Waste & Disposal Research Unit at SCK•CEN to support ONDRAF/NIRAS in the preparation of the Safety and Feasibility Case 1, is aiming at achieving two main objectives:

1) to demonstrate a sound understanding of the mechanisms and processes ("confidence building") that govern the radionuclide migration and retention behaviour in the clay host rock, and
2) to derive site specific (Mol site) migration/retention parameters for the waste relevant radionuclides that serve as an input for performance assessment (PA) calculations.

The following parameters are currently used to describe the diffusion-driven transport of the waste-relevant radionuclides in Boom Clay:

- C_{max} [mol/L]: maximum radionuclide concentration which is able to move in the pore water of the host rock (solubility limited).
- R [-]: a retardation factor describes how the transport is slowed down due to sorption effects.
- D_{pore} [m^2/s]: Diffusion coefficient of a dissolved or suspended species in the pore water of a porous medium.
- η [-]: Diffusion accessible porosity, part of the porosity which is accessible for diffusion (radionuclide dependent).

As it is difficult to study all waste relevant radionuclides, due to safety reasons, budget and time limitations, the strategic approach consists of subdividing all radionuclides into groups [9] based on analogous chemical characteristics (e.g. oxidation state, place in periodic table) and behaviour, such as their hydrolysis/complexation behaviour, their sorption behaviour (affinity and mechanism), etc. Each group is represented by one or two "reference elements", which are studied in detail and for which so-called "phenomenological models" are elaborated. These phenomenological or geochemical models represent the cornerstone of the developed methodology, describing in a qualitative and quantitative way the RN retention/migration processes under the specific conditions of the considered disposal site. The described methodology relies on a huge experimental programme, which is supported by geochemical calculations as well as transport modeling.
Currently, four main groups of RN's have been differentiated, characterized by an increasing complexity in interaction mechanisms, which coincides with their charge and valence/oxidation state:

Group I) Mono- and multivalent anions (non-metallic and metallic):
↳ I, Cl, Se, C, and Mo

Group II) Mono- and divalent cations (alkali and alkaline earth metal group):
↳ Cs, Rb, Ca, Sr and Ra

Group III) Mono- and multivalent transition metals and metalloids:
↳ Ag, Ni, Tc, Zr, Sn and Be

Group IV) Tri- and tetravalent lanthanides and actinides (+ pentavalent protactinium):
↳ Ac, Am, Cm, Eu, Sm, Pu, U, Th, Np, and Pa

INORGANIC SPECIATION

This subdivision has been – among others - based on thermodynamic speciation calculations [10], which were performed using the geochemical code Geochemist's Workbench [11] and the in-house compiled thermodynamic database MOLDATA [12]. It is well known that aqueous complexation and redox[2] reactions determine the species distribution and stability [13]. Based on the pore water composition, its ionic strength and pH, the predominant oxidation states of the radionuclides and their aqueous speciation can be identified. In Table I, the reference inorganic Boom Clay pore water composition (Mol site) is summarized and in Table II, the dominant calculated (inorganic) species of all safety-relevant radionuclides are displayed.

Under BC conditions, the halogens Cl and I form simple anionic species, while Mo and C occur as oxyanions, i.e. molybdate MoO_4^{2-} and bicarbonate (HCO_3^-). In case of Se, two oxidation states are retained, Se(II) in form of HSe^- representing the predicted stable species under the reducing BC conditions, and selenate (SeO_4^{2-} /Se(VI)) is also retained as it may persist in reducing pore waters due to redox disequilibrium and/or slow reduction kinetics [14]. Niobium (Nb) has been also added to this group. All mono- and divalent cations belonging to the alkaline and alkaline earth metals are characterized by a high stability of the simple cation. Except for possible complexation with CO_3^{2-} in the case of alkaline earth metals, no other complexation reactions (including hydrolysis) are relevant for these metals. The transition metals and metalloids comprise a very diverse group. They occur either as hydrolysed species (Pd, Tc, Zr, Sn), or as complexes (Ag, Ni) with inorganic ligands. The actinides and lanthanides belonging to group IV are mainly forming negatively charged carbonate species, neutral hydrolysis species and/or mixed hydroxyl-carbonate complexes. Besides these species, lanthanides and (especially) actinides are also known to form "eigencolloids"[3], which cannot be thermodynamically described.

Table I: BC pore water composition at 25°C [10]

Species	[mol/L]
Al	4.6×10^{-8}
Si	2.0×10^{-4}
Mg	6.6×10^{-5}
Ca	5.0×10^{-5}
Fe	3.7×10^{-6}
K	1.9×10^{-4}
Na	1.6×10^{-2}
Cl^-	7.3×10^{-4}
SO_4^{2-}	2.4×10^{-5}
pCO_2(g), atm	$10^{-2.4}$
E_h, mV	-281
pH	8.35

[2] REDOX: REDuction/OXidation

[3] eigencolloid is a term derived from the German language (*eigen*: own) and used to designate colloids of pure phases

Table II: Predominant aqueous species of the different radionuclide groups

GROUP I: Anions	GROUP II: Mono- and divalent cations	GROUP III: Transition metals	GROUP IV: Tri- and tetravalent La/Ac (+ pentavalent Pa)
I^-	Cs^+	AgHS(aq)	$Am(CO_3)_2^-$
Cl^-	Rb^+	AgCl(aq))	$Ac(CO_3)_2^-$
HCO_3^-	Ca^{2+}	$Ni(CO_3)_2^{2-}$	$Cm(CO_3)_2^-$
HSe^- / SeO_4^{2-}	Sr^{2+}	$Pd(OH)_2$(aq)	$Sm(CO_3)_2^-$
$Nb(OH)_6^-$	Ra^2	$TcO(OH)_2$(aq)	$Pu(CO_3)_3^{3-}$ / $Pu(OH)_4$(aq)
MoO_4^{2-}		$Sn(OH)_5^-$	$Np(OH)_4$(aq)
		$Zr(OH)_4$(aq)	$ThCO_3(OH)_3^-$
			$PaO_2(OH)$(aq)
			$UO_2(CO_3)_3^{4-}$ / $U(OH)_4$(aq)

Based on the description given above, it is obvious that the radionuclide grouping can not be premised only on the calculated inorganic aqueous speciation. Further substantiation for grouping is indeed obtained from sorption, solubility and transport experiments in Boom Clay. The interaction of the different RN's with NOM plays a crucial role in these processes, as will be shown hereafter.

ORGANIC SPECIATION

In order to systematically check for the possible change in solution speciation due to the presence of DOM, the geochemical Humic Ion-binding model VI [15] introduced into the Phreeqc code [16] was used for scoping calculations. The Humic Ion-binding model VI is able to simulate metal-humic substance binding in a wide range of pH, ionic strength and competing cation conditions. In the scoping calculations, generic humic substance acid-base and metal-DOM complexation parameters were used [15]. Application of these generic parameters to an extended in-house experimental data set has shown that these generic parameters provide a good general description of the radionuclide speciation in BC pore water [16]. Results indicated that all metal ions with $\log K_{MA}$ (HA) values ≤ 1.1 are relatively uncomplexed by humic substances, while for the other metals, DOM may have a profound influence on the aquatic speciation in the slightly alkaline pH range of Boom Clay. Humic substance complexation is therefore an issue restricted to heavy metal cations and multivalent lanthanides/actinides (Group III and IV). The speciation of anions as well as alkali and alkaline earth metals under BC conditions is considered not to be influenced by DOM [17, 18, 19].

SOLUBILITY/CONCENTRATION LIMIT

Solubility determines the concentration of a radionuclide released from the radioactive waste and therefore is one of the key parameters influencing the mobile fraction within and out of the disposal system. Consequently, taking into account solubility/concentration limits into transport calculations represents an important feature when assessing the performance of the disposal system and safety of the environment. With respect to solubility in the far-field, the parameter/phase selection strategy has been mainly based on thermodynamic calculations [7]. The difficulty of such calculations is/was to select the most relevant phase from the thermodynamically possible phases and the kinetically most likely/favourable phases to form. Experimental data – if available - as well as expert judgement were therefore integrated into the selection procedure. Concerning elements for which thermodynamic (and experimental) data were lacking, parameter ranges were derived by chemical analogy (e.g. for Ac, Cm – were treated as chemically similar to Am). It should be mentioned, that the calculations were performed without considering an initial concentration constraint. In other words, the radionuclide inventory of the waste was not used to calculate or estimate an "inventory-limited pore water concentration" for the far-field. The formation of "eigencolloids" and/or "pseudo-colloids" was also not integrated into the solubility assessment, although it is well known that this process may increase the amount of actinides/lanthanides in the groundwater and lead to "apparent solubilities", which may exceed the thermodynamic solubilities by several orders of magnitude. The formation of mixed solid phases or aqueous solid solutions (Aq-SS), co-dissolution and co-precipitation were also not taken into account in the thermodynamic calculations. The latter implies that the role of dissolution and precipitation kinetics was neglected also in the applied approach. However, phase selection was based on the so-called Ostwald principle/rule [20, 21, 22], meaning that concentration limit through formation of an amorphous phase was considered to be generally kinetically favored compared to the precipitation of a crystalline solid.

SORPTION

Sorption is one of the most important processes able to attenuate RN release from the disposal system. Illite and smectite are the principal adsorptive constituents of Boom Clay. They have a similar structure, with both belonging to the group of three-layer clay minerals, consisting of a regular sequence of octahedral Al-O-OH layers sandwiched in between two tetrahedral Si-O layers. The layer or permanent charge of the clay minerals is created by isomorphous substitution of higher charged cations by cations of lower valency in the crystal lattice. Usually in the tetrahedral sheets, Si^{4+} is replaced by Al^{3+}, whereas in the octahedral layer Mg^{2+} or Fe^{2+} may substitute for Al^{3+}. The unbalanced negative charges may be compensated and neutrality restored through interlayer cations on the one hand and adsorption/coordination of cations to the external surface sites via pH independent cation-exchange processes. While for illite the interlayer cation (predominantly potassium) is generally not exchangeable, within the expandable smectite structure, the interlayer cation(s) can be exchanged. Cation exchange reactions are generally fast, stoichiometric and reversible.

Besides the permanent charge, clay minerals also possess a variable charge, which develops mainly along broken edges of the clay platelets as function of pH due to the protonation or deprotonation of the surface hydroxyl- (OH-) groups. These groups are able to form stable surface complexes with cations at mostly neutral-to-alkaline pH.

Based on a large amount of in-house performed experimental data, it could be demonstrated that monovalent alkali metals and alkaline earth metals are mainly sorbed via cation exchange on illite (e.g. Cs) and smectite (e.g. Sr, Ca), while transition metals, metalloids as well as lanthanides and actinides are mainly sorbed via surface complexation onto illite and smectite.

Although different surface complexation models are available to describe/model metal uptake by clay minerals, at SCK•CEN mainly two approaches are used:

1) the so-called "generalized 3-sites cation exchange model" developed by [23, 24], and
2) the 2 Site Protolysis Non Electrostatic Surface Complexation and Cation Exchange (SPNE SC/CE) sorption model developed by [25, 26].

These models have been successfully used to describe adsorption of Ni^{2+}, Eu^{3+}, $Am^{3+}$and Th^{4+} onto BC. The reader is referred to [16, 27, 28] for further details.

RETARDATION

Although the above-mentioned models are used to describe the sorption data in a "mechanistic way", in performance assessment calculations a much simpler approach is used to represent RN retention. Sorption is frequently quantified by so-called distribution/partition coefficients (K_d-values), representing the partitioning of a contaminant between the aqueous phase and a solid surface:

$$K_d \text{ [l/kg]} = \{C\}_{solid,eq} / [C]_{liquid,\,eq}$$

with $\{C\}_{solid,\,eq}$ the concentration of a contaminant adsorbed on the solid phase (mol/kg) at equilibrium, and $[C]_{liquid,eq}$ the corresponding concentration of the contaminant in the liquid phase (mol/l);

Based on these distribution/partition coefficients, the retardation factor for a certain radionuclide can be calculated according to:

$$R = 1 + \frac{\rho \cdot K_d}{\eta}$$

with ρ the bulk dry density (1.7 kg/m³) and η the diffusion accessible porosity (~0.35).

In order to derive transport parameters and properties, different types of migration experiments have been performed at SCK•CEN in the last decades [29]: in-diffusion, through-diffusion, percolation, and impulse injection. In Figure 5, a schematic representation of a percolation-type migration experiment is illustrated. These experiments have been typically used to investigate the transport of RN's with mobile DOM colloids.

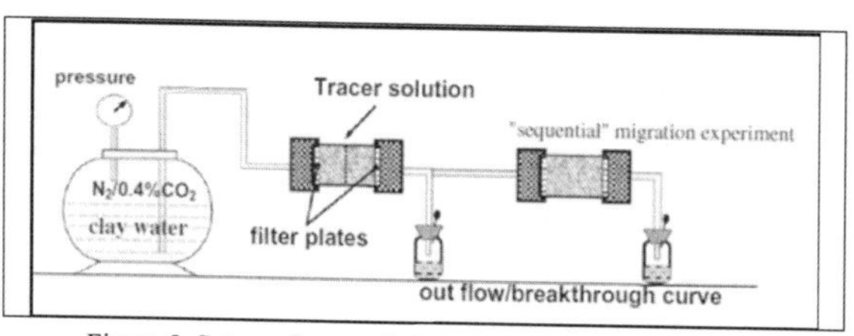

Figure 5: Set-up of a percolation type migration experiments

In these migration experiments, for each radionuclide to be investigated, separate stainless steel cells were filled with 2 BC cores of equal length (total length: 7×10^{-2} m) and in between the clay cores, a cellulose filter, spiked with the radionuclide under consideration, was placed. The migration cells were then connected to a pressure vessel containing Real Boom Clay Water (RBCW) at a pressure of 1 MPa. The RBCW was percolated through the clay cores and sampled at the outlet. The radionuclide concentration/activity was monitored over time resulting in a so-called break-through curve. After several years, when a constant RN concentration was reached in the outlet solution, the outflowing water was coupled to the inlet of a second cell containing another BC core of 3×10^{-2} m length. Through this coupling, a constant concentration boundary condition was provided. At the end of the experiments, the clay cores were extracted from the migration cells and cut in thin slices of 0.1-1 mm and the isotope activity profile determined via gamma analysis. In Figure 6, the measured outlet concentrations of migration experiments performed with Tc, Cm, Pu, Np and Pa over time are illustrated. It can be seen that the RN concentration profiles of the 1st and 2nd core show very similar features. After an initial rapid break-through, a constant outlet concentration was reached. However, the concentration profiles of the 2nd core are generally lower than the outlet concentrations of the 1st core. The latter observation reveals that the measured constant concentrations are not solubility controlled, otherwise both profiles would be superposable. Equal profiles would be also observable, if the radionuclides were behaving as conservative tracers, which is also not the case.

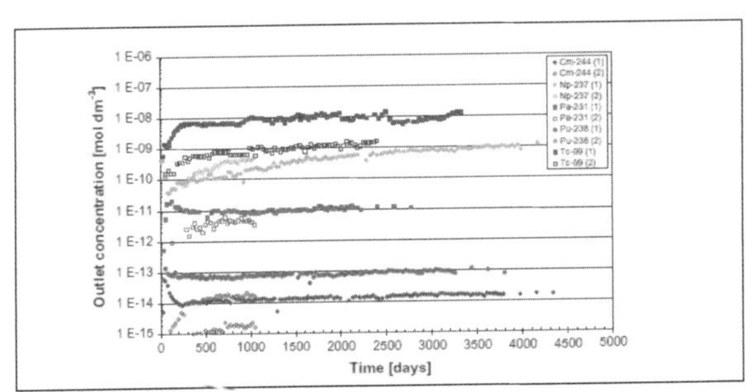

Figure 6: Evolution of outlet concentrations measured in sequential/column migration experiments

Instead, another interpretation was put forward. As outlined before, group III and IV elements show a strong interaction with DOM, which led to the description of the RN migration by an organic matter facilitated transport model [30].

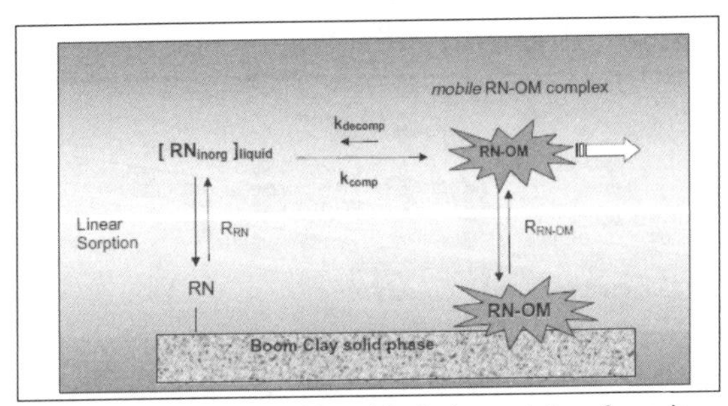

Figure 7: Conceptual model used for the interpretation of organic matter linked radionuclide migration in BC

Within this transport model (see Figure 7), the number of parameters remains limited and most of them can be obtained from independent measurements (batch complexation/solubility experiments, batch sorption experiments, OM transport experiments). In this model, two components per radionuclide are allowed to migrate: 1) the dominant (as revealed by the speciation calculations) inorganic aqueous species under Boom Clay conditions ([RN_{inorg}]) and 2) the species (*i.e.* radionuclide) associated with the organic matter colloid ([RN-OM]). Both components are described by the classical advection-diffusion-reaction with linear sorption equation. The transport of the RN-OM mobile complex is described by:

$$(\eta+\rho_b K_{d,RNOM})\frac{\partial(c_{RNOM})}{\partial t}-\nabla\cdot\left[\eta D_{p,RNOM}\nabla c_{RNOM}+u_{Darcy}c_{RNOM}\right]=-\lambda(\eta+\rho_b K_{d,RNOM})c_{RNOM}+\eta Q_{sol-OM}$$

and the transport of the free RN-species in solution is described by:

$$(\eta + \rho_b K_{d,RN})\frac{\partial(c_{RN})}{\partial t} - \nabla \cdot \left[\eta D_{p,RN} \nabla c_{RN} + u_{Darcy} c_{RN}\right] = -\lambda(\eta + \rho_b K_{d,RN})c_{RN} - \eta Q_{sol-OM}$$

where c_{RN}, c_{RNOM}, c_{OM} are the concentrations of free inorganic radionuclide species in solution, concentration of the mobile RN-OM complex and the concentration of the mobile OM, respectively. $D_{p,RNOM}$ and $D_{p,RN}$ are the pore diffusion coefficients (D_p) of the RN complexed to the mobile OM and of the free inorganic RN species in solution, respectively.

The exchange of the radionuclide between $[RN_{inorg}]_{liquid}$ and the RN-OM complex is described by a complexation constant which is the ratio between the association and dissociation kinetics. The mass transfer (Q_{sol-OM}) of the RN between the OM complexed form and the "free" RN in solution is given by:

$$Q_{sol-OM} = k_{comp} c_{RN} c_{OM} - k_{decomp} c_{RNOM}$$

$$\left[RN_{inorg}\right] + \left[OM\right] \underset{k_{decomp}}{\overset{k_{comp}}{\rightleftarrows}} [RN - OM] \text{ and } K_{RN \cdot OM} = \frac{k_{comp}}{k_{decomp}} = \frac{c_{RNOM}}{c_{RN} \cdot c_{OM}}$$

where k_{comp} and k_{decomp} are the kinetic rate constants for the RN-OM complexation and decomplexation reactions, respectively; and K_{RNOM} is the equilibrium constant.
This model lead to the successful simulation of the break-through curves of different radionuclides belonging to groups III and IV (Tc, Cm, Np, …), indicative for their very similar (geochemical) migration behaviour in Boom Clay.

CONCLUSIONS

The Belgian radioactive waste management organisation ONDRAF/NIRAS is preparing its first Safety and Feasibility Case (SFC-1) to demonstrate the long-term safety of its deep disposal concept for Category B and C wastes. In order to support O/N, SCK•CEN has developed a research strategy to characterize the migration/retention behaviour of the waste relevant radionuclides within Boom Clay at Mol under present-day geochemical conditions. It was shown that through a combination of laboratory experiments, thermodynamic and kinetic transport calculations, different radionuclide groupings, characterized by similar speciation (organic and inorganic), solubililty and sorption behaviour, could be identified. Detailed phenomenological studies enabled to understand the underlying mechanisms of the former processes and the derivation of consistent transport parameters, that are necessary as input for performance assessment calculations. Recently obtained results of long-term running percolation experiments, enabled also the development of a new conceptual model with respect to the colloid facilitated RN transport in BC.

ACKNOWLEDGEMENTS

This work is performed in close cooperation with, and with the financial support of ONDRAF/NIRAS, the Belgian Agency for Radioactive Waste and Fissile Materials, as part of the programme on geological disposal of high-level/long-lived radioactive waste that is carried out by ONDRAF/NIRAS.

REFERENCES

[1]ONDRAF/NIRAS, Plan Déchets pour la gestion à long terme des déchets radioactifs conditionnées de haute activité et/ou de longue durée de vie et aperçue de questions connexes Rapport NIRONS 2011-02 F (2011).
[2]ONDRAF/NIRAS, Aperçu technique du rapport SAFIR 2, Safety Assessment and Feasibility Interim Report 2, NIROND 2001-05 F (2001).
[3]L. Yu, M. Gedeon, I. Wemaere, J. Marivoet, and M. De Craen, Boom Clay hydraulic conductivity, A synthesis of 30 years of research, External Report, SCK•CEN-ER-122 (2011).
[4]M. Honty, CEC of Boom Clay – a review, External Report, SCK•CEN-ER-134 (2010).
[5] C. Bruggeman, D. J. Liu, and N. Maes, Influence of Boom Clay organic matter on the adsorption of Eu^{3+} by illite - geochemical modeling using the component additivity approach, *Radiochimica Acta*, **98**, 597-605(2010).
[6] ONDRAF/NIRAS, Geological Disposal Programme, Feasibility Strategy and Feasibility Assessment Methodology for the Geological Disposal of Radioactive Waste, SFC-1 Level 4 report, NIROND-TR 2010-19 E (2011).
[7]ONDRAF/NIRAS, Overview of SFC-1 quality measures and activities - V2 May 2007, NIROND note 2006-1103, Working Document v. 1.4 (2006).
[8]ONDRAF/NIRAS, The Long-term Safety Assessment Methodology for the Geological Disposal of Radioactive Waste – SFC-1 Level 4 Report: Second Full Draft, NIROND-TR 2009-14 E (2009).
[9]C. Bruggeman, S. Salah, N. Maes, L. Wang, A. Dierckx, and M. Ochs, Outline of the experimental approach adopted by SCK•CEN for developing radionuclide sorption parameters, External Report SCK•CEN-ER-73 (2008).
[10]S. Salah and L. Wang, Speciation and solubility calculations for waste relevant radionuclides in Boom Clay, External Report, SCK•CEN-ER-198, First Full Draft (2011).
[11]C. M. Bethke and S. Yeakel, The Geochemist's Workbench Release 8.0 (four volumes), Hydrogeology Program. University of Illinois, Urbana, Illinois (2010).
[12]L. Wang, S. Salah, and H. De Soete, MOLDATA: A Thermochemical Data Base for phenomenological and safety assessment studies for disposal of radioactive waste in Belgium, External Report SCK•CEN-ER-121 (2011).
[13]W. Runde, The chemical interactions of actinides in the environment. Los Alamos Science, N° 26, 392-411 (2000).
[14]P. De Cannière, A. Maes, S. Williams, C. Bruggeman, T. Beauwens, N. Maes, and M. Cowper, Behaviour of Selenium in Boom Clay, External Report SCK•CEN-ER-120 (2010).
[15]E. Tipping, Humic ion-binding model VI: An improved description of the interactions of protons and metal ions with humic substances, *Aquatic Geochemistry*, **4**(1), 3-48 (1998).
[16]D. J. Liu, C. Bruggeman, and N. Maes, The influence of natural organic matter on the speciation and solubility of Eu in Boom Clay pore water, *Radiochimica Acta*, **96**(9-11), 711-720 (2008).
[17]N. Maes, S. Salah, C. Bruggeman, M. Aertsens, and E. Martens, Caesium retention and migration behaviour in Boom Clay, Topical report – First Draft, External Report, SCK•CEN-ER-153 (2011).
[18]N. Maes, S. Salah, C. Bruggeman, M. Aertsens, E. Martens, and L. Van Laer, Strontium retention and migration behaviour in Boom Clay, Topical Report – First Full Draft, External report SCK•CEN-ER-197 (2011).
[19]C. Bruggeman, M. Aertsens, N. Maes, and S. Salah, Iodine retention and migration behaviour in Boom Clay. Topical report – First Full Draft, External report SCK•CEN-ER-119 (2010).

[20]H. C. Helgeson, W. M Murphy, and P. Aargard, Thermodynamic and kinetic constraints on reactions among minerals and aqueous solutions. II. Rate constants, effective surface area, and the hydrolysis of feldspar, *Geochimica et Cosmochimica Acta*, **48**, 2405-2432 (1984).
[21]W. Stumm and J. J. Morgan, *Aquatic Chemistry*, 3rd ed., John Wiley and Sons (1996).
[22]C. I. Steefel and P. Van Cappellen, A new kinetic approach to modeling water-rock interaction: The role of nucleation, precursors, and Ostwald ripening, *Geochimica et Cosmochimica Acta*, **54**, 2657-2677 (1990).
[23]M. H. Bradbury and B. Baeyens, A generalised sorption model for the concentration dependent uptake of caesium by argillaceous rocks, *Journal of Contaminant Hydrology*, **42**(2-4), 141-163 (2000).
[24]C. Poinssot, B. Baeyens, and M. H. Bradbury, Experimental and modeling studies of caesium sorption on illite, *Geochimica et Cosmochimica Acta*, **63**(19-20), 3217-3227 (1999).
[25]M. H. Bradbury and B. Baeyens, Sorption modeling on illite Part I: Titration measurements and the sorption of Ni, Co, Eu and Sn, *Geochimica et Cosmochimica Acta*, **73**(4), 990-1003 (2009).
[26]M. H. Bradbury and B. Baeyens, Sorption modeling on illite. Part II: Actinide sorption and linear free energy relationships, *Geochimica et Cosmochimica Acta*, **73**(4), 1004-1013 (2009).
[27]N. Maes, S. Salah, D. Jacques, M. Aertsens, M. Van Gompel, P. De Cannière, and N. Velitchkova, Retention of Cs in Boom Clay: Comparison of data from batch sorption tests and diffusion experiments on intact clay cores, *Physics and Chemistry of the Earth*, **33**, 149-155 (2008).
[28]N. Maes, M. Aertsens, S. Salah, D. Jacques, M. Van Gompel, Cs, Sr and Am retention on argillaceous host rocks: comparison of data from batch sorption tests and diffusion experiments, Updated version of the PID1.2.18 delivered to the FUNMIG project, External Report SCK•CEN-ER-98 (2009).
[29]M. Aertsens, Migration in Clay: experiments and models, External Report SCK•CEN-ER-165 (2011).
[30]N. Maes, C. Bruggeman, J. Govaerts, E. Martens, S. Salah, and M. Van Gompel, A consistent phenomenological model for natural organic matter linked migration of Tc(IV), Cm(III), Np(IV), Pu(III/IV) and Pa(V) in the Boom Clay, *Physics and Chemistry of the Earth*, **36**, 1590-1599 (2011).

Green Technologies for Materials Manufacturing and Processing

RECLAIMING FIBROUS MATERIAL IN MANUFACTURING PROCESSES

Kevin D. Baker
USDA, ARS, Southwestern Cotton Ginning Research Laboratory
Mesilla Park, New Mexico, U.S.

ABSTRACT

In the manufacture of faced, fibrous materials, separation of fibrous materials from the facing material would allow the fibrous material to be recycled – in the case of fiberglass, melted and reused in the plant – while some facing materials, such as oil-impregnated paper, could be used as a fuel for a cogeneration facility. In all cases, material delivered to the landfill would be significantly reduced. A machine was developed that used a 22 kW (30 hp) fan to pull a vacuum on a rotating perforated round screen to hold the material in place while a series of a bar, air knife and rotating saw blades removed the fibrous material from its facing. The machine was tested with faced, fiberglass insulation. Optimum machine feed rate equivalent was 20 rpm rotational speed of the perforated screen. Minimum clearance between the bar and the perforated roller was 9.5 mm (0.375 in") with the saw operating and 12.7 mm (0.5 in) without the saw operating. At least 80 % of the fibrous material must be removed before the saw or air knife can be used. The bar and air knife combination was not much better than the bar alone. The saw improved both the feeding of the material and the effectiveness of material removal. Percent fiberglass removed averaged 98.0 % for paper and aluminum facing when the saw was set at a 550 rpm saw speed. The effectiveness of the hand/machine fiberglass removal was improved to 99.5 % when the insulation was fed through the machine five times per each piece. Use of all three devices (bar, saw and air knife) resulted in the best material removal – averaging 99.5 % for paper and aluminum facing and 99.9 % for polyethylene facing. There was no facing material observed in any of the fiberglass that was removed.

INTRODUCTION

In the manufacture of faced fibrous materials, defects may occur or scraps may be produced that cause the manufacturer to reject or have surplus material. In the case of fiberglass insulation manufacturing, from 36 to 54 tonnes (40 to 60 tons) of material may be rejected each month. This material is currently hauled to a landfill, placing both a financial and an environmental burden on the company. It is thought that some cotton ginning equipment could be modified so that it could be used to cleanly separate the fibrous material from its facing material. This separation of materials would allow the fibrous material to be recycled – most commonly to be reused in the plant – while the facing material could be reused as well. An example of facing material reuse would be oil-impregnated paper from the fiberglass insulation, with a heat value of 28,000 kJ/kg (12,000 Btu/lb), which could be used as a fuel for a cogeneration facility. (There are estimated to be 29 fiberglass insulation manufacturing plants across the U.S. and others worldwide that would benefit from this technology.) By reusing some or all of the materials, the amount delivered to the landfill would be significantly reduced, thereby reducing manufacturing costs.

The objective of this project is to utilize concepts from cotton cleaning equipment to design and develop a machine to remove fibrous material from a thin laminar facing material. The machine should meet the following specifications:

- processing throughput of 1.8 tonnes (2 T) of raw material per 8-hour day
- manual feed of raw material (paper-backed insulation or other)

- manual removal of backing material from machine
- fibrous material separated and directed to materials handling equipment
- capable of working with raw material lengths ranging from 0.3 to 15 m (1 to 50 ft)
- capable of working with raw material widths of 0.3 to 1.2 m (12 to 48 in)
- capable of working with raw material thicknesses of approximately 0.025 to 0.02 m (1 to 8 in)
- machine will be powered by electricity at 240 to 480 volts, 3 phase
- machine will separate materials with less than 1 percent (by weight in randomly selected samples) of foreign material in either the fibrous material output or the backing material output.

MATERIALS AND METHODS

A machine was developed that used a vacuum roller to hold material in place while a stationary bar, a series of rotating saw blades, and an air knife removed the fiberglass from the backing. An option of a brush against the roller was also tried. The vacuum roller used a 22 kW (30 hp) centrifugal fan to pull a vacuum of 2 kPa (8 inches w.c.) on a cylinder made of perforated metal that was 0.5 m (20 in) in diameter and had 20 % open area (figure I). The length of the roller was 1.4 m (54 in), enough to easily accommodate material that could be as wide as 1.2 m (4 ft).

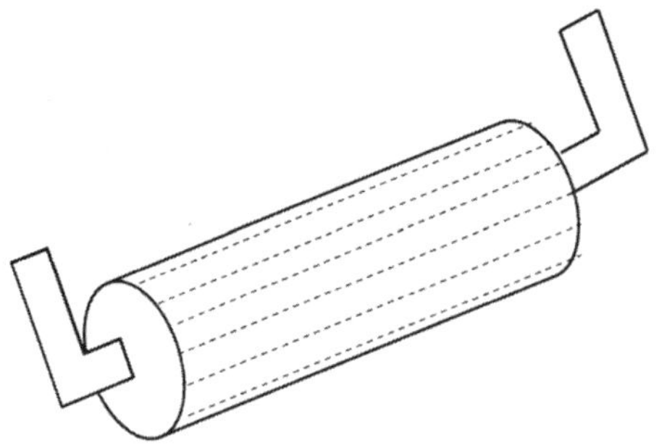

Figure I. Sketch of a perforated roller. Ducts in each of the two ends connect to a fan that will pull a vacuum on the roller.

The stationary bar was mounted along the length of the perforated roller. (figure II) It was 0.078 m (3 in) wide and 9.5 mm (0.375 in) thick, with a 15 degree taper on the feed end. The gap between the bar and the roller could be adjusted from 0 to 0.025 m (0 to 1 in).

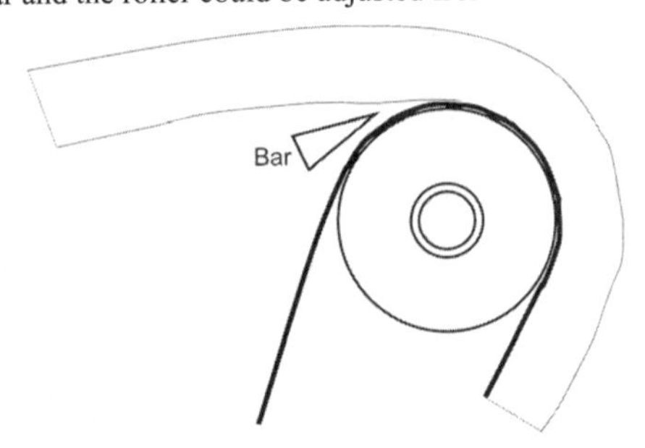

Figure II. Sketch of perforated roller with bar for removal of fibrous material.

The air knife was mounted along the length of the perorated roller (figure III). It released an air jet through a 0.05 mm (0.002 in) gap against the facing material and was used to remove whatever fibrous material remained on the facing material after other devices had been incorporated. A pressure regulator was used to adjust the pressure up to 690 kPa (100 psi), with air flow rate varying up to 140 L/min per m (1.5 standard ft^3/min per foot) of air knife width, depending upon the air pressure.

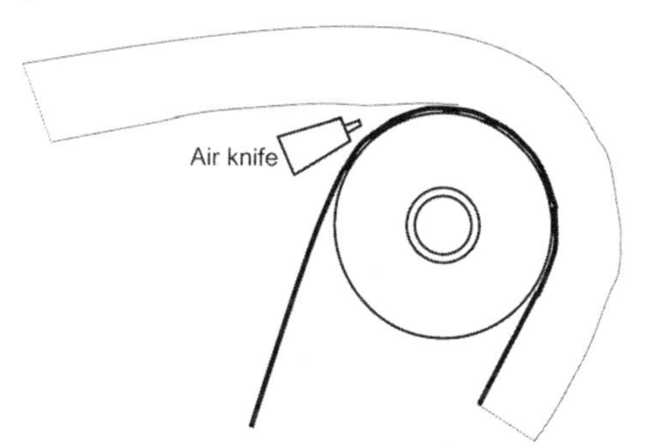

Figure III. Sketch of perforated roller with air knife for removal of fibrous material.

The saw was a long, continuous saw blade wrapped around a 0.3 m (12 in) diameter solid cylinder that was mounted along the perforated roller with 1.6 mm (0.0625 in) clearance between the saw and the roller (figure IV). The saw had approximately 2.4 teeth per cm (6 teeth per inch) that projected up from the cylinder by 6.4 mm (0.25 in). The rows of teeth were 3.2 mm (0.125 in) apart. Saw teeth were aggressive in removal, being pointed in a forward rotating direction at an angle of 30 degrees. Saw speed could be varied from 127 to 550 rpm. Material removed by the saw could be doffed using an air blast or by brushing the saw with a cylindrical series of steel-wire brushes. Brush speed was fixed in proportion to saw speed at 2.25 times the saw speed.

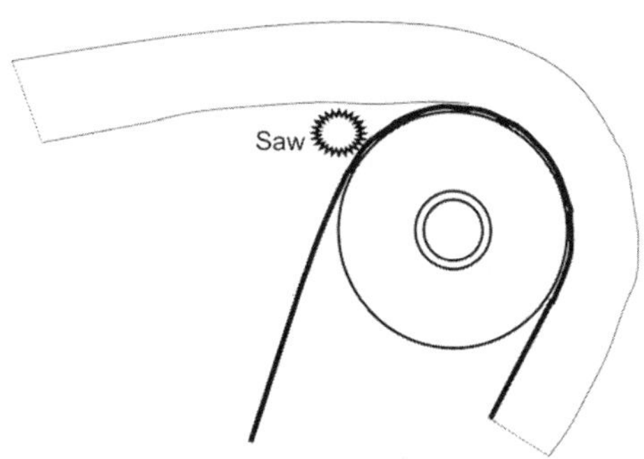

Figure IV. Sketch of perforated roller with saw for removal of fibrous material.

A fourth method of separating the fibrous material from its facing material was the brush method (figure V). This method was only tested by hand with a steel-wire brush and was not automated.

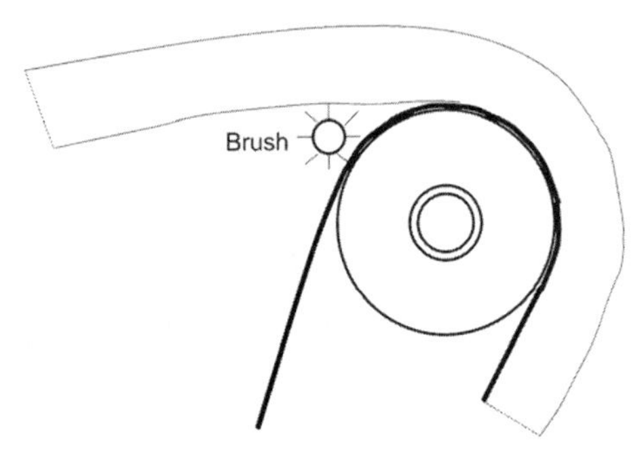

Figure V. Sketch of perforated roller with brush for removal of fibrous material.

Materials tested were all faced, fiberglass insulation rolls, some of which were manufacturing defects and some of which were new materials with no defects. Three facing materials were tested: oil-impregnated paper, aluminum foil, and polyethylene film. Material dimensions ranged from 0.3 to 2.4 m (1 to 8 ft) in length, from 0.3 to 1.2 m (12 to 48 in) in width, and from 0.05 to 0.25 m (2 to 10 in) thick (R-11 to R-30). Material with polyethylene facing was covered on both sides of the fiberglass. Hand separation of one side of the polyethylene material was needed before the any components of the machine could be used. Materials were also evaluated to determine the effectiveness of hand separation of fibrous material and facing material.

In each test, the material was manually-fed into the machine and the bulk of the fiberglass was manually removed by the same person that was feeding the material. A second person was on the other side of the machine to manually remove the facing material. Effectiveness was determined using weights from a lab scale before and after the separation processes. Material remaining on the facing material was removed manually by brushing with a steel-wire brush.

RESULTS AND DISCUSSION

Hand separation of the material resulted in an average of 80 % fiberglass removal for material with paper and aluminum facing and an average of 90 % fiberglass removal for material with polyethylene facing. The amount of material to be removed is greater than the saw capacity and interferes with the blast from the air knife, therefore either hand separation or use of the bar is necessary before the saw or air knife can be used.

The vacuum pulled on the roller could be up to 2 kPa (8 inches w.c.). In nearly all cases, this amount of vacuum was sufficient to hold the materials flat against the roller so the other devices could remove the fibrous material. The only exception to this was when using the saw with material having polyethylene facing. The saw would pull the material through at a feed rate that was too fast to be effective. A greater vacuum may have prevented this, but equipment was not available to produce a greater vacuum.

The bar was tested alone with a 6.4, 9.5 and 12.7 mm (0.25, 0.375 and 0.50 in) gap between the perforated roller and the bar. None of the materials tested would feed uniformly through the machine with a 6.4 mm (0.25 in) gap. Material could feed uniformly through the 9.5 mm (0.375 in) gap with the saw operating at low speed (127 rpm), but not without the saw operating unless there was manual assistance. Material would feed uniformly through the bar with a 12.7 mm (0.5 in) gap without any additional assistance. The percent of fiberglass removed using a 9.5 mm (0.375 in) bar gap was a little higher than was removed using a 12.7 mm (0.5 in) bar gap with 95.5 % removed versus 93.7% removed for paper and aluminum facings and 99.1 % removed versus 97.9 % removed for polyethylene facing (Table I).

Table I. Effectiveness of fiberglass removal from faced, fiberglass insulation batts.

Device used	Paper and aluminum facing (% fiberglass removed)	Polyethylene facing (% fiberglass removed)
Hand separation	80	90
Bar with 12.7 mm gap	93.7	97.9
Bar with 9.5 mm gap*	95.5	99.1
Bar (12.7 mm gap) with air knife	94.3	98.1
Bar (9.5 mm gap) with saw at 300 rpm	96.5	Not operable
Bar (9.5 mm gap) with saw at 550 rpm	98.0	Not operable
Bar (9.5 mm gap) with saw (127 rpm) and air knife	99.5	99.9

*Material feed was manually augmented for testing purposes.

The bar plus air knife was tested with a 12.7 mm (0.5 in) gap between the perorated roller and the bar and the air knife operating at 620 kPa (90 psi). Flow rate through the knife was about 85 L/min (3 standard ft^3/min). Because the material would not feed through the system, this combination could not be tested with a 9.5 mm (0.375 in) gap. The air knife offered only a slight improvement over the bar alone, increasing the percent of material removed to 94.3 % for paper and aluminum facings and to 98.1 % for polyethylene facing. However, the material removed is lower for the bar and air knife combination with the bar set at a 12.7 mm (0.5 in) gap than for the bar alone set at a 9.5 mm (0.375 in) gap.

The bar plus saw was tested with a 9.5 mm (0.375 in) gap between the perforated roller and the bar and a 1.6 mm (0.0625 in) clearance between the saw and the roller. The saw operated at 300 and 550 rpm. Because the saw would pull the plastic-faced material through the machine very rapidly, the plastic-faced material could not be tested with this combination. This condition may have been remedied by pulling a greater vacuum on the perforated roller, but our equipment did not allow this. The saw was more effective when operated at 550 rpm than at 300 rpm, with 98.0 % of fiberglass removed with the saw operating at 550 rpm and 96.5 % material removed with the saw operating at 300 rpm. When operating this combination and the saw set at 550 rpm, the effectiveness of material removal could be increased by feeding material through a repeated number of times. Effectiveness increased from 98.0 % of fiberglass removed to 98.5, 98.9, 99.2 and 99.5 percent fiberglass removal with a second, third, fourth and fifth pass of the facing material through the machine, respectively. Maximum removal was 99.5 %, despite further passes through the machine.

The bar plus saw plus air knife combination was tested with a 9.5 mm (0.375 in) gap between the perforated roller and the saw, the saw operating at 127 rpm and the air knife set at 620 kPa (90 psi). This combination resulted in the best fiberglass removal. The percent of material removed averaged 99.5 % for paper and aluminum facings and to 99.9 % for polyethylene facing. This combination achieved the fibrous material objective for the project.

CONCLUSION

A machine was developed that used a 22 kW (30 hp) fan to pull a vacuum on a rotating perforated round screen to hold the material in place while a series of a bar, air knife and rotating saw blades removed the fibrous material from its facing. The machine was tested with faced, fiberglass insulation. Optimum machine feed rate equivalent was 20 rpm rotational speed of the perforated screen. Minimum clearance between the bar and the perforated roller was 9.5 mm (0.375 in) with the saw operating and 12.7 mm (0.5 in) without the saw operating. At least 80 % of the fibrous material must be removed before the saw or air knife can be used. The bar and air knife combination was not much better than the bar alone. The saw improved both the feeding of the material and the effectiveness of material removal. Percent fiberglass removed averaged 98.0 % for paper and aluminum facing when the saw was set at a 550 rpm saw speed. The effectiveness of the hand/machine fiberglass removal was improved to 99.5 % when the insulation was fed through the machine five times per each piece. Use of all three devices (bar, saw and air knife) resulted in the best material removal – averaging 99.5 % for paper and aluminum facing and 99.9 % for polyethylene facing. There was no facing material observed in any of the fiberglass that was removed.

DISCLAIMERS

Some of the material discussed in this manuscript comprises the subject matter of a patent application currently pending with the US Patent and Trademark Office. If you are interested in licensing the technology described herein, please contact Dr. Kevin Baker. Dr. Baker can direct you to the appropriate US Department of Agriculture (USDA), Agricultural Research Service (ARS) personnel who can answer your questions and provide you with further information regarding licensing opportunities. In the event that Dr. Baker is not immediately available, please contact the Technology Transfer Coordinator for the USDA-ARS Southern/Northern Plains Areas at the Northern Plains Area Research Center in Fort Collins, Colorado.

Mention of trade names or commercial products in this article is solely for the purpose of providing specific information and does not imply recommendation or endorsement by the U.S. Department of Agriculture.

DIRECTIONAL DRIVERS OF SUSTAINABLE MANUFACTURING: THE IMPACT OF SUSTAINABLE BUILDING CODES AND STANDARDS ON THE MANUFACTURERS OF MATERIALS

Amy A. Costello, PE, LEED-AP
Marsha S. Bischel, Ph. D.
Armstrong World Industries, Inc.
Lancaster, PA 17603

ABSTRACT

One of the driving forces behind the environmental sustainability movement remains the push towards "green" buildings. As a movement, environmental sustainability has experienced rapid growth and development. In the past five years, there has been a broad shift away from voluntary systems, largely focused on energy and carbon reductions associated with building operations, towards holistic, mandated ones. As part of this movement, there are now green building codes such as CALGreen and the International Green Construction Code, legislation, executive orders and updated green building schemes that will directly impact the manufacturing of materials used in the built environment. Many of these initiatives are beginning to converge towards a common set of requirements for manufacturers. Among the items which will impact manufacturers are the drive for increased recycling and recycled content, limits on chemicals of concern, on-site energy production and storage facilities, a preference for bio-based materials, and a move towards life cycle assessments.

EVOLUTION OF REQUIREMENTS FOR SUSTAINABLE BUILDINGS

In 1990, the Building Research Establishment (BRE) in the United Kingdom developed a voluntary tool to measure the sustainability of new non-domestic buildings in the UK which they called the BRE Environmental Assessment Method (BREEAM). Little did they realize that they would start a phenomenon that would result in over 260 copy cat rating systems around the world. Today, the United States Green Building Council's (USGBC) Leadership in Energy and Environmental Design (LEED) is the world's most widely used rating system. First launched in 1998, the USGBC has certified over 2 billion square feet of building space using the LEED system.[1]

In the more recent past, there has been a shift away from totally voluntary systems to ones that are mandated, or legislated. As an intermediate step, many state and local governments in the US began requiring LEED certification of their own buildings as early as 2005. As of Dec. 2011, LEED was required in some way, from resolution, executive order or legislation, by 384 local governments, 34 state governments and 14 federal agencies or departments.[1] In 2007, the California Building Standards Commission was tasked with adopting green building standards for residential, commercial, and public buildings. This effort led to the creation of the California Green Building Codes or "CALGreen" which became effective on January 1, 2011, and was the first "green building code" in the US. This code is mandatory for all new construction in the state of California.

In 2009, the International Codes Council began developing the International Green Construction Code (IgCC), which was designed as an overlay code which jurisdictions could select to use in addition to any of the other existing ICC's codes, such as the International Building Code or International Residential Code. The IgCC was finalized in March 2012 and as of June 2012, seven states or jurisdictions within those states had adopted the code, with more expected to follow suit.

The voluntary and mandated systems all have requirements that are applicable to materials, such as low volatile organic compound (VOC) emissions or recycled content. These requirements have a direct impact on raw material and process selection for building product manufacturers. Other requirements such as requirements for the use of on-site renewable energy sources or the avoidance of chemicals of concern also have a direct impact on materials manufacturers. A comparison of the sustainable building rating schemes and building codes in Table I shows that overall, there is a group of common needs for materials that are encouraged and rewarded by the systems used in North America. The first five of these will be addressed further.

Table I. Material needs encouraged by existing codes and building rating systems.

Material Criteria	**LEED**[3,4]	**CALGreen**[2]	**IgCC**[5]
Recycled Content	X	X	X
Recycling of Materials/Waste diversion	X	X	X
Chemical Restrictions	X	X	X
On-site Energy generation/storage	X	X	X
Bio-based Content	X	X	X
Low VOC Emitting Materials	X	X	X
Regional Materials	X		X

RECYCLED CONTENT, RECYCLABILITY AND WASTE DIVERSION LINKS

Significant emphasis is placed in all of the sustainable building systems on recycled content, the ability to recycle materials, and requirements to divert waste from landfills. In fact, these three items are intimately linked.

Recycled Content

Recycled content is perhaps the most commonly recognized single attribute associated with sustainability. Globally, recycled content is typically calculated using guidance provided in ISO 14021, Environmental labels and declarations — Self-declared environmental claims (Type II environmental labeling). This standard defines the term "pre-consumer" (also known as "post-industrial") material as "Material diverted from the waste stream during a manufacturing process. Excluded is reutilization of materials such as rework, regrind or scrap generated in a process and capable of being reclaimed within the same process that generated it."

ISO 14021 defines post-consumer recycled content as "Material generated by households or by commercial, industrial and institutional facilities in their role as end-users of the product which can no longer be used for its intended purpose. This includes returns of material from the distribution chain."

It is important to note that per the ISO definition, recycled content cannot include scrap material generated during the manufacturing process that can be re-used in that process. So, material generated in-process, such as internal glass cullet, does not count toward the recycled content value.[6]

The LEED 2009 rating system references ISO 14021 but offers credits for projects which select materials with recycled content, such that the sum of 100% of the post-consumer recycled content plus 50% of the pre-consumer content constitutes at least 10% or 20% (for 1 or 2 points, respectively), based on cost, of the total value of the materials in the project that contain recycled content. The USGBC allows twice as much credit for post-consumer recycled content because they believe that there are already sufficient economic drivers for the recycling of industrial waste materials, and that this system will therefore promote the use of post-consumer goods. CALGreen uses the same formulation in calculating recycled content. It should be noted that this methodology puts certain industries, such as gypsum or glass making,[7] at a competitive disadvantage since they do use significant amounts of pre-consumer waste during manufacturing. For example, drywall made of 100% synthetic "flu gas" gypsum will be credited as having only 50% recycled content.

In some cases, these schemes have targeted specific materials for having recycled content. For example, CALGreen cites the use of fly ash or slag in concrete, thus encouraging the recycling of these materials into concrete. Additional requirements in CALGreen are to use at least 10% recycled materials. As the tier level increases, these amounts increase.[8]

While many may view recycled content as simply a waste diversion credit, it is critical to note that recycled content replaces natural resources that would have been used to manufacture a product, thereby decreasing the burden on the environment, and lengthening the lifespan of existing reserves. Substituting recycled content for virgin natural materials may also eliminate or reduce processing energy associated with extracting and harvesting virgin natural resources. For example, it takes less energy to recycle glass into new glass than it takes to make glass from sand, a virgin natural resource: for each 10% of recycled material added, the energy to melt the total batch is reduced by 2 – 3%.[7,9] For metals, the difference may be even greater:[10] there is a 95% reduction in energy and greenhouse gas emissions when making new aluminum ingot from recycled material as compared to using virgin bauxite.[10,11]

When materials are recycled, solid waste is eliminated and so are the environmental impacts associated with the disposal of these materials. Table II compares key environmental impacts associated with both landfill and incineration disposal options and demonstrates the environmental trade-offs associated with each disposal option; these impacts are based on European data[12] and assume methane gas capture at the landfill. For example, when 1000 kg of municipal waste are recycled, 715 to 1,265 kg of carbon dioxide equivalents are avoided, significantly decreasing greenhouse gas emissions to the air.

Table II. Environmental Impacts associated with final disposal of 1000 kg of municipal waste.[12]

Environmental Impact Category (CML – Nov 2010)	**Final Disposal (per 1000 kg)**	
	Landfill	**Incinerate**
Acidification Potential (kg SO_2-Equivalent)	0.49	0.29
Eutrophication Potential (kg Phosphate-Equivalent)	0.09	1.27
Global Warming Potential, 100 years (kg CO_2-Equivalent)	1265	715
Ozone Layer Depletion Potential, steady state (kg R11-Equivalent)	8.63E-08	1.95E-07
Photochemical Ozone Creation Potential (kg Ethene-Equivalent)	0.03	0.17

There are also many instances when the inclusion of a recycled material also has a direct, positive impact on the environmental footprint of the new material in which it is being included. Examples of several different building materials are shown in Table III; impacts were calculated using data in the NIST BEES database,[13] and show the impacts associated with using recycled components back into their original products as part of a closed-loop system. These impacts extend beyond waste diversion. For example, the recycling of a ton of vinyl siding results in quantifiable reductions in water use and global warming potential; these reductions result from the elimination of the processes required to produce the virgin materials.

Table III: Impact of using 1 ton of recycled material in lieu of virgin raw materials in a closed loop system

Material Recycled in Closed-loop system	Global warming potential avoided, kg CO_2 equivalents	Water saved (gallons)
Generic Vinyl CompositeTile	882	2,670
Generic Gypsum*	28	2,817
Generic Vinyl Siding*	2,070	516
Generic Nylon Carpet Tile*	8,423	122,918

* Includes the impacts associated with needed installation components

Construction Waste Management

Construction waste management requirements incentivize the diversion of materials from landfills and incinerators. Most voluntary rating systems (i.e., LEED, Green Globes) and some building codes (i.e. IgCC, CALGreen) offer credits or include mandates for diverting materials while simultaneously offering credits for selecting products that use the diverted materials. These systems have created an ideal situation that creates both a push and a pull that encourages waste diversion behavior on both the supply and the demand sides of the issue.

On the supply side, the construction and demolition industry is eager to divert waste and are investing in infrastructure to do so; in some instances, such in California, they are required to do so. On the demand side, manufacturers are using several approaches to this issue, including: re-engineering their products to displace virgin materials with recycled materials; creating new products based on these recycled streams; or re-designing products such that they can be easily disassembled and recycled by component. Some building product manufactures are taking responsibility of their products at the end of life by offering product specific recycling programs.

Some materials with recycled content can be easily reused again, often as the same material such as aluminum, steel, paper and glass. However, less traditional materials from buildings such as ceiling tiles, vinyl composition tile flooring, and gypsum can also be reused in "closed loop" systems - meaning that they can be recycled back into the original product. Other waste materials can be recycled into other products: for example, crushed concrete can be used as aggregate in road beds or in fresh concrete.[14]

Currently, CALGreen mandates a 50% construction waste diversion for all building construction; the IgCC also requires a minimum or 50% diversion, but offers jurisdictions the choice of selecting higher diversion rates. As higher levels of waste diversion are mandated, the supply of diverted materials available for recycling will increase. As a steady stream of recycled materials become available, manufacturers will have to find innovative new ways to use them. Thus, the interaction of waste diversion and recycling creates a perpetual sustainable system that in an ideal world eliminates all waste.

Recyclability of Materials

Because of waste diversion mandates, the pressure to recycle materials into new materials or products is increasing. These mandates also assume that materials are recyclable. The ability to recycle a product is known as recyclability.

As noted, some materials are easily recycled: glass, metals, etc. Other materials or products are less easily recycled. There may be any number of reasons for this difference in recyclability, including the use of multilayer or multi-component structures (composites, layered glasses), impurities (laminated glass), chemical properties (alloying elements), and the degradation of physical properties (polymers). The recyclability of a specific material may be impacted by other regulations or requirements related to so called "chemicals of concern" or hazardous waste classifications. These restrictions may enhance recyclability or hinder it.

For example, in the United Kingdom, gypsum wallboard is no longer allowed to be disposed of in standard landfills; its high sulfur content means it must be considered a hazardous waste. As a result, there is a need to recycle gypsum wallboard that is removed during renovations or demolition, as well as fresh cut-offs from new construction. A study by The Waste and Resources Action Programme (WRAP) states that in 2010, 50% of waste gypsum wallboard was targeted to be recycled in the UK; by 2020 this needs to be 70%. The total estimate of available waste wallboard per year is 1.75 million tonnes, which is approximately three times the capacity of UK wallboard manufacturers to use recycled material. This capacity is limited in part by technical limitations that indicate that 25% recycled content is the maximum level that will achieve the required performance.[15] Thus, in order to use all the waste material that is available, the gypsum manufacturers will need to develop new technologies or products, or other manufacturers will need to develop products that can use this as a raw material.

Recycling may be hindered by the presence of chemicals of concerns in materials. For example, while there are various efforts to recycle plastics, there are also regulations and systems that prohibit the use of certain heavy metals and chemicals that are used in many plastics. Antimony, bis-phenol A, halogenated flame retardants, and some phthalate-based plasticizers are examples of chemicals that all considered to be hazardous by some organizations.[16,17] Their presence in materials may limit the ability of those materials to be recycled.

Another potential conflict with recyclability and mandated waste diversion is with regards to bio-based materials. While some materials, such as wood, can easily be recycled, albeit often as a down-cycled material such as paper, the recyclability of other materials is not as obvious. Many of these are new materials (composite bamboo flooring), and have not yet been introduced into the waste stream in large numbers. While many bio-based materials may have significant end-of life value as bio-mass energy sources, this is not always widely available.

BIO-BASED MATERIALS:

Materials using rapidly renewable, plant-based ingredients are highly prized by some of the sustainable building systems. For example, CALGreen and LEED 2009 stipulate that 2.5% of all materials, based on value, be from rapidly renewable (bio-based) sources. The materials selection section of IgCC requires that at least 55 percent of the total building materials used in the project, based on mass, volume or cost, comply with a menu of requirements including "used materials," recycled content, recyclable materials, bio-based materials, or "indigenous" materials. In cases where a material complies with more than one section, the value shall be multiplied by the number of sections that it complies with, thus encouraging the use of materials that meet multiple requirements, such as reclaimed wood.

It should be noted that single attributes, such as a percent content requirement, do not consider impacts associated with a product's entire life cycle. This is particularly noteworthy for bio-based materials. For example, such content-based requirements do not take into consideration the herbicides, pesticides or fertilizers that may be applied to bio-based feed stocks during the growing process. They also do not consider or compare the durability of a bio-based product versus a non-bio-based product. If a bio-based product lasts only three years, but a comparable petroleum based product lasts 100 years, only considering whether the resource is renewable may not be enough information to make an informed decision regarding the overall environmental preferability of a material.

MATERIALS FOR ON-SITE ENERGY PRODUCTION AND STORAGE FACILITIES:

The use of on-site renewable energy systems to offset building energy has been traditionally rewarded by the voluntary sustainable building schemes, such as LEED and BREEAM.[18] For example, LEED currently offers points based on the percentage of renewable energy that a building uses, based on energy cost.

When CALGreen was introduced, California began mandating such systems: 1% of the electrical needs of the building (based on cost) must be provided by on-site, renewable energy sources such as photovoltaic systems, wind, biomass, etc.

Both the performance and prescriptive paths in IgCC Section 6, Energy Conservation, Efficiency and COe Emission Reduction Section, include renewable energy requirements. For the performance path, buildings must be equipped with one or more renewable energy systems that have the capacity to provide not less than 2 percent of the total calculated annual energy use. For the prescriptive path, IgCC offers a complex formulation that addresses the use any single or combination of such renewable energy generation systems.

These requirements for on-site energy generation, and the associated need for storage systems, will drive the renewable energy and battery industries and the materials used in them, providing significant new markets and encouraging manufacturers to continue to innovate to provide compact energy sources that can be used within relatively small spaces.

However, due to the issues related to chemicals of concern, some of the basic materials used in these systems potentially become problematic. For example, cadmium telluride based solar cells have been considered to be cost effective options, and are theoretically capable of being recycled into new panels;[10] however, cadmium is classified by IARC as a Group 1 human carcinogen.[19] Other photovoltaic systems rely on gallium arsenide;[10] arsenic is also a carcinogen,[19] and is has been placed on "red-lists" of hazardous chemicals used by many architects.[16,17] The latest generations of battery materials vary with regards to chemicals of concern: $LiTiS_2$ and $LiFePO_4$ are considered to be benign;[10] $LiCoO_2$ contains cobalt, which is considered to be a possible human carcinogen.[19]

Nano-based materials are being actively researched for use in energy capturing and storage materials, and may hold the promise of higher efficiencies,[10] and for reducing some environmental effects.[10,20] However, there are concerns in some quarters that the health effects of nanomaterials have not been adequately investigated.[20]

Many of the basic components used in these renewable systems are not easily recycled,[21] nor are they made of recycled materials. In addition to hazardous materials, photovoltaics contain valuable materials that should be recycled, including copper;[21] however, this is costly and does not generally occur.[21,22] The blades of wind turbines are made of fiber reinforced composite materials;[10,23] it is extremely difficult to separate the constituent parts from such structures.[23] The most common way to reuse FRC's currently is to shred them and use as fill; however, there are many competing materials for this use.[23] It is clear that as mandated waste diversion increases, manufacturers will need to develop energy capture systems that: are made of homogeneous, recyclable materials; are designed such that they can be easily and efficiently be disassembled into their constituent raw materials; use novel, low impact methods of recovering the materials; and/or find new uses for using these materials in other products.

MOVE TOWARDS LIFE CYCLE ASSESSMENTS

The existing requirements for materials in the sustainable building systems and codes are largely based on a series of "single attributes," such as a minimum requirement for percent recycled content, etc. Under such schemes, the more "checkboxes" a single material can meet, the more desirable it is. However, as discussed earlier, these single attributes do not consider all aspects of a product. In the future, more holistic, multi-attribute assessments will be needed that address all the environmental impacts of a product; these will most likely be based on Life Cycle Assessments (LCA).

Life Cycle Assessment is a tool for quantifying all the environmental impacts associated with the life of a product. These are detailed "cradle-to-grave" analyses and include such things as the mining and refining of raw materials, processing, transportation, installation and use phases, and end-of-life (landfilling, recycling, etc.) The process for conducting an LCA is described in ISO 14044. The output from an LCA is a list of environmental impacts; in general, in addition to the five factors shown in Table I, an LCA will also consider embodied energy and water usage, among other things. Because these analyses cover all aspects of a product's life, it is important to note that the impact of materials in the procurement chain will need to be understood. Manufacturers of materials will therefore be under pressure to conduct their own LCA's or to provide input for their customers.

LCA's can be used internally by a manufacturer to determine the areas of greatest environmental impact, allowing action plans to be developed for reducing certain impacts. They can also be used to aid in the design of products that have smaller environmental impacts. They are also the basis for developing Environmental Product Declarations (EPD's), a type of environmental dossier for a given product.

There is a definite trend towards the inclusion of LCA's in the sustainable building systems and codes. For example, EPD's are required in certain European countries for all construction products, and will be required throughout the European Union as the requirements of Section 56 of the Construction Products Regulation of 2011 are phased in.[24] CALGreen has an option for performing a whole-building LCA that incorporates the impacts from all the materials used in the building. LCA requirements were included in early drafts of IgCC 2012; these were removed from the final document, but are expected to be included in the 2015 version. Draft versions of LEED Version 4 included credits for life cycle assessments of building products and for whole buildings. In all cases, these requirements have been proposed as a way to transform the market by making the holistic impacts associated with a product available to consumers, allowing them to make informed decision about the products they buy.

For example, while recycling of materials is definitely a good practice since it diverts material from the landfill; however, a better understanding the ramifications of this action in terms of additional quantifiable impacts reductions can be determined through the use of LCA. Such analyses may suggest, as was seen in Table III, that it is more impactful to recycle some materials as compared to others. There could theoretically even be instances where recycling of a material has one or more negative environmental impacts such as requiring more energy to recycle the material than the ~~original~~ energy required to manufacture ~~the~~ using virgin materials.

While many may view LCA's as complicated, the widespread use LCA's will enable consumers and manufacturers to make educated decisions regarding the true impacts associated with a material, product or building.

SUMMARY OF SPECIFIC MATERIALS REQUIREMENTS

As can be seen in the previous sections, although there are common elements for materials among the major sustainable building systems, there are significant differences in the details. This lack of truly common requirements presents a challenge to manufacturers who would prefer to design to a single common baseline.

The major requirements for materials as outlined in LEED, CALGreen and the IgCC are shown in Table IV.

Table IV: Summary of Key Material Requirements in Major NA-Based Sustainable Building Systems and Codes

Material Attribute	**LEED 2009**[3,4]	**CALGreen**[2]	**IgCC**[5]
Voluntary or Mandatory System?	Voluntary, except where mandated by local governments	Mandatory Code for all new buildings	Optional add-on to other ICC-based codes
Recycled Content	• 10% - 1 point • 20% - 2 points	Minimum 10%	Part of a menu of options for materials content
Waste Diversion	• 50% - 1 point • 75% - 2 points	50% minimum	• 50% minimum • Can be raised by local authorities
Bio-based Materials	2.5% of all materials	2.5% of all materials	Part of a menu of options for materials content
On-site Energy Generation	• Can purchase renewable energy or have on site • Points vary according to amount and system • 1% of energy cost – 1 point • Up to 13% of energy cost for 7 points	Minimum of 1% of energy	Minimum of 2% of energy

TRADE-OFFS AND CONFUSION ASSOCIATED WITH SUSTAINABILITY STANDARDS

As can be seen from the previous discussion, there are very often complex trade-offs when designing, using or selecting "sustainable materials" for use in the various sustainable building systems. A product or material that meets some criteria may not meet other criteria, or may present other challenges. It may meet the requirements of some systems, but not others.

For example, there were several examples cited of the conflict between recycling of materials and simultaneously meeting criteria around materials of concern.

Many of these conflicts can be summarized at a very high level in Table V below. This table shows potential issues that need to be addressed in a holistic way by manufacturers as they attempt to create products that meet the needs of the new sustainable building codes.

Even when a life cycle assessment is used, trade-offs may exist in determining the "ideal" product. Because LCA's consider a variety of environmental impacts, it will be possible to weigh the importance of specific categories, such as water or energy use, as well as consider the overall impacts; individual concerns or preferences will help drive material selection in this case.

CONCLUSIONS:

The requirements for sustainable buildings and the materials that are used in them are currently transitioning from voluntary systems to mandatory ones. This change will impact both producers and consumers of materials, since meeting the requirements will no longer be an optional, value-added proposition, but rather will be a basic necessity of doing business with the construction industry. Manufacturers who are the first to reformulate, redesign or reposition their products according to these mandates will likely have a competitive advantage. In addition, holistic requirements for materials are becoming the norm, and the market is shifting away from single attributes, such as percent recycled content. However when looking at groups of attributes, it is clear they are often inconsistent and can conflict with one another. The advancement of life cycle assessments, which consider all impacts of a product during its life, will add another layer of complexity to these systems, but will ultimately help consumers and manufactures make better choices when it comes to environmental actions.

A common set of requirements is needed to meet the requirements of both the voluntary and mandatory systems, so that manufacturers and consumers will have a single baseline to use. At a high level, many of the key attributes and needs seem to be shared among the major systems; however, the details still vary, making for confusion and duplication within the market.

Table V: Trade-offs among materials-related needs associated with sustainable building codes

	Recycling/Waste Diversion	Recycled Content	Chemicals of concern	On-site energy	Bio-based materials
Recycling/Waste Diversion	• Mandated? • Has the product been designed for disassembly?	• Can the material be used in a closed-loop system? • Can the material be up-streamed? • Will the materials be down-streamed?	• Some materials are listed as hazardous and can't be recycled • Some materials can't be easily land filled because they have been listed as hazardous	• Can components of solar cells and fuel cells be recycled?	• Can bio-based materials be re-used? If not, can they be composted or used as a bio-mass source?
Recycled Content		• Is there a maximum recycled content that can be used before properties are adversely impacted? • Is there a minimum content specified by the system?	• Some materials contain trace amounts of COC's • Post-consumer sources may be difficult to characterize	• Can systems be devised that use more recycled content?	• For some bio-based materials, using recycled content may be an issue
Chemicals of concern			• There are many lists of COC's that must be tracked	• Can systems be developed that use less hazardous materials?	• Some materials need biocides, etc., which could be of concern • Some materials naturally contain or emit COC's
On-site energy				• In some systems the use of on-site or renewable energy is mandated	• Are there more opportunities for using bio-based materials in these systems?

REFERENCES

[1]US Green Building Council, www.usgbc.org
[2]CALGreen, Guide to the (Non-Residential) California Green Building Standards Code (1/2012), http://www.documents.dgs.ca.gov/bsc/CALGreen/MasterCALGreenNon-ResGuide2010_2012Suppl-3rdEd_1-12.pdf
[3]US Green Building Council, LEED 2009 for Schools, New Construction and Major Renovations (2009).
[4]US Green Building Council, LEED 2009 for Schools, New Construction and Major Renovations (2009).
[5]International Code Council, 2012 International Green Construction Code, (2012)
[6]PPG, Recycled Content Claims for Glass Cullet (3/2010).
[7]Glass for Europe, Recyclable waste flat glass in the context of the development of end-of-waste criteria (6/2010).
[8]S. Mann, CalGreen: California's Building Code (5/1/2011) http://www.homeenergy.org/show/article/nav/greenbuilding/id/797/viewFull/yes
[9]St. Gobain, Sustainable Development Report 2011 (5/2012).
[10] D. S. Ginely and D. Cahen, eds., Fundamentals of Materials for Energy and Environmental Sustainability, Materials Research Society, Cambridge University Press (2012).
[11]ALCOA, 2011 Sustainability at a Glance (2012).
[12]PE International (2011)
[13]NIST BEES database
[14]Concrete Network, Recycling Concrete (2012) http://www.concretenetwork.com/concrete/demolition/recycling_concrete.htm
[15]Waste and Resources Action Programme, Waste Protocols Project: Gypsum, Partial Financial Impact Assessment of a Quality Protocol for the production and use of gypsum from waste plasterboard (2008).
[16]Cascadia Region Green Building Council, Living Building Challenge 2.1 (5/2012).
[17]Google Healthy Materials Program
[18]BRE, BREEAM
[19]International Agency for Research on Cancer, Monographs on the Evaluation of Carcinogenic Risks to Humans (2012), http://monographs.iarc.fr/ENG/Classification/index.php
[20]US EPA, Nanotechnology, http://www.epa.gov/ncer/nano/questions/index.html
[21]N. Weadock, Recycling Methods for Used Photovoltaic Panels, *University of Maryland Watershed Program* (1/9/2011), http://2011.solarteam.org/news/recycling-methods-for-used-photovoltaic-panels
[22]E. Geis, Solar Panel Recycling Gears Up, *The Daily Green* (8/12/2010), http://www.thedailygreen.com/environmental-news/latest/solar-panel-recycling-460810
[23]K. Larsen, Recycling wind, *Reinforced Plastics* (1/31/2009), http://www.reinforcedplastics.com/view/319/recycling-wind/
[24]Official Journal of the European Union, Regulation (EU) No 305/2011 of The European Parliament And Of The Council of 9 March 2011laying down harmonised conditions for the marketing of construction products and repealing Council Directive 89/106/EEC (4/4/2011)

DEVELOPING THERMAL PROCESSES WITH ENERGY EFFICIENCY IN MIND

Brian Fuller, Bruce Dover and Tom Mroz
Harper International Corporation
Lancaster, New York, USA

ABSTRACT

Thermal processing is typically the most energy intensive portion of a materials manufacturing process. Opportunities to conserve energy not only reduce ecological impact, but can result in significant cost saving, as thermal processing is a critical cost driver in direct equipment and installed utility requirements. To make the best use of possible opportunities, it is important to consider the form of the commercial process as early as during laboratory development to ensure it will be suitable for application in the intended continuous form. In particular, continuous thermal processes offer multiple opportunities for energy conservation over batch, including reduced residence time, minimizing thermal requirements to heat ancillary process components, and opportunities for use of byproduct materials and heat for preheating. Detailed discussion on opportunities for continuous, energy efficient thermal processing will be provided with examples of commercially relevant processes to illustrate the concept and provide a guide for future considerations.

INTRODUCTION

Material refinement and processing can involve dozens of unit operations to finally arrive at a saleable, high-value advanced material. Thermal processing is typically the most energy intensive portion of a materials manufacturing process. Thermal processing has many opportunities to conserve energy for both cost savings and ecological impact.

Often the link between initial process investigation and the commercial scale production plant is made too late in the process development. For an existing process, every time you want to save energy, you have to put equipment in place perhaps with an unfavorable return on investment. With demands for decreasing time-to-market, realizing a more energy efficient process exists, or worse yet is required for commercial success, needs to occur up front in development. The affect of scaling the process to meet the short term demand of sales and marketing for supplying test quantities of final product to potential customers leads to the trend of batch furnaces with increasing residence time proportional to production size. Frequently these steps are repeated several times and producers tailor their processes to the particular characteristics of the batch product. The residence time should not be ignored; the furnace construction and operation should be considered for each scale up step with the commercial end in mind.

With steady, high volume material demand the goal should be continuous over batch processing. Significant energy conservation advantages of continuous thermal processing over batch processing are discussed broadly. Example processes are presented for future consideration.

COMMERCIAL PROCESS

Process Types

The specific starting material and target product place the thermal process into one of several broad types: Calcining and Pyrolysis, Sintering and Synthesis. These process types each have their own thermal requirements which dictate furnace type and construction, and also opportunities for heat recovery. Each type can exist in a wide range of temperature and atmosphere requirements.

Calcining and Pyrolysis processes involve the heat treatment of a solid to produce both solid and gas reaction products. Key processing concerns efficient heat transfer into the solid bed and gas management to control the atmosphere and extract condensable gases. An example of Calcination is the removal of bound water and carbon dioxide as in formula (1) below, whereas Pyrolysis can be the decomposition of a polymer in an inert atmosphere leaving behind carbonaceous solid and condensable gases.

$$Nd_2(CO_3)_3 \cdot 2H_2O \rightarrow Nd_2O_3 + 3CO_2 + 2H_2O \quad (1)$$

Sintering processes most often involve the heat treatment of a part, but may also apply to loosely agglomerated powders, with the goal of increasing the density and strength of the part. Key processing concerns are uniform temperature control and atmosphere control.

Synthesis processes can be solid-solid or gas-solid complex reactions to produce a new solid material. Key processing concerns efficient heat transfer into the solid bed and gas management to control the process atmosphere or volatiles.

Material Flow

For each of these processing types, the flow of the material through the thermal process will gauge which furnace type is possible. How will the material be transported through the thermal processing units? If the material is a powder, low energy advantages exist if the bulk powder is free flowing throughout the process. If the powder is sticky it prevents the use of such high heat transfer and uniform processing units such as rotary tube or vertical tube furnaces. For processing parts, how can they be transferred through the furnace while allowing uniform heating, gas diffusion and shrinkage while optimizing the furnace volume? In both cases where a bulk powder is free flowing and a process part can be arranged in smaller loads with more direct paths to heat transfer and gas diffusion, the benefits are often shorter processing times and a more uniform product.

Commercial Volume

Based on the potential applications for new advanced materials, the markets need to be sized for future commercial plant capacities. The volume of material required at the commercial scale can strongly influence the form of the commercial process. For example, a high temperature process where at the pilot production scale a ceramic tube rotary furnace is suitable for the production capacity, at the commercial production scale a whole new furnace type will need to be implemented. The reason is scalability. The ceramic rotary tube furnace will be limited in diameter and length due to the materials of construction. Knowing the final production rate required will enable thoughtful decisions regarding the furnace configuration.

Maximum Temperature

The maximum design temperature of the furnace, independent of whether or not it is ever operated there, means the manufacturer will use materials capable of continuous operation in that high temperature. The higher the temperature rating of an insulating material is the higher the cost and the lower the insulating properties will be. This is one of the reasons insulating walls are typically thermally graded, with varying composition as you get farther from the hot face. The design of the furnace insulation is not as simple as adding thicker insulation. This increases the furnace volume and cost and also raises the average temperature across a piece of hot face insulation, limiting its useful life. Operating at a relatively high heat flux through the insulation wall gives strength to the refractory, but may require the use of a water-cooled steel shell to maintain the integrity of the furnace. Water-cooling is frequently required for high temperature graphite furnaces. The added treatment of cooling and quality in a closed-loop water cooling system adds to the plant's operating cost. These trade-offs should be considered with the need for operational flexibility and processing various grades of product when specifying the maximum capabilities. A furnace designed for a higher temperature will be less energy efficient.

Product Quality

The source and quality of the raw material has an impact on the number and type of processing steps. Take the extraction, calcination, and separation of rare earth oxides for example. An ore sample with but a trace amount of the target oxide, for example 6 wt-% light rare earth oxides in the Bayan Obo, China deposit[1], is an intensive series of wet chemistry and thermal steps. The wet chemistry steps including the preparation of aqueous solutions, removal of bulk and trace water, and handling of waste water are costly. In addition, these material sources often lead to low yields at thermal treatment. For example a 67% recovery in the ore calcinations from formula (1) above, or even 18-20% recovery in the pyrolysis of rayon fibers[2], means each thermal step is sized for the large volume of incoming feed. If the volatiles are removed at the front of the furnace then only a small volume fraction of the remaining furnace is utilized.

Related, the final target purity and avoidance of detrimental elements for products such as battery and electronic materials quickly limits the process environment. Consider materials for electronic and energy storage applications which demand very low impurities. Trace, large, metallic elements such as iron, nickel and zinc have a detrimental effect by changing bulk electrical properties, promoting side reaction, or changing crystal morphologies. However, the major concern is safety of the battery and avoiding the formation of internal short circuits often caused by crystal growth promoted by the foreign metal deposit from the anode surface to the cathode, penetrating the separator[3]. Limiting the process to non-alloy materials of construction quickly leads to long cycle times limited by slow ramp rates of high mass kiln furniture, or multiple smaller capacity furnaces which are inherently inefficient. A rotary tube furnace with non-alloy process tube has limited tube sizes due to manufacturing materials and techniques. For commercial scale plants making 10 tons per day or larger, this leads to dozens of small rotary furnaces or fewer large footprint tunnel furnaces. Advancements in manufacturing and furnace designs can lower the risk associated with larger diameter quartz process tubes, but ceramic tubes may never be available in larger diameters.

The final purity should be dictated by the end-use application, and sometimes the requirements are unavoidable. Likewise the process atmosphere is often not flexible and dictated by the reaction. A majority of the time a process requires either an oxidizing, inert, or reducing atmosphere. Within each category the specific utility gas to use is important to the overall plant energy footprint. For example, consider a process which needs an inert cover gas: the preferable gas to use is nitrogen, but nitrogen is not always inert. So a safer position may be the use of argon. However, at an average atmosphere composition of 0.93 vol-% compared to nitrogen's 78 vol-%, argon is much more energy intensive to produce[4]. The use of inert, noble gases has their specific uses, but should not be the default.

CONTINUOUS PROCESSING

Residence time

The scale-up of capacity from lab to test production, pilot production, and then commercial production often introduces new process limitations that were not observable at the previous scale: kinetic rate limitations due to temperature uniformity, peak temperature, and mass diffusion. These all can, if taken without optimization, lead to a significant residence time requirement in the heating chamber for the process. The volumetric rate times the residence time then leads to a larger furnace size with higher capital and installed utility costs, higher surface area for wall heat loss, and higher gas consumption. With operational consideration, this can even drive groups towards batch furnaces which are inherently inefficient, cycling the entire thermal mass of the heating chamber.

The residence time also helps frame in possible furnace types. At very short residence times less than 1 minute a dilute-phase vertical drop reactor may be suitable. With short residence times less than 2 hours a rotary tube furnace or dense-phase moving bed vertical reactor would be ideal. For longer residence times, the processes are usually limited to tunnel furnaces.

Non-optimized, long residence times may also have a negative impact for product quality of powder processing. Grain growth and agglomeration of particles can lengthen, or add entirely, a milling step downstream, which can be very energy intensive depending on the process material. The residence time requirement specified may be the single largest impact on energy efficiency on the thermal unit.

While not directly an energy efficiency issue, the amount of material held within the thermal process is proportional to the residence time. At start-up several of these volumes must pass through the process before steady state is established. At shut-down another volume of potentially off-spec material is produced. All or part of the value added by upstream operations can be wasted.

Ancillary Process Components

As mentioned above, sometimes the imposed long cycle times and material flow properties can quickly limit useable furnace types. The operation which heats only the process material and atmosphere as required will be lower than any other in terms of kWh/kg of product. With traditional heating methods, electrical resistance and hydrocarbon-fired, a furnace with no product carrier is optimal: rotary tube or vertical tube furnace for example. The material must be in powder, pellet or agglomerate format and be free flowing through the entire temperature range. Adding a carrier could double the energy input for sensible heats. Complex kiln furniture for stacking parts in a large furnace volume can range from 50-95+% of the load thermal mass.

In a batch furnace, the complex kiln furniture is typically required. This in addition to the entire thermal mass of the furnace shell, insulation, process containment, and process load is heated up and cooled down every cycle. This is in addition to the energy requirement that is seen on a continuous furnace when adding kiln furniture, described above. Now, the heating elements and connected load to the furnace are sized not for the energy going into the process, but the energy being thrown away heating and cooling the furnace itself. Energy and total operating costs are always higher with batch furnaces.

Atmosphere and Byproducts for Reuse

With a continuous furnace it is possible to configure the gas handling systems in complex, well integrated ways to recover heat. The key feature of a continuous furnace that enables the following solutions is an uninterrupted gas flow with steady gas composition. The steady gas flow and composition enables gas recycling, for example the recycling of hydrogen gas in a reduction process, or energy recovery, for example by oxidizing volatile organic compounds followed by a heat exchanger to preheat the furnace process atmosphere. In addition the system design can take advantage of the fact that in a continuous furnace all points in the process are occurring simultaneously but at different locations in the furnace. Consider a

Pyrolysis process which evolves fuel-rich streams at the high-heat soak temperature of 750°C. The majority of the required heat input is in the preheating zones to bring the process material and atmosphere to 750°C. To take advantage, the fuel-rich exhaust can be externally piped and oxidized to provide heat to the preheating zones. However in a batch furnace the fuel is never around when it is needed, so the fuel-value of the process byproduct is wasted.

Wisely handling the incoming process atmosphere in a continuous furnace can also lead to energy benefits. The default way to manage the flow of the solid feed and gas feed in a thermal system is to introduce them counter-current to each other. And with good reason; the cold gas is preheated by the discharged product, and the hot product is cooled by convective heat transfer from the gas, often more efficient compared to conduction from water cooling. It is most effective when the summation of each material's mass rate times its heat capacity (effectively watts per Kelvin, W/K) in the opposing streams are matched, just like a well-designed heat exchanger. This arrangement is not possible in a batch furnace and no energy benefits will be gained.

MODELS AND EXAMPLES

Counter-current Gas Enhanced Heat Transfer Model

The following model was developed to expand on the above enhanced heat transfer concept of counter-current process gas. The model uses a rotary tube furnace 1 meter in diameter, 10 meters heated length, operated at 1150°C with 1000 kg/h of solids with a specific heat of 1000 J/kg-K. A coarse particle size of approximately 200 micrometers was considered. Particles of this size allow significant through flow of the process gas within the bed of material, enhancing gas-solid heat transfer. An overall gas to solid heat transfer coefficient of 50 W/m^2-K was considered. This would be expected to be significantly lower with a fine, not-free-flowing powder.

Figure 1 and 2 below illustrate the case of balanced and unbalanced mass rate times heat capacities for the opposing solid and gas streams. Figure 2 has twice the gas requirement of the balanced case. The temperatures of the furnace setpoint, solid and gas are plotted against the distance from the furnace entrance.

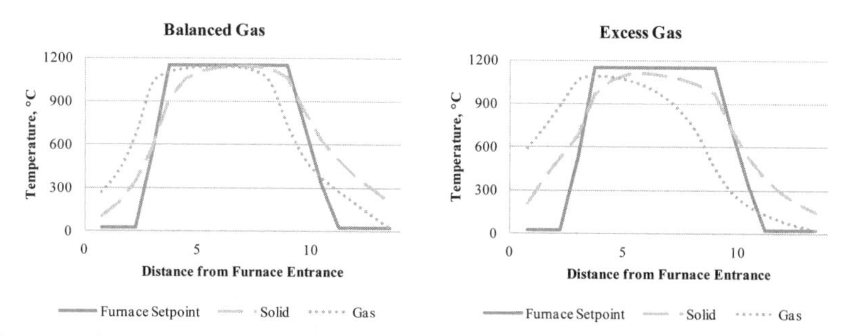

Figure 1 and 2. Plots illustrate the difficulty in controlling the temperature profile of the solid material in the furnace when the reaction is operated with significant excess mass rates of the gas.

A significant excess mass rate times heat capacity of the gas will enable faster cooling and preheating of the solid product. However, the energy required for the reaction is higher at

382 kW in Figure 2 compared to 170 kW in Figure 1. It also diminishes the temperature control of the solid material along the length of the furnace. The loss of temperature control is even more pronounced in situations where processes are developed with complex temperature profiles.

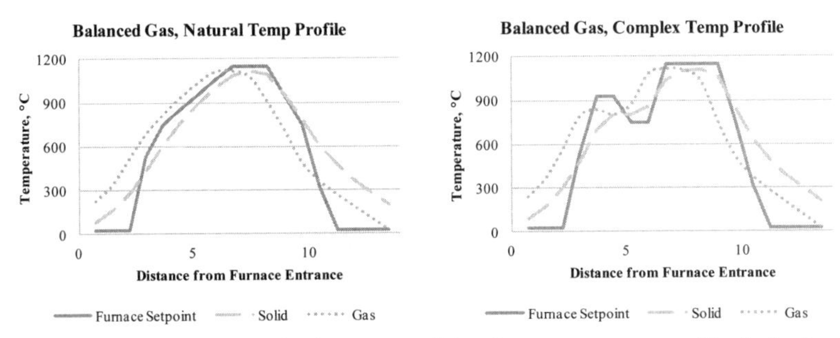

Figure 3 and 4. Plots illustrate the impact on solid and gas temperature profiles in the furnace with a natural temperature ramp rate and soak compared to an imposed complex temperature profile.

The counter-current strategy in Figure 3 works very well. The solid material is more closely held to the furnace setpoint. However, as seen in Figure 4, complex ramp rates with many zones of temperature control where you ramp, hold, ramp and hold may be ill-conceived with high counter-current gases. The sensible heat of the counter current gas will carry heat to the low temperature zones at the beginning of the furnace. Correcting the situation and lowering the counter-current gas rate will cause an unbalanced W/K in opposing streams. In addition, the scenario in Figure 4 requires approximately 10% more power at 156 kW compared to 142 kW in Figure 3.

This sample model shows the benefits of a counter-current solid-gas heat exchange within a continuous furnace. While developing your process, consider the optimal process gas rates for both reaction kinetics and furnace heat exchange. In addition, be careful of scaling up complex time-temperature profiles from batch to continuous for the commercial production.

Metal Oxide Reduction Sample

For a small production operation of a particular metal oxide reduction, the reaction time and hydrogen gas requirements were so high that it was decided the process should be produced in a batch fluid bed furnace. In scale-up testing, alternative ways to reduce the oxide were investigated. Gas-solid interaction was critical so a rotary tube furnace with lifters was tested with similar gas flow rates. The material flowed uniformly off each lifter, mixing with the counter-current process atmosphere, but the residence time requirement pushed the limits of a rotary tube furnace at more than two hours. In addition, the quantity of reaction gas caused low yields due to significant powder entrainment in the gas stream. To improve, alternative reductants were investigated. Methane was chosen for its carbothermal reduction properties. Upon cracking it generates both carbon and two molecules of hydrogen.

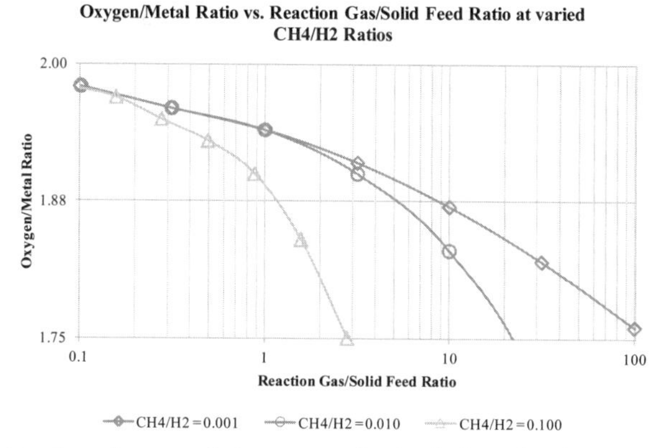

Figure 5. The target reaction completion was a reduction in the ratio of oxygen/metal atoms in the product from 2 to 1.75.

A small addition of methane improved the reaction time to have a uniform process and reduce the hydrogen consumption. Almost two orders of magnitude of gas quantity was saved by this addition. The gas utility cost and the energy to heat the gas to temperature could be significantly reduced. This discovery could allow the process to economically scale up to commercial production volumes.

CONCLUSION

The thermal process is a major consumer of energy in a commercial plant. The form of the commercial scale plant must be explored earlier in the development process. Even at the lab scale, where influences of gas quantities, temperature ramp rates, and materials of construction may not be critical, they should be considered with the end in mind. Optimizing the thermal process may also eliminate unit operations all together, such as powder milling and closed-loop water systems. Similarly, in consideration of furnace volume utilization it may be that an additional upstream step is required for eliminating low-temperature volatiles such as water in a dryer or high-temperature condensable solids in a wet-chemistry reactor. The benefits of a continuous furnace operation over a batch operation are increased up-time, product quality and consistency, decreased residence time, reduced power requirements, and integration for complex heat recovery or utilization from process byproducts. Every step of a new production scale will present a whole new set of factors that may not have been apparent. As you consider your process development, keep energy efficiency in mind.

REFERENCES

[1]C. K. Gupta and N. Krishnamurthy, *Extractive Metallurgy of Rare Earths,* Boca Raton, FL: CRC Press (2005), 61.

[2]V. Immanuel and K. Shee Meenakshi, Synthesis of Carbon Cloth by Controlled Pyrolysis of Rayon Cloth for Aerospace and Advanced Engineering Applications, *Indian J. Science and Technology*, **4**(7), 759 (2011).

[3]S. Sriramulu et al., Implantation, Activation, Characterization and Prevention/Mitigation of Internal Short Circuits in Lithium-Ion Cells, 2012 DOE Annual Merit Review Meeting, 14-18 May 2012.

[4]F. R. Field, J. A. Isaacs and J. P. Clark, Life-cycle Analysis of Automobiles: A Critical Review of Methodologies, *J. Minerals, Metals and Materials Soc.*, **46**(4), 12-16 (1994).

JOINING OF SILICON NITRIDE CERAMICS BY LOCAL HEATING TECHNIQUE - STRENGTH AND MICROSTRUCTURE

Mikinori HOTTA*, Naoki KONDO, Hideki KITA**, and Tatsuki OHJI

Advanced Manufacturing Research Institute, National Institute of Advanced Industrial Science and Technology (AIST)
2266-98 Shimo-Shidami, Moriyama-ku, Nagoya 463-8560, Japan

ABSTRACT

The joint strength was examined in conjunction with the microstructural development in details for silicon nitride ceramics joined by a local-heating joining technique. Commercially available silicon nitride ceramic pipes sintered with Y_2O_3 and Al_2O_3 additives were used for parent material, and powder slurry of Si_3N_4-Y_2O_3-Al_2O_3-SiO_2 system was brush-coated on the rough or uneven end faces of the pipes. Joining was carried out by locally heating the joint region at different temperatures from 1500 to 1650°C for 1 h with a mechanical pressure of 5 MPa in N_2 flow, and four-point bending tests were conducted for measuring the joint strength. Microstructural observations using a scanning electron microscopy (SEM) clarified that grains were fine and equiaxed and substantial amounts of voids and pores were left in joint layers of the specimens joined at 1500 and 1550°C, which showed the low strength of 412 MPa. On the contrary, the specimens joined at 1600 and 1650°C which indicated the relatively high strength of about 680 MPa had fully densified joint layers with well developed fibrous grains. Fracture surface observations also revealed that intergranular fractures were predominant when the fracture origins were in joint regions while those failing from parent regions showed a feature of mixed intergranular and transfranular fractures.

INTRODUCTION

Silicon nitride ceramics are expected to be used as industrial components for saving the energy for production and improving the quality of products in manufacturing industries, because of their excellent heat resistance, wear resistance, good corrosion resistance, high specific elastic modulus, and lightweight.[1] These components are very often required to have large scale of more than 10 m. Joining is one of the most important technologies to obtain such huge ceramic components, because it is critically difficult to fabricate them as single units.[2] When producing large ceramic components, however, it is essentially important to join ceramic units by locally heating the joint region in order to reduce the energy consumption and cost in the production process. Furthermore, the ceramic components employed in industries are most often exposed to high-temperature and/or corrosive environments, and excellent sealing performance is required against fluid gas and liquid. In order to meet these requirements, the joint layers need to have sufficient mechanical, thermal and chemical properties which are equal to those of the parent parts. Most ideally this can be attained by making the microstructure and chemical composition identical to each other.

There have been a substantial number of reports on joining of ceramics via heat treatments.[3-20] A variety of adhesive inserts such as silicon nitride and/or oxides powder mixture,[3-6,15-20] deformable SiAlON plate,[7] and active braze alloy[11] have been used for joining of silicon nitrides. In these studies, however, high joining temperatures (close to sintering temperatures of the parent silicon nitrides),[15] high mechanical pressures,[18] and/or long post heat-treatments (at temperatures higher than joining ones)[6,19] have been required for obtaining high joint strength of silicon nitrides. In addition, the joined bulk samples of these studies were generally small and the whole of the samples were heated in conventional furnaces, which does not fit to the above-stated joining of large components via local heating where joining temperatures and applied pressures are significantly restricted.

The authors recently reported researches on a local heating technique using electric furnace equipment specially developed for joining long silicon nitride pipes.[21,22] Powder slurry of Si_3N_4-Y_2O_3-Al_2O_3-SiO_2 system was brush-coated on the end faces of silicon nitride pipes; the advantage of this technique was easily joining pipes even with rough or uneven end faces for fabricating long silicon nitride tubular components. The pipes were joined by locally heating the joint region using the furnace at 1500 to 1650°C, and the joint strength was investigated as a function of joining temperatures in relation to microstructures and chemical compositions, resulting in relatively high strength of about 680 MPa for the specimens joined at 1600 and 1650°C. This paper investigated in more details the microstructure evolution in the joint regions in conjunction with the strength properties. In addition, fractographic studies were carried out for the facture surfaces to clarify the fracture behaviors.

EXPERIMENTAL PROCEDURE

The local heating equipment for joining ceramic pipes is illustrated in Fig. 1.[22] One side of the ceramic pipes is fixed to the equipment. The end faces of the pipes are fixed to each other, and the joint region is subsequently heated by an electric furnace to produce a joined ceramic long pipe. The heating element made of graphite in the furnace is about 300 mm long, and inert gas is passed into the heating furnace and inside the pipes during heating. Commercially available silicon nitride ceramic pipes sintered with Y_2O_3 and Al_2O_3 (SN-1 grade, Mitsui Mining & Smelting Co., Ltd., Tokyo, Japan) were used as the joined pipes. The pipes were 1 m in length, and had outer and inner diameters of 28 and 18 mm, respectively. A powder mixture of Si_3N_4–Y_2O_3–Al_2O_3–SiO_2 system in a composition of oxynitride glass was used as the insert material. The composition of the mixture was 30.2 mass% Si_3N_4, 43.3 mass% Y_2O_3, 11.7 mass% Al_2O_3, and 14.8 mass% SiO_2; this was previously reported for the joining of silicon nitride using an oxynitride glass insert by Xie *et al.*.[6] α-Si_3N_4 (SN-E10 grade, Ube Industries, Ltd., Japan, average particle size: 0.5 μm), Y_2O_3 (RU-P grade, Shin-Etsu Chemical Co., Ltd., Japan, 1.3 μm), Al_2O_3 (AL-160SG-4 grade, Showa Denko K. K., Japan, 0.6μm), and SiO_2 (Kojundo Chemical Laboratory Co., Ltd., Japan, 0.8μm) were weighted to realize the composition and then mixed by ball-milling in a plastic pot with Al_2O_3 balls and ethanol for 48 h. The mixed slurry was dried and passed through a sieve. A small amount of ethanol was added to the mixture to prepare a homogeneous slurry for the insert material. The slurry was brush-coated to the rough or uneven joint surface of the silicon nitride pipes and then dried. Subsequently, the two silicon nitride pipes were fixed to the local heating equipment, and the joint region of the pipes was inserted into the electric furnace in the equipment. The silicon nitride pipes were joined at temperatures from 1500 to 1650°C for a holding time of 1 h in N_2 gas flowing at a rate of 5 L/min at a heating rate of 10°C /min. During the joining, a mechanical pressure of 5 MPa was applied to the joint surface, and the pipes were rotated at a rate of 3 rpm. The temperature of the pipe surface around the joint region was measured using an optical pyrometer. After the joining of two silicon nitride pipes that were each 1 m in length, the obtained 2 m long pipe was joined with a third 1 m long pipe to subsequently fabricate a 3 m long silicon nitride pipe by the same joining process as previously described.

The joined silicon nitride pipes were cut perpendicular to the joint surface in order to observe the microstructure at the joint region and measure the flexural strength of the joined pipes. The surfaces of the specimens were polished with a 0.5 μm diamond slurry and then etched by plasma in CF_4 gas. The etched and fractured surfaces were observed by scanning electron microscopy (SEM). The flexural strength of the specimens was measured at room temperature using a four-point bending method (Sintech 10/GL, MTS Systems Corp., USA); the specimens had dimensions of 3 mm x 4 mm x 40 mm, and the joint region was at the center of the bending bar. The outer and inner spans were 30 and 10 mm, respectively, and the crosshead speed was 0.5 mm/min.

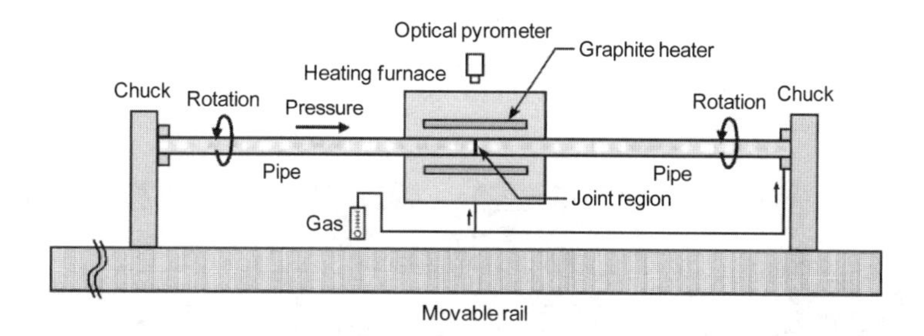

Figure 1. Schematic diagram of local heating equipment for joining ceramic pipes.

RESULTS AND DISCUSSION

Figure 2 shows the room-temperature flexural strength and fracture sites of the specimens joined at 1500 to 1650°C and the parent silicon nitride bulks. The specimens joined at 1600 and 1650°C exhibited high strength values of about 680 MPa, which is relatively close to the average strength of the parent silicon nitride bulks, 737 MPa. Out of six tested specimens joined at 1600 and 1650°C, four and three were fractured from the parent region, respectively. For the specimens joined at 1500 and 1550°C, the strength was both 412 MPa, which is substantially lower than those at 1600 and 1650°C, and all the specimens failed from the joint region.

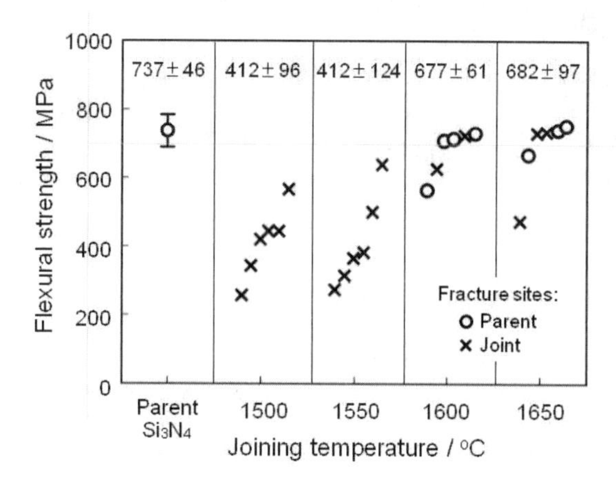

Figure 2. Room-temperature flexural strength and fracture sites of the specimens joined at 1500 to 1650°C and parent Si_3N_4 bulks.

Figures 3 and 4 show SEM micrographs of the etched surface of the joint regions of the specimens joined at 1500 and 1550°C, respectively. The microstructure consisted mainly of equiaxed grains with a small number of fibrous grains, indicating that the phase transformation from α to β-silicon nitride grains already occurred in the joint layers. However, the grain size was substantially small and typically less than one micron. In addition, substantial amounts of small pores (particularly that joined at 1500°C, Fig. 3 (a)) and voids (arrows in Figs. 3 (b) and 4 (b)) were observed in the joint regions, indicating that the densification were are completed. The relatively low joint strength of the specimens joined at 1500 and 1550°C are due to these pores and voids as well as fine, equiaxed microstructure.

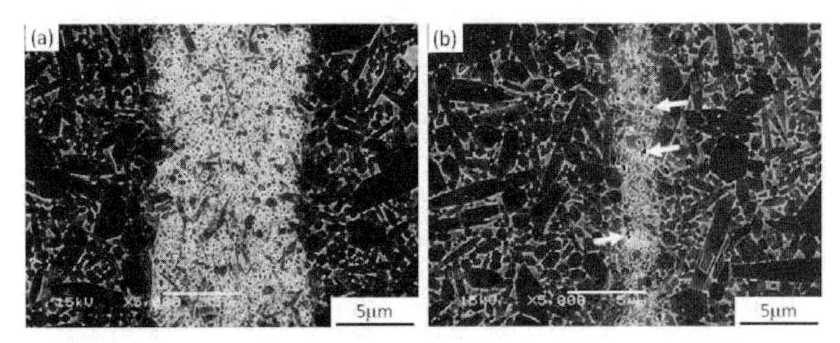

Figure 3. Scanning electron micrographs of the joint regions joined at 1500°C.

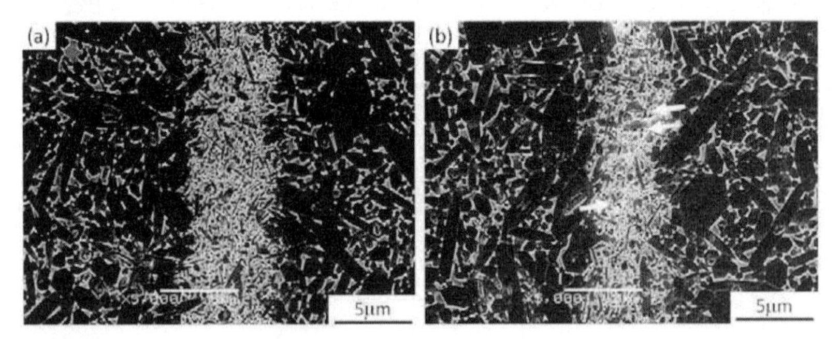

Figure 4. Scanning electron micrographs of the joint regions joined at 1550°C.

Figure 5 shows SEM micrographs of the etched surface of the joint regions of the specimens joined at 1600°C. Such pores and voids as observed in the cases of 1500 and 1550°C (Figs. 3 and 4) were hardly identified, indicating completed densification. In addition, many fibrous grains were formed indicating similar grain morphology to the parent region, though the grain size was substantially smaller than that of the parent silicon nitride. When the joint layer was thin (Fig. 5 (b)), the border with the parent region was not clearly distinguishable due to relatively large fibrous grains formed in the layer. Figure 6 shows SEM micrographs of the etched surface of the joint regions of the specimens joined at 1650°C. Similarly to the case at 1600°C, few voids or pores were observed, indicative of the fully densified layer. However, the grain growth was further enhanced in the joint region and some fibrous grains appear to interpenetrate between the parent and joint regions, resulting in the less distinct border compared to the case at 1600°C.

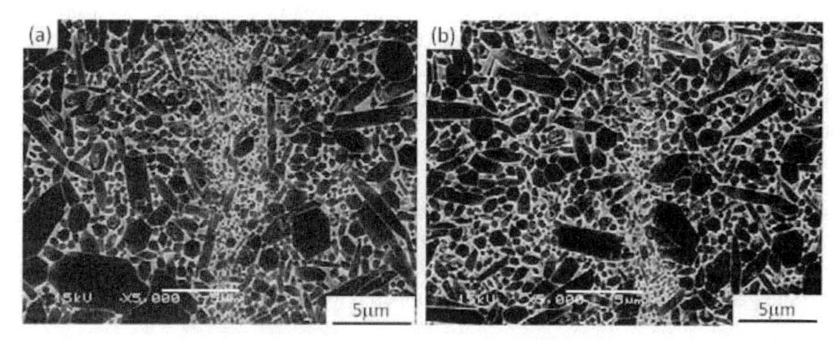

Figure 5. Scanning electron micrographs of the joint regions joined at 1600°C.

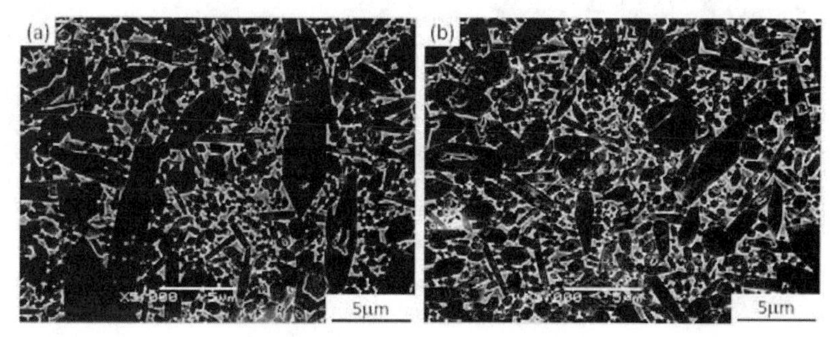

Figure 6. Scanning electron micrographs of the joint regions joined at 1650°C.

Compared to the specimens joined at 1500 and 1550°C, those at 1600 and 1650°C showed relatively high strength of about 680 MPa, which can be attributed to several factors. First, there are almost no voids and pores, compared to the cases at 1500 and 1550°C due to the full densification of the joint layers. Second, a lot of fibrous grains of β-silicon nitride were grown substantially and some fibrous grains appeared to interpenetrate the border between the parent and joint regions, indicating strengthening of both the joint layer itself and the border due to crack shielding effects of fibrous grains. Third, the chemical composition of the joint region became close to that of the parent one by the diffusion, leading to reduction of thermal expansion mismatch between the two regions.[22] The diffusion of the chemical elements also enhances formation of the strong border. These speculations are supported by the fact that more than half of the specimens joined at 1600 and 1650°C failed from the parent region while all joined at 1500 and 1550°C did from the joint one.

Xie *et al.*[3-6] studied joining of silicon nitrides with the ceramic adhesive based on the system Si_3N_4–Y_2O_3–SiO_2–Al_2O_3, which is the same as this study, and realized the similarity in chemistry and microstructure between the parent and joint layers. When the joining was conducted at 1600°C for 30 min with an applied pressure of 5 MPa, the joint strength was

550 MPa at room temperature. The post-joining hot-isostatic pressing at 1780°C for 2 h at 150 MPa led to increased strength of 668 MPa. Xie *et al.*[6] also reported that the applied pressure during joining showed significant influence on the joint strength. For example, for the as-joined material, the joint strength with 0 and 2 MPa was as low as about 90 and 200 MPa, respectively. The joint strength of the specimens joined at 1600 and 1650°C in this work was higher than that of the as-joined specimens and was almost the same as that with the post-joining hot-isostatic pressed ones in their work. This is most likely because of the strong border between the parent and joint regions, which is caused by well developed fibrous grains bridging these two regions formed in the long holding time of 1 h (Fig. 6).

Figure 7 shows SEM micrographs of the fracture surfaces of specimens joined at 1600°C. Intergranular fractures were dominant when the fracture origins were in the joint regions (Fig. 7 (a)), while those failing from the parent regions indicated a feature of mixed integranular and transgranular fractures. Actually, the specimen fractured from the parent region showed higher strength of 727 MPa, compared to that from the joint region (Fig. 2).

Figure 7. Scanning electron micrographs of fracture surfaces of specimens joined at 1600°C. Fractures from joint region (a) and parent region (b).

CONCLUSIONS

Silicon nitride ceramics were joined by a local-heating joining technique, and the joint strength was examined in conjunction with the microstructural development in details. Commercially available silicon nitride ceramic pipes sintered with Y_2O_3 and Al_2O_3 additives were used for parent material, and powder slurry of Si_3N_4-Y_2O_3-Al_2O_3-SiO_2 system was brush-coated on the rough or uneven end faces of the pipes. Joining was carried out by locally heating the joint region at different temperatures from 1500 to 1650°C for 1 h with a mechanical pressure of 5 MPa in N_2 flow, and four-point bending tests were conducted for measuring the joint strength. The specimens joined at 1500 and 1550°C showed the low strength of 412 MPa, and the microstructural observations using an SEM clarified that the grains were fine and equiaxed and substantial amounts of voids and pores were left in the joint layers. On the contrary, those joined at 1600 and 1650°C showed the relatively high strength of about 680 MPa; and the joint layers were fully densified with well developed fibrous grains. Fracture surface observations also revealed that intergranular fractures were predominant when the fracture origins were in joint regions while those failing from parent regions showed a feature of mixed intergranular and transfranular fractures.

ACKNOWLEDGMENT

This work was supported by the Project for "Innovative Development of Ceramics Manufacturing Technologies for Energy Saving" from Ministry of Economy, Trade and Industry (METI), Japan and New Energy and Industrial Technology Development Organization (NEDO), Japan.

FOOTNOTES

*Corresponding Author. e-mail: mikinori-hotta@aist.go.jp
**Present address: Nagoya University, Nagoya 464-8603, Japan

REFERENCES

[1]H. Kita, H. Hyuga, N. Kondo and T. Ohji, "Exergy Consumption through the Life Cycle of Ceramic Parts", *Int. J. Appl. Ceram. Tech.*, **5** [4] 373-381 (2008).
[2]H. Kita, H. Hyuga and N. Kondo, "Stereo Fabric Modeling Technology in Ceramics Manufacture", *J. Eur. Ceram. Soc.*, **28**, 1079-1083 (2008).
[3]R. -J. Xie, L. -P. Huang, Y. Chen and X. -R. Fu, "Evaluation of Si_3N_4 Joints: Bond Strength and Microstructure", *J. Mater. Sci.*, **34**, 1783-1790 (1999).
[4]R. Xie, L. Huang, Y. Chen and X. Fu, "Effects of Chemical Compositions of Adhesive and Joining Processes on Bond Strength of Si_3N_4/Si_3N_4 Joints", *Ceram. Int.*, **25**, 101-105 (1999).
[5]R. Xie, L. Huang, Y. Chen and X. Fu, "Bonding Silicon Nitride Using Y_2O_3-Al_2O_3-SiO_2 Adhesive", *Ceram. Int.*, **25**, 535-538 (1999).
[6]R. -J. Xie, M. Mitomo, L. -P. Huang and X. -R. Fu, "Joining of Silicon Nitride Ceramics for High-Temperature Applications", *J. Mater. Res.*, **15** [1] 136-141 (2000).
[7]R. -J. Xie, M. Mitomo, G. -D. Zhan, L. -P. Huang and X. -R. Fu, "Diffusion Bonding of Silicon Nitride Using a Superplastic β-SiAlON Interlayer", *J. Am. Ceram. Soc.*, **84** [2] 471-473 (2001).
[8]M. Singh, "Microstructure and Mechanical Properties of Reaction-Formed Joints in Reaction-Bonded Silicon Carbide Ceramics", *J. Mater. Sci.*, **33**, 5781-5787 (1998).
[9]M. Singh and R. Asthana, "Joining of Zirconium Diboride-Based Ultra High-Temperature Ceramic Composites Using Metallic Glass Interlayers", *Mater. Sci. Eng. A*, **460**, 153-162 (2007).
[10]M. Singh and R. Asthana, "Joining of ZrB_2-Based Ultra-High-Temperature Ceramic Composites to Cu-Clad-Molybdenum for Advanced Aerospace Applications", *Int. J. Appl. Ceram. Tech.*, **6** [2] 113-133 (2009).
[11]M. Singh, R. Asthana, F. M.Varela and J. Martinez-Fernandez, "Microstructural and Mechanical Evaluation of a Cu-Based Active Braze Alloy to Join Silicon Nitride Ceramics", *J. Eur. Ceram. Soc.*, **31**, 1309-1316 (2011).
[12]K. P. Plucknett, "Joining Si_3N_4-Based Ceramics with Oxidation-Formed Surface Layers", *J. Am. Ceram. Soc.*, **83** [12] 2925-2928 (2000).
[13]J. Q. Li and P. Xiao, "Joining Alumina Using an Alumina/Metal Composite", *J. Eur. Ceram. Soc.*, **22**, 1225-1233 (2002).
[14]S. M. Hong, C. C. Bartlow, T. B. Reynolds, J. T. McKeown and A. M. Glaeser, "Ultrarapid Transient-Liquid-Phase Bonding of Al_2O_3 Ceramics", *Adv. Mater.*, **20**, 4799-4803 (2008).
[15]M. Gopal, M. Sixta, L. D. Jonghe and G. Thomas, "Seamless Joining of Silicon Nitride Ceramics", *J. Am. Ceram. Soc.*, **84** [4] 708-712 (2001).
[16]M. A. Sainz, P. Miranzo and M. I. Osendi, "Silicon Nitride Joining Using Silica and Yttria Ceramic Interlayers", *J. Am. Ceram. Soc.*, **85** [4] 941-946 (2002).
[17]N. Kondo, H. Hyuga and H. Kita, "Joining of Silicon Nitride with Silicon Slurry via Reaction Bonding and Post Sintering", *J. Ceram. Soc. Japan*, **118** [1] 9-12 (2010).
[18]N. Kondo, H. Hyuga, M. Hotta and H. Kita, "Semi-Homogeneous Joining of Silicon Nitride

with a Silicon Nitride Powder Insert", *J. Ceram. Soc. Japan*, **119** [4] 322-324 (2011).
[19]N. Kondo, M. Hotta, H. Hyuga and H. Kita, "Semi-Homogeneous Joining of Silicon Nitride using Oxynitride Glass Insert Containing Silicon Nitride Powder and Post-Heat Treatment", *J. Ceram. Soc. Japan*, **120** [3] 119-122 (2012).
[20]N. Kondo, H. Hyuga, H. Kita and K. Hirao, "Joining of Silicon Nitride by Microwave Local Heating", *J. Ceram. Soc. Japan*, **118** [10] 959-962 (2010).
[21]M. Hotta, N. Kondo and H. Kita, "Joining of Silicon Nitride Long Pipe by Local Heating", *Ceram. Eng. Sci. Proc.*, **32** [8] 89-92 (2011).
[22]M. Hotta, N. Kondo, H. Kita and T. Ohji, "Joining of Silicon Nitride by Local Heating for Fabrication of Long Ceramic Pipes", *Int. J. Appl. Ceram. Tech.*, (in press).

DEVELOPMENT OF THERMAL SPRAYING AND PATTERNING TECHNIQUES BY USING THIXOTROPIC SLURRIES INCLUDING METALS AND CERAMICS PARTICLES

Soshu Kirihara, Yusuke Itakura and Satoko Tasaki
Joining and Welding Research Instite, Osaka University
Ibaraki, Osaka, Japan

Yusuke Itakura
Graduate School of Engineering, Osaka University
Suita, Osaka, Japan

ABSTRACT

Thermal nanoparticles coating and microlines patterning were newly developed as novel technologies to fabricate fine ceramics layers and geometrical intermetallics patterns for mechanical properties modulations of practical alloys substrates. Nanometer sized alumina particles were dispersed into acrylic liquid resins, and the obtained slurries were sputtered by using compressed air jet. The slurry mists could blow into the arc plasma with argon gas spraying. On stainless steels substrates, the fine surface layers with high wear resistance were formed. In cross sectional microstructures of the coated layers, micromater sized cracks or pores were not observed. Subsequently, pure aluminum particles were dispersed into photo solidified acrylic resins, and the slurry was spread on the stainless steel substrates by using a mechanical knife blade. On the substrates, microline patterns with self similar fractal structures were drawn and fixed by using scanning of an ultra violet laser beam. The patterned pure metal particles were heated by the argon arc plasma spray assisting, and the intermetallics or alloys phases with high hardness were created through reaction diffusions. Microstructures in the coated layers and the patterned lines were observed by using a scanning electron microscopy.

INTRODUCTION

Thermal nanoparticles spraying had newly developed as a novel coating technique to create fine ceramic layers without structural defects on practical alloys substrates [1,2]. The nanometer sized alumina ceramic particles were dispersed into liquid resins. Fluid viscosity of the mixt pastes were measured and optimized to realize homogenized particles dispersions and smooth slurry flows. The obtained thixotropic slurry of rheological flow was sputtered by using compressed air jet, and the slurry mist blew into the arc plasma with an argon gas spraying. On a stainless steel substrate, the ceramic layers were formed. The formation of micro cracks and pores in the coated layers will be discussed by scanning electron microscope observations. Subsequently, thermal microline patterning to realize metal phases drawing on the alloys substrates had been developed [3,4]. Micrometer sized pure aluminum particles were dispersed into the photo solidified liquid resins. The slurry pastes were spread on the stainless steel substrate, and micro patterns were drawn and fixed by using an ultra violet laser scanning. The patterned pure metal particles were heated by the argon arc plasma spraying, and the intermetallics and alloys phases with high hardness were created through reaction diffusions. Microstructures and composite distributions in the vicinity of formed alloy and metal interfaces were observed and analyzed by using an electron microscope. Load dispersion abilities of the network were evaluated by using conventional mechanical tests and compared with simulated and visualized profiles by using a numerical analysis simulation. The modulation effects of stress distributions on the material surfaces by design and fabrication of these geometric fractal patterns will be discussed.

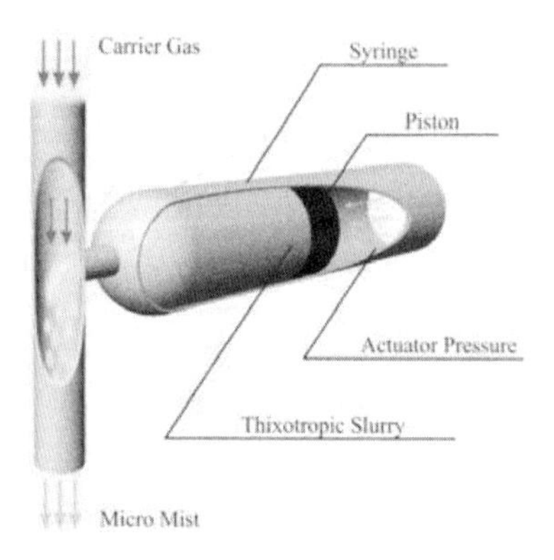

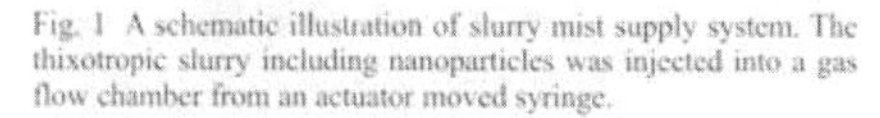
Fig. 1 A schematic illustration of slurry mist supply system. The thixotropic slurry including nanoparticles was injected into a gas flow chamber from an actuator moved syringe.

Fig. 2 An appearance photograph of the slurry mists supplying system mounted on a plasma torch. The slurry syringe piston is moved by a mechanical actuator in micromater order accuracy.

EXPERIMENTAL PROCEDURE

The nanometer and micrometer sized alumina and aluminum particles of 200 nm and 10 μm in average diameters were dispersed into the acrylic liquid resins at 40 % and 60 % in volume fractions, respectively. These obtained slurries were mixed in airtight containers through planetary configuration of rotation and revolution movements. Fluid characteristics of the slurry pasts were evaluated by using a viscosity and viscoelasticity measuring instrument to realize thixotropic fluid flows. The thixotropic slurry was injected into a gas flow chamber perpendicularly from a syringe nozzle of 40 mm in inner diameter, and the fluid viscosity was decreased by shear stress loading as schematically illustrated in Fig. 1. The slurry mists were sprayed toward a glass plate to measure and observe droplet diameters and spattering patterns. The micro sized mists were formed through the compressed air jet of 2 atm in gas pressure, and introduced into the into the arc plasma with an argon gas spray of 50 slpm in flow late through the stainless still guide nozzle of 4 mm in caliber size. Figure 2 shows the mist supplying equipment mounted on the plasma spraying apparatus. The alumina coated layer was formed on the SUS-304 stainless steel substrate of 50×50×1 mm in size placed at 140 mm in distance from the plasma gun. Crystal phase of the coated layer was analyzed by X-ray diffraction, and cross sectional micro-structures and compositional distributions were observed by using an optical and scanning electron microscopy. Subsequently, self similar patterns of Hilbert curve with stage numbers 1, 2, 3 and 4 of the fractal line structures were designed by using a computer graphic application, and these graphic images were converted into the numerical data sets by computer software. These patterns of 25×25 mm in whole size were composed of arranged lines of 400 μm in width. These graphic models were transferred into the processing apparatus as operating data sets. The pure aluminum particles were patterned on the SUS-304 stainless steel substrate by a stereolithographic pattering system as schematically illustrated in Fig. 3. The mixed resin paste with metal particles was spread with 100 μm in layer thickness on the substrate of 30×30×2 mm in size by a mechanically moved knife edge. An ultraviolet laser beam of 355 nm in wavelength and 100 μm in beam spot was scanned on the resin surface. The solid pattern was obtained by light induced photo polymerization. Figure 4 shows the appearance of stereolithographic system. After the removing uncured resin by ultrasonic cleaning in ethanol solvent, the patterned sample was heated above the reaction temperature by using the argon gas arc plasma spraying. The microstructures and composite distributions were observed by using the scanning electron microscopy and energy dispersive X-ray spectroscopy. The stress distributions on the patterned samples were simulated by using the finite element method of the numerical simulation.

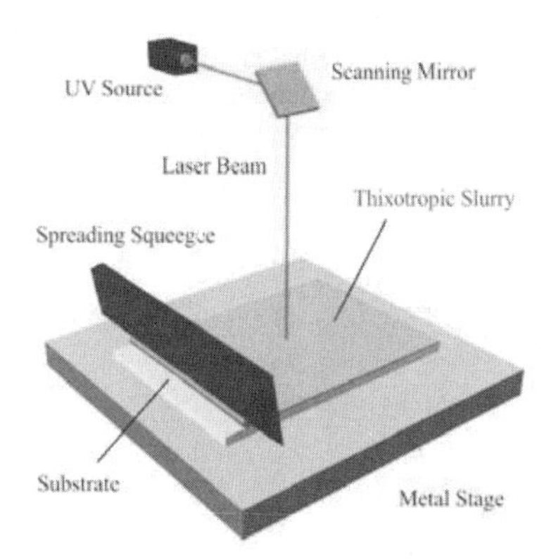

Fig. 3 A schematic illustrated stereolithograpy system using laser scanning processes. The thixotropic slurry can be spread smoothly on a metal substrate by using a mechanically movedsqueegee.

Fig. 4 The appearance photograph of stereolithograpy equipment with the slurry supplying syringe. An ultra violet laser beam was irradiated from the top position of this apparatus for a metal stage.

RESULTS AND DISCUSSIONS

The acrylic liquid resin with the alumina nanoparticles of 200 nm in diameter at high volume contents exhibited the thixotropic characteristics of a non-Newtonian fluid. The formed slurry with the particles dispersion above 42 % in volume contents could not flow smoothly. For the stable materials supplying and the continuous mist creation, the maximum volume content of the nanoparticles can be optimized at 40 %. In this slurry fluid, the nanoparticles are considered theoretically to spatially disperse without coagglutinations. The obtained slurry blew toward the glass plate to observe and measure the mist droplets shapes and sizes. These droplet diameters became finer as shown in Fig. 5, and the variations could be reduced by decreasing the slurry supply rate. The minimum diameter size was 50 μm. Figure 6 shows the fine alumina coated layer formed on the stainless steel substrate. The ceramic layers of 50 μm in thickness were formed at 300 gpm in supply rate. The structural defects of cracks or pores were not observed through microscopic observations. The coated ceramics layer and alloy substrate were joined successfully without the defects of voids or exfoliations in the interfaces. Moreover, the carbon element contaminations in the created alumina layers derived from the acrylic liquid resins were not detected though the X-ray diffraction analysis. Subsequently, the iron aluminide micro pattern with the fractal structure of Hilbert curve was formed successfully on the stainless steel substrate. Figure 7 shows the formed fractal polyline of the number 3 in fractal stage. The micrometer order geometric structure was composed of fine intermetallics lines of 450 μm in width. The part accuracy of these microline patterns were estimated as 10 % approximately. The iron aluminide composite was formed widely comparing with the designed line width, though the reaction diffusion between the aluminum and stainless steel substrates. The defects of crack or pores could not observed in the cross sectional microstructure of the formed intermetallics. And the intermetallic layer and the alloy substrate were joined successfully without void formation through the reaction diffusion. The composite phase could be identified as Al-rich Fe_3Al. The stress distributions on the patterned surfaces were visualized for the Hilbert curve of stage number 3 through the numerical simulation. The required mechanical properties of Young's modulus were defined along the compositional analysis and the phase identifications. The stress intensities concentrate into the vicinity of fixed edge and are distributed along the patterned lines and the corners with the higher hardness. The fractal patterns can include the more numbers of sides and nodes in limited aria comparing with the periodic arrangements of polygon figures.

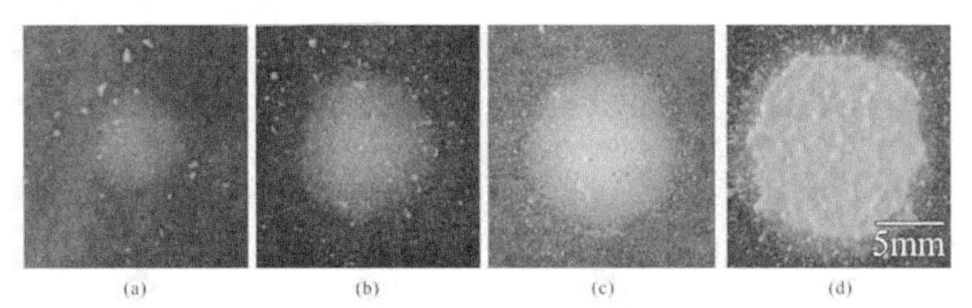

Fig. 5 Micro mist spraying patterns formed by blowing of the thixotropic resin slurry including the alumina nanoparticles. The slurry supplying rate was varied systematically to create the fine mist droplet. The mist droplet sizes in these spraying patterns (a), (b), (c) and (d) are measured by using a digital optical microscope as about 50, 100, 200 and 500 μm, respectively.

Fig. 6 A microstructure of an alumina coated layer formed on a stainless steel substrate. A side vew of the coated sample was observed to show the thermal sedimention of nanoparticles.

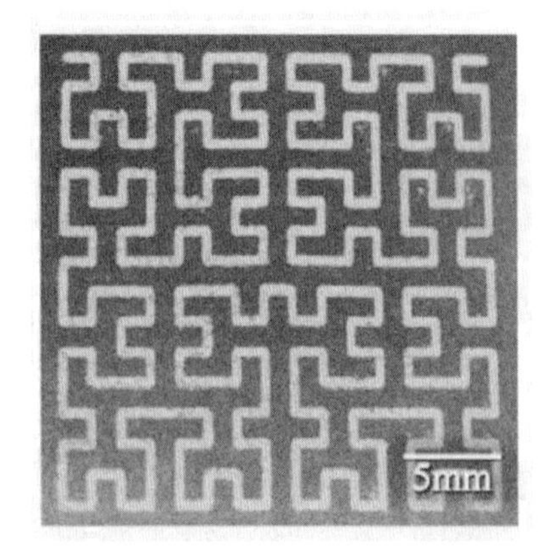

Fig. 7 An iron aluminide micro pattern with Hilbert curve of a fractal structure formed on the stainless steel substrate through reaction diffusion assisted by the plasma spray irradiation.

CONCLUSION

Fine alumina coated layers and aluminide composite lines could be created successfully on stainless steel substrates by thermal nanoparticles spraying and microlines patterning to improve thermal and mechanical properties on components surfaces. As raw materials using these techniques, thixotropic slurries including nanometer and micrometer sized particles were formulated systematically. The slurry handling methods are key techniques to realize the nanoparticles arrangements create useful functional surfaces on various components. In the near future, the investigated coating and patterning techniques will become the candidate for efficient processes to realize effective thermal barrier coating without structural defects and mechanical properties improvement on various mechanical components

References

[1] S. Kirihara: Thermal Spraying Technology, 30 (2010) 44-50.
[2] S. Kirihara: Journal of Japan Welding Society, 80 (2011) 6-9.
[3] Y. Uehara, S. Kirihara: Journal of Smart Prosessing, 1 (2012) 186-189.
[4] Y. Uehara, S. Kirihara: Joural of Functionally Graded Materials, 22 (2012) 36-42.

NOVEL JOINING METHOD FOR ALUMINA BY SURFACE MODIFICATION AND REDUCTION REACTION

Ken'ichiro Kita[1], Naoki Kondo[1]

[1]National Institute of Advanced Industrial Science and Technology (AIST)
2266-98 Shimo-shidami, Moriyama-ku, Nagoya, 463-8560, JAPAN

ABSTRACT

This paper describes a novel ceramic joining method containing surface modification by polymer and direct reaction between the modified surface and metal foil. By using polymer, the surface of alumina can be modified to SiO_2 and SiOC. In the case of the combination of SiO_2 and aluminum foil, the components of SiO_2 and aluminum could be diffused and transformed into uniform alumina silicate joining layer. On the other hand, the combination of SiOC surface and aluminum could be resulted metal silicon near the center of joining layer and carbon substance with cracks in the peripheral of joining layer. It was considered the reduction producing metal silicon was preferentially carried out in the center and the by-product of carbon substance would bleed out to the peripheral of joining layer.

1. INTRODUCTION

Energy saving is one of the global problems and we should approach this problem from multilateral phases. From the point of view of material engineering, the weight loss is one of the solutions for the problem. For example, the weight loss of the machine such as a stage for processing of semiconductor which needs rapid and complicated action in operation is very effective, because such machines usually consist of heavy metal because of stability and hardness. Therefore, we thought that the substitute from heavy metal to ceramic is one of the good works for the solution of the problem. Ceramic is lighter than heavy metal and the toughness and stability of ceramic is not inferior to that of heavy metal [1,2].

Ceramics consist of strong covalent bond or ionic bond, so that ceramics have lots of great characters such as heat resistance, high hardness, abrasive resistance, etc [3-5]. Therefore, the making of the huge and complex architecture which consists of ceramics is very difficult. If such architecture should be made by traditional method, a huge electronic furnace, a large amount of energy, and a lot of cost will be required. We must investigate a novel process for making such ceramic architecture.

We paid attention to joining method of ceramics. This method has the following features; a huge electronic furnace is not needed, complicated shaped ceramic architecture can be made, and partial heating enables to join ceramics [6,7]. Many researches of ceramic joining method have been reported [8-11]. Metallization and glass joining are famous joining methods for ceramics. These methods can make ceramic joining by low-temperature heating and apply to a lot of kinds of ceramics. However, the components of most adhesive agent is different from that of joined ceramic, so that joined ceramics by using adhesive agent has sometimes poor thermal resistance and quality which restricts usage of the joined ceramic.

Based on the above facts, precursor polymer is considered as one of good adhesive agent for ceramic joining [12,13]. However, there was a problem that the polymer generates many cracks in joining layer derived from shrinking of polymer during ceramization.

In this study, we tried joining ceramics by direct reaction between pure metal foil and the ceramized membrane derived from the polymer. We thought that we could obtain a novel joining method without the generation of cracks and poor quality in joining layer through this study.

2. EXPERIMENTAL PROCEDURE

The experimental procedure in this study was shown in Fig.1. Bulk alumina whose purity was more than 99.9 % was cut into alumina pieces (length: 20 mm, width: 30 mm, height: 20 mm). A surface with an area of 600 mm^2 was abraded by a grinder, and the following investigations were carried out on the abraded surface.

Two types of polymers were prepared in this study. One is polycarbosilane (PCS) containing Si-CH_2-Si of main chain, methyl group and hydrogen as side chains. The reason why this polymer would be used is that PCS is one of the best polymers of modification of ceramic surface. By using PCS, the modified surface on ceramic can be obtained easily without exfoliation [14]. The other is polymethylphenylsiloxane (PMPhS) which is one of siloxanes including phenyl groups in the side chain. Polysiloxane includes large amount of oxygen and carbon, so that SiOC ceramic can be obtained easily [15]. Besides, the solubility between PCS and PMPhS is very good [16]. Commercial PMPhS (KF-54, Shin-Etsu Chemicals Co. Ltd., Japan) was blended with PCS (NIPUSI-Type A, Nippon Carbon, Japan) at a PMPhS to PCS blend ratio of 30 mass%. Hereafter, this blended polymer was referred to as "PS30".

Bulk aluminas were dipped into PCS or PS30 solution whose density was fixed to 0.1mol/L. After drying, samples were cured in air flow. The heating rate of samples in curing was fixed at 8K/min up to 473 K and it kept for 1h. After the curing, these samples were pyrolyzed at 1273 K for 1 h under Ar atmosphere and re-pyrolyzed at 1473K for 2 h under Air flow. The ceramization of these polymers was completed by the first pyrolysis and the oxidation of ceramics membranes from these polymers was completed by re-pyrolysis.

An aluminum foil with a thickness of approximately 11 μm was placed between ceramics membranes on the abraded surfaces of the samples, and these were heated at 1073 K for 2 h in vacuum as joining. Hereafter, the joined sample by using PCS polymer was referred to as "PCS sample" and the joined sample by using PS30 was referred to as "PS30 sample".

After the above-mentioned process, these samples were cut into the shape of sticks for the bending test of JIS R1601. The joining area in the pyrolyzed surface was observed by scanning electronic microscopy (SEM; JEM-5600, JEOL, Japan), energy dispersive X-ray spectroscopy (EDS; JEM-2300, JEOL, Japan), and X-ray diffraction (XRD; RINT2500, Rigaku Corporation, Japan). After the observation, a four-point bending test was performed along JIS R 1601.

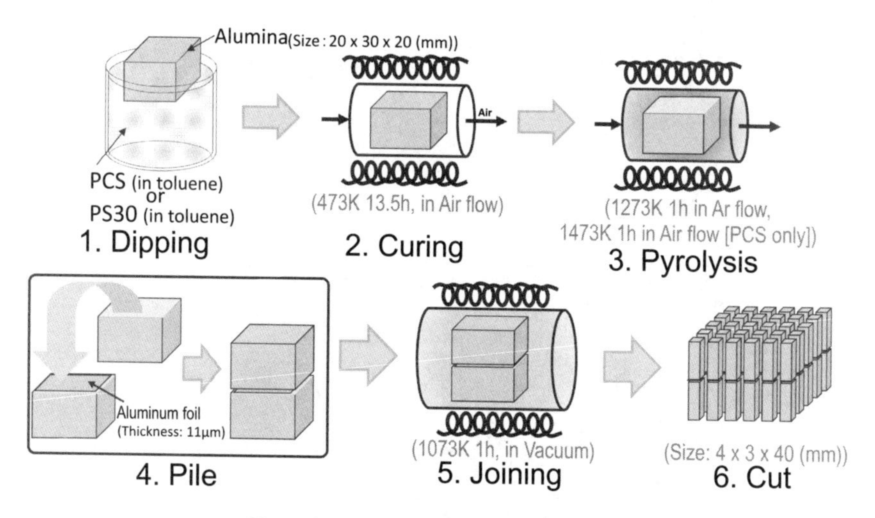

Figure 1. The outline of this experiment

3. RESULTS AND DISCUSSION

3-1. Comparison between PCS sample and PS30 sample

We tried observing the cross section of joining areas of these samples to clarify the composition. Figure 2 shows the SEM image and EDS mappings of the cress section of joining area of PCS sample. The color of the joining area was lighter than that of aluminum and the width of the area was about 20μm. This SEM image shows that there was no crack in the interface between joining area and alumina. The EDS mappings show that the joining area contained aluminum, silicone, and oxygen. Moreover, these elements uniformly existed in this area. It was considered that SiO_2 on alumina and metal aluminum were integrated and uniformly diffused in this joining area. On the other hand, the SEM image and EDS mappings of the cress section of joining area of PS30 sample was shown in Fig.3. The width of the joining area was about 30μm. In this case, there was also no crack in the interlayer between joining area and alumina. However, the EDS mappings show that the spectra of elements were strongly observed the spectrum of silicone element in the joining layer and any other spectra could not be found. Silicon element existed in the modified membrane only on alumina. Therefore, this joining area was occupied by the elements derived from polymer.

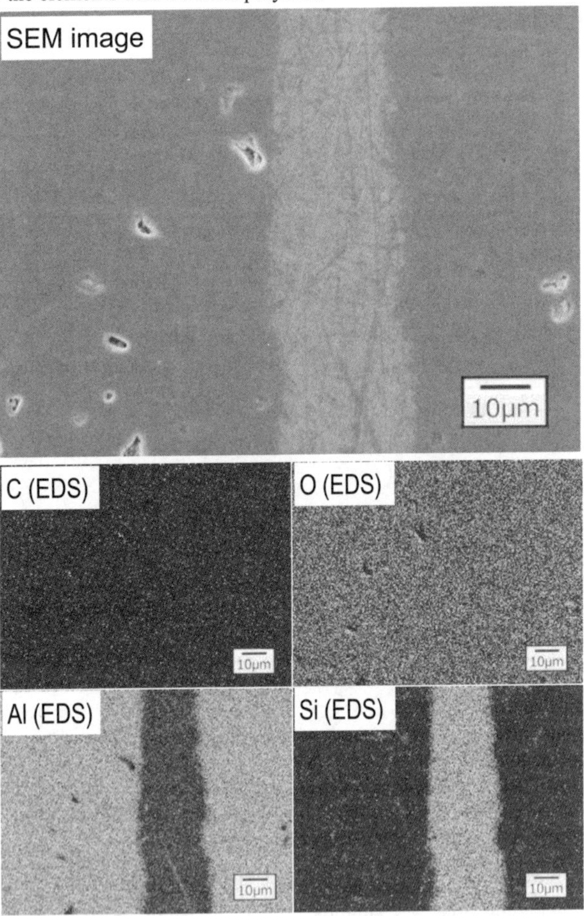

Figure 2 SEM image and EDS mappings of the cross section of the joining area of PCS sample

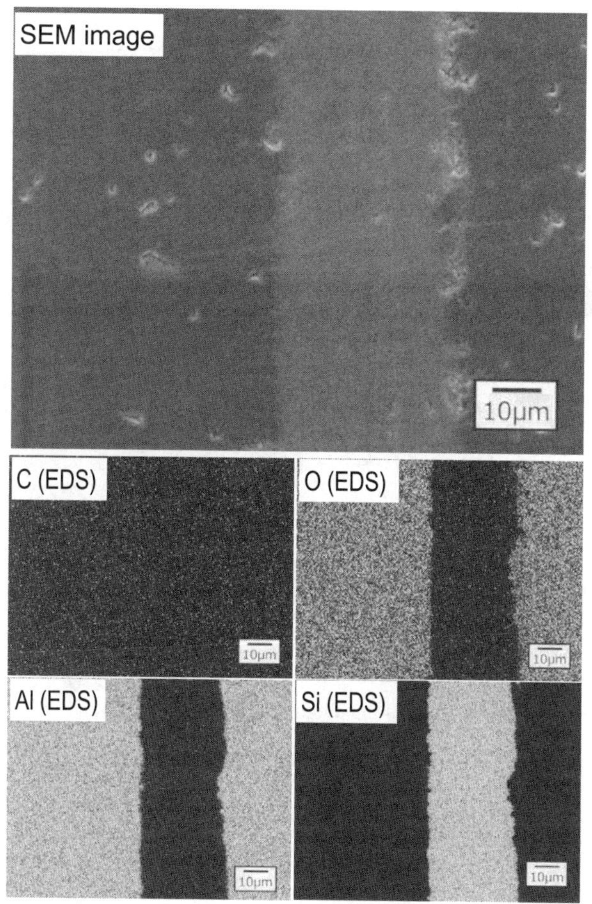

Figure 3 SEM image and EDS mappings of the cross section of the joining area of PS30 sample

To identify the substance in these joining layers, XRD patterns of these joining layers were investigated and these spectra are shown in Fig.4. Both patterns show that alumina, metal aluminum, silicon, and carbon as the background. Moreover, there was a wide and continuous peak for θ = 9.0°–13.0°. Such peak shows the existence of an amorphous compound and the compound was considered to be derived from amorphous of aluminum silicate or metal Si [17]. In the case of PCS sample, the peaks of aluminum silicate are mainly shown in 9.7°, the set of small peaks around 20°, 31.8°, 41.6° and 51.2°. These peaks are shown in the pattern of the joining layer of this sample. Therefore, taking account of the existence of amorphous, it can be seen that the joining area of PCS sample consists of aluminum silicate. On the other hand, in the case of PS30 sample, the remarkable peaks of aluminum silicate were hardly observed. However, the remarkable peaks of crystal silicon were gained compared with that of PCS sample. This result is agreeable because the result of EDS mappings suggest the existence of metal silicon in the joining layer, and it is considered that the continuous peak of amorphous compound shows amorphous metal silicon.

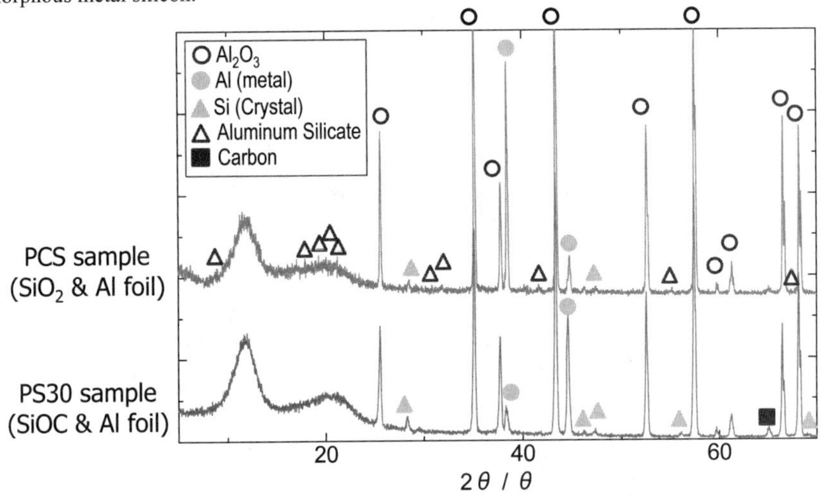

Figure 4 XRD patterns of the joining area of PCS sample and PS30 sample

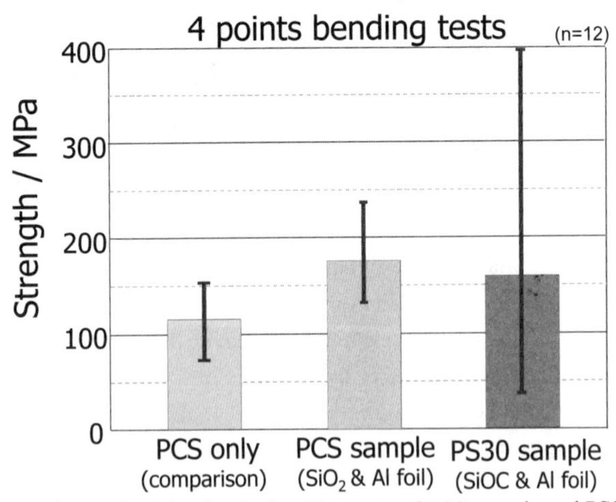

Figure 5 The result of the 4 point bending tests of PCS sample and PS30 sample

The 4 point bending test was carried out, and the results are shown in Fig.5. As a comparison, the 4 point bending strength of alumina sample joined by using PCS only was also shown in this figure [18]. The number of these samples for this test was fixed to 12. In the case of PCS sample, the average strength was about 177MPa and maximum strength was 239 MPa. On the other hand, in the case of PS30 sample, although maximum strength was 395 MPa, the average strength was only 155MPa. In addition, the error bar of this sample was too wide. Therefore, we re-investigated this sample to clarify the result of wide error bar.

3-2. Comparison between PS30-C sample and PS30-P sample

From the above-mentioned result of PS30, it could not be found where carbon existed in spite of there was a large number of carbon in SiOC membrane. To clarify where carbon existed, PS30 sample was distinguished by PS30-C sample and PS30-P sample. PS30-C sample was defined as the sample near the center of PS30 sample, and PS30-P sample was defied as the sample near the perimeter of PS30 sample. We re-investigated these samples.

In the case of PS30-C, SEM image shows that there was no crack in the joining area, EDS mappings showed that very strong peak of silicon spectrum was observed, XRD pattern shows the many metal silicon peaks and the curve of pattern of amorphous. It was considered that PS30-C was almost the same as PS30.

On the other hand, in the case of PS30-P, the difference compared with PS30-C could be clearly observed. Figure 6 shows the SEM image and EDS mappings of the cress section of joining area of PS30-P. The joining area had been sunk in and made into a porous shape. Moreover, the strong spectrum of carbon could be observed from the area. It reveals that the joining area of PS30-P consisted of carbon component and porous shape.

Figure 7 shows the XRD pattern of the joining area of PS30-P. As a comparison, the XRD pattern of PS30-S is also shown in Fig.8. The remarkable differences compared with PS30-S are the lack of the curve of pattern of amorphous and the increase of the peaks derived from carbon components. The curve revealed the metal silicon because of the result of Fig2-4, so that it was reasonable that the joining area was mainly occupied by carbon contents.

Figure 8 shows the strength of PS30-P and PS30-C measured by the 4-point bending test. In the case of PS30-S, the average strength of bending test was 252 MPa and the maximum strength was 395 MPa. Most samples broke off from the alumina area in the bending test and this result suggests that the center of the joining area was very strong. On the other hand, in the case of PS30-P, the average strength of bending test was only 58 MPa. The broken point in this all samples was joining area and this result was acceptable because the joining layer consisted of carbon components and porous shape.

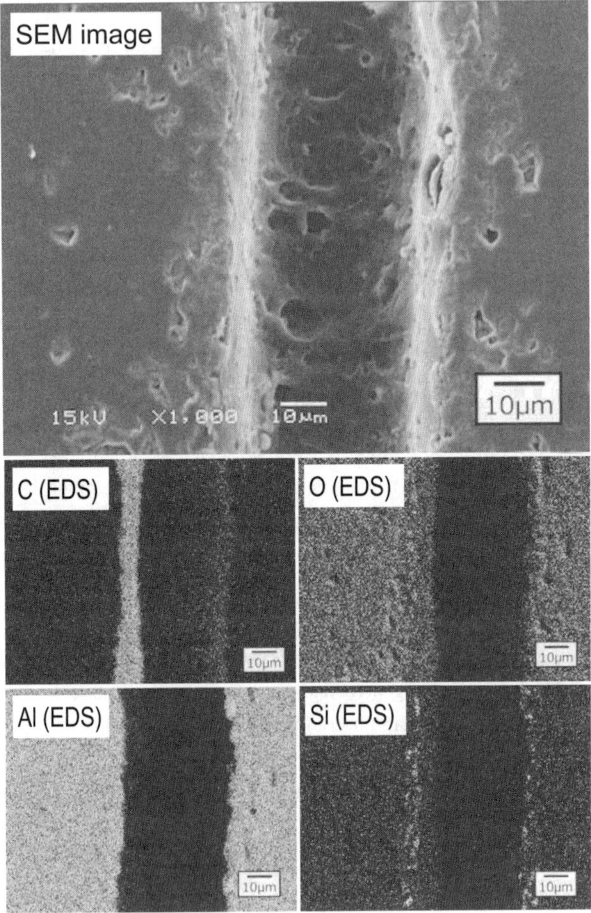

Figure 6 SEM image and EDS mappings of the cross section of joining area of PS30-P sample

From the above-mentioned result of PS30, the center of the joining layer consisted of metal silicon. The considerable formula between metal aluminum and SiOC which can be transformed into metal silicon was the following;

$$SiO\text{-}2C\ (s) + 10/3\ Al\ (l) \rightarrow 1/3\ Al_2O_3\ (s) + 2/3\ Al_4C_3\ (s) + Si\ (s) \qquad (1)$$

Al_4C_3 formation is permitted because aluminum moves rapidly at low temperature region, if there is no barrier on aluminum diffusion. Therefore, the carbon content was considered as the aluminum carbide derived from the direct reaction between metal aluminum and SiOC membrane. However, the peak of aluminum carbide mainly appears lower than 15° and no remarkable peaks in the pattern of PS30-P sample [19]. The peaks suspected as carbon are similar to that of aluminum oxycarbide; therefore it is possible that the joining layer was aluminum oxycarbide [20].

These results reveal that following; metal silicon is made and remains the center of the PS30, the by-product of carbon bleeds out to the peripheral of joining layer, and the carbon consisted of the carbon-rich and porous joining layer during the direct reaction by heating.

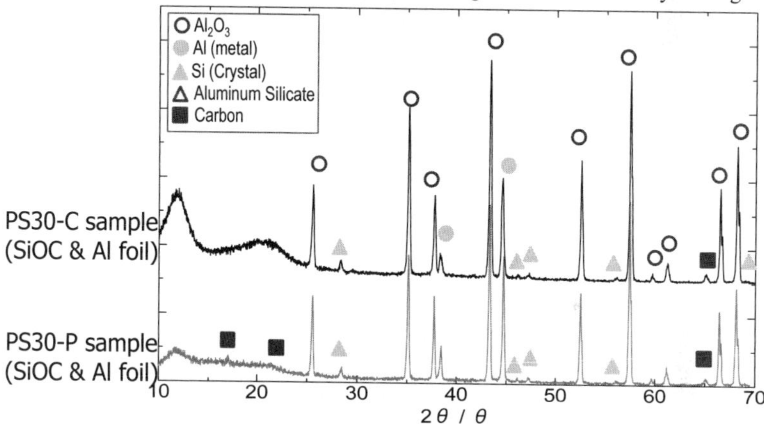

Figure 7 XRD patterns of the joining area of PS30-C sample and PS30-P sample

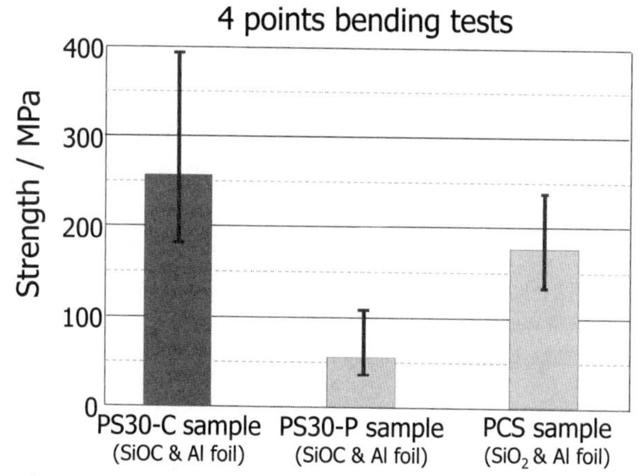

Figure 8 The result of the 4 point bending tests of PS30-C, PS30-P, and PCS sample

4. CONCLUSION

SiO_2 surface on alumina and aluminum foil can be transformed into aluminum silicate in joining layer, and the strength of the joined sample was about 177 MPa average. It was considered that this layer was derived from direct reaction and uniform diffusion between SiO_2 surface and metal aluminum.

On the other hand, SiOC surface on alumina and aluminum foil can be transformed into metal silicon near the center of joining layer and carbon substance with cracks in the peripheral of joining layer. The strength of the sample near the center of joining layer was about 252 MPa average, and that of sample in the peripheral of joining layer was about 58 MPa average. It was considered that this difference of joining layers were derived from reduction reaction between SiOC surface and metal aluminum. The reduction would product metal silicon and the by-product of carbon substance would bleed out to the peripheral of joining layer.

REFERENCES

[1] W. B. Hillig, R. L. Mehan, C. R. Morelock, V. J. de Carlo and W. Laskow, "Silicon/ silicon carbide composites", *Am. Ceram. Soc. Bull.*, **54** (1975), 1054.

[2] S. Prochazka and R.M. Scanlan, "Effect of Boron and Carbon on Sintering of SIC", *J. Am. Ceram. Soc.*, **58** (1975), 72.

[3] R. L. Crane and V. J. Krukonis, "Strength and fracture properties of silicon carbide filament" *Am. Ceram. Soc. Bull.*, **54** (1975), 184-188.

[4] I. A. Aksay, D. M. Dabbs, and M. Sarikaya, "Mullite for Structural, Electronic, and Optical Applications", *J. Am. Ceram. Soc.*, **74** (1991), 2343-2358.

[5] M. Fukushima, Y. Zhou, Y. Yoshizawa, and K. Hirao, "Oxidation behavior of porous silicon carbide ceramics under water vapor below 1000 °C and their microstructural characterization", *J. Ceram. Soc. Jpn.*, **114** (2006), 1155-1159.

[6] H. Kita, H. Hyuga, and N. Kondo, "Stereo Fabric Modeling Technology in Ceramics Manufacture", *J. Euro. Ceram. Soc.*, **28** (2008), 1079-1083.

[7] H. Kita, H. Hyuga, N. Kondo, and T. Ohji, "Exergy Consumption Through the Life Cycle of Ceramic Parts", *Int. J. Appl. Ceram. Technol.*, **5** (2008), 373-381.

[8] M. Nicolas, "The strength of metal/alumina interfaces", *J. Mater. Sci.*, **3** (1968), 571-576.

[9] J. T. Klomp and T. P. J. Botden, "Sealing Pure Aluminia Ceramics to Metals", *Ceram. Bull.*, **49** (1970), 204-211.

[10] L. S. D. Glasser, J. A. Gard, and E. E. Lachowski, "The reaction of zinc oxide and zinc dust with sodium silicate solution" *J. Appl. Chem. Biotechnol.*, **28** (1978), 799-810.

[11] M. Itoh, E. Sugimoto, and Z. Kozuka, "Solid reference electrode of SO_2 sensor using

alumina solid electrolyte" *Trans. J. Inst. Metal.*, **25** (1984) , 504-510.

[12]E. Anderson, S. Ijadi-Maghsoodi, O. Ünai, M. Nostrati, and W. E. Bustamante, "Ceramic Joining," *Ceramic Transactions Vol.77*, ed. I. E. Reimanis, C. H. Henager and A. P. Tomsia (Westville, OH: The American Ceramics Society, 1997), 25-40.

[13]P. Colombo, V. Sglavo, E. Pippel, and A. Donato, "Joining of reaction-bonded silicon carbide using a preceramic polymer", *J. Mater. Sci.*, **33** (1998), 2405-2412.

[14]K. Kita, N. Kondo, Y. Izutsu, H. Kita, "Investigation of the properties of SiC membrane on alumina by using polycarbosilane", *Mater. Lett.*, **75** (2012), 134-136.

[15]K. Kita, M. Narisawa, H. Mabuchi, M. Itoh, M. Sugimoto, M. Yoshikawa, "Synthesis of SiC Based Fibers with Continuous Pore Structure by Melt-Spinning and Controlled Curing Method", *Adv. Mater. Sci.*, **66** (2009), 5-8.

[16]K. Kita, M. Narisawa, A. Nakahira, H. Mabuchi, M. Sugimoto, and M. Yoshikawa, "Synthesis and properties of ceramic fibers from polycarbosilane/polymethylphenylsiloxane polymer blends", *J. Mater. Sci.*, **45** (2010), 3397-3404.

[17]X. Yuan, S. Chen, X. Zhang, and T. Jin, "Joining SiC ceramics with silicon resin YR3184", *Ceram. Int.*, **35** (2009), 3241-3245.

[18]K. Kita, N. Kondo, Y. Izutsu, H. Kita, "Joining of alumina by using organometallic polymer", *J. Ceram. Soc. Jpn.*, **119** (2011), 658-662.

[19]C. Ji, Y. Ma, M. –C. Chyu, R. Knundson, and H. Zhu, "X-ray diffraction study of aluminum carbide powder to 50 GPa", *J. Appl. Phys.*, **106** (2009), 083511.

[20]J. H. Cox, and L. M. Pidgeon, "THE X-RAY DIFFRACTION PATTERND OF ALUMINUM CARBIDE Al_4C_3 AND ALUMINUM OXYCARBIDE AL_4O_4C", *Can. J. Chem.*, **41** (1963), 1414-1416

IONIC LIQUIDS USED AS CLEANING SOLVENT REPLACEMENTS

Melissa Klingenberg, Ph.D. and Janelle Yerty
Concurrent Technologies Corporation
Johnstown, PA, USA

Elizabeth Berman, Ph.D.
Air Force Research Laboratory
Wright-Patterson Air Force Base, OH, USA

Natasha Voevodin, Ph.D.
University of Dayton Research Institute
Dayton, OH, USA

ABSTRACT

Some industrial solvents used in cleaning operations have been designated by the United States (U.S.) Environmental Protection Agency (EPA) as producing greenhouse gases (GHGs) and/or containing volatile organic compounds (VOCs), hazardous air pollutants (HAPs), and/or ozone-depleting substances (ODSs). Although many ODS concerns have been addressed previously and "green" solvents were implemented in cleaning operations, some replacements still contain HAPs, VOCs, or GHGs. Ionic liquids (ILs) have been studied as replacements for common organic solvents used in cleaning operations.

Three candidates were identified and have the potential to be used as cleaning agents: 2-ethylhexyl lactate (2ehl), 1-ethyl-3-methylimidazolium (EMIM) acetate and EMIM ethyl sulfate. Candidates were evaluated according to quality testing procedures currently performed at U.S. Air Force Air Logistics Centers and military cleaning specifications. Test results on cleaning efficiency, corrosion properties, hydrogen re-embrittlement of steel substrates, as well as other effects of candidate cleaners on integrity of substrate properties will be presented, and recommendations will be provided on use of ILs for cleaning operations.

INTRODUCTION

Many solvents used in cleaning operations have been designated by the United States (U.S.) Environmental Protection Agency (EPA) as producing greenhouse gases (GHGs) or containing volatile organic compounds (VOCs), hazardous air pollutants (HAPs), and/or ozone-depleting substances (ODSs). Although many ODS issues were addressed previously, some issues involving HAPs, VOCs, or GHGs still remain, including concerns related to worker health and safety. Ionic Liquids (ILs) have been investigated previously for coatings deposition, energy harvesting, deicing, and as green solvents for munitions applications. Recently, ILs have been considered for cleaning applications because of their negligible vapor pressure, high thermal and electrochemical stability, and low melting points (less than 100 degrees Celsius (°C))[1]. Typically, the cation determines the viscosity and conductivity, which ultimately affects the mass transport properties of the IL[2]. Anion chemistry can determine the reactivity with water, coordinating ability, and hydrophobicity, though not all ILs are hydrophobic[2]. The toxicology of ILs also depends on the specific anion and cation pairing. For example, 1-ethyl-3-methylimidazolium chloride (EMIM [Cl]) is non-toxic, while a close derivative, 1-butyl-3-methylimidazolium chloride (BMIM [Cl]) is toxic[2,3]. Due to the paired nature, ILs can be designed to suit a specific requirement; however, the myriad of combinations also results in reduced availability, potential for high synthesis costs, and limited life cycle knowledge, which may prove to be an obstacle for using ILs as "green" solvents[4].

Through literature searches, vendor surveys, and discussions with chemical suppliers regarding specific cleaning requirements, IL chemistries were identified as having potential for cleaning applications. Vendor discussions led to the selection of 1-ethyl-3-methylimidazolium (EMIM) acetate, EMIM methane sulfonate, EMIM ethylsulfate, and triethylsulfonium bis(trifluoromethylsulfonyl)imide as potential cleaning agents[5]. The literature review indicated that 2-ethylhexyl lactate (2ehl) had been tested as a cleaner and had performed similarly to hydrofluorinated ether (HFE) 7100, which is currently used in vapor degreasing operations at U.S. Air Force Air Logistics Centers (ALCs)[3] for cleaning typical parts, including 4000 series steel C-130 barrels, C-130 lever support sleeves, C-5 pivot shafts, stainless steel C-5 pivot brackets, as well as challenge parts, including 4000 series steel C-130 barrel bolts, C-130 dome retainer nuts, C-130 thrust rings, and Al sheet metal skins. Although 2ehl was determined to be an organic solvent with low vapor pressure rather than a traditional IL (having considerably lower melting point in the mixed form than in each individual component, similar to traditional ILs), it was selected as an alternative cleaner candidate.

Given the lack of performance data for most candidates, the cleaners were selected for testing largely based on evaluation of environmental health and safety (EH&S) risks in terms of Health Materials Information System (HMIS) ratings and any issues indicated on the associated material safety data sheet (MSDS) and/or technical data sheet (TDS). EMIM acetate, EMIM ethylsulfate, and 2ehl (shown in Table 1 with HMIS ratings) were selected for evaluation.

Table 1. Cleaner Candidate Characteristics

Cleaner/Chemical Abstract Service (CAS) Number	Chemical Structure	Indication of Danger/HMIS Rating*	Issues
2ehl 186817-80-1	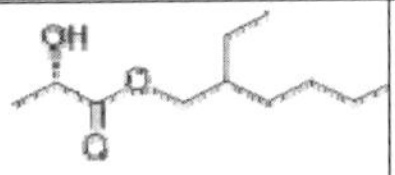	Xi**	Irritant to skin and eyes.
EMIM acetate 143311-17-4	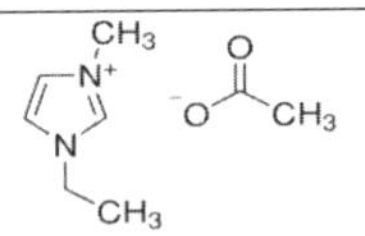	Health – 0 Flammability – 1 Reactivity – 0	Irritant; harmful if swallowed; acute toxicity; stable under normal conditions
EMIM ethylsulfate 342573-75-5	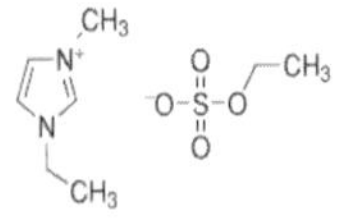	Health – 1 Flammability – 0 Reactivity – 0	Considered hazardous by the Occupational Safety and Health Administration Hazard Communication Standard; irritant

EXPERIMENTAL DETAILS

EMIM acetate was procured from Government Scientific Source, EMIM ethylsulfate was procured from Expotech USA, Inc., and 2ehl was procured from Purac America, Inc. Test requirements were derived from technical orders and military specifications related to cleaning operations that were supplied by ALCs paint/depaint and plating facilities[7-17]. Testing and evaluation included cleaning efficiency and materials compatibility testing. All testing followed industry and federal standards, including ASTM test methods, U.S. EPA standard test methods, and ALC military specification test methods.

For cleaning efficiency pre-weighed 4130 steel and 2024-T3 aluminum (Al) panels 10.2 centimeter (cm) x 15.2 cm x 0.8 cm were coated with 200 milligrams (mg) of molybdenum disulfide grease (containing 50 mg of free soil) and baked in an oven at 100 °C for 60 minutes (min) per MIL-PRF-87937D, Section 4.5.21. Each cleaner was tested in the undiluted form or diluted in a 1:1, 1:2, 1:3, and 1:9 ratio (i.e., cleaner to water) on 9 panels of each alloy to determine optimum performance. Five wear cycles, using each cleaner or dilutions thereof, were performed on each of three panels using an in-house fabricated apparatus equivalent to Gardner wear tester. Percent (%) cleaning efficiency was calculated as a comparative value to that of the control solution [% by weight (wt.) d-limonene, 5% by wt. diethanolamine, 5% by wt. Triton X-100, and 60% by wt. distilled water per ASTM D 1193, Type IV] using equation 1[13].

$$\text{Cleaning Efficiency} = \left(\frac{\left(\frac{(A-B)}{(A-C)} \right)}{\left(\frac{(X-Y)}{(X-Z)} \right)} \right) \times 100 \tag{1}$$

In Equation 1, A, B, C, X, Y, and Z are defined as follows:

- A = the weight of the soiled panel before cleaning with product
- B = the soiled panel after cleaning with product
- C = the unsoiled panel used in product cleaning test
- X = the weight of the soiled panel before cleaning with the control formula
- Y = the weight of the soiled panel after cleaning with the control formula
- Z = the weight of the unsoiled panel used in the control formula cleaning test.

MIL-PRF-87937D, Section 4.5.21, does not specify acceptance criteria beyond calculating the average cleaning efficiency of the sample set. To derive pass/fail results, cleaning efficiency values were statistically compared to the control formulas at a 95% confidence level.

Compositional analysis was performed on panels tested at the 1:9 dilutions (i.e., MIL-Spec recommended dilutions for cleaning efficiency), using energy dispersive spectroscopy (EDS) via scanning electron microscopy (SEM) at 20.0 kilo-electron volt (keV) (Falcon Camscan Model# MV2300) per ASTM E 1508. Visual examination was performed using white and ultraviolet (UV) light. Each panel was subjected to a gentle, dry air stream to remove any dust. White light inspection via conventional laboratory lighting revealed any evidence of rust scale, dirt, paint, preservative or organic materials such as grease, oil, ink, or dye. A 365 nanometer Blak-Ray Lamp, Model UVL-56 black light (e.g., ultraviolet [UV] spectrum) was used to reveal the presence of any remaining hydrocarbons. Water break tests were conducted in accordance with MIL-STD-1359B Sections 5.2.1.1 and 5.2.1.3. The water break test was performed to uncover any contaminants (e.g., grease, oil, or other hydrocarbons remaining) not revealed by white or UV light inspection. In water break testing, panels were immersed in distilled water, raised horizontally when removed from the water, and kept flat. Water behavior on the panel was evaluated for 10 seconds (s) (with use of a stopwatch), including the amount of water remaining on the panel surface at this time to determine if there was sufficient contamination remaining on the panel to cause the water to break. It should be noted that the cleaning efficiency test included a post-bake for 10 min at 100°C which has the potential to leave a residue of any remaining cleaner due to their "negligible" volatility. Evidence of water droplets or a "water break" would signify sufficient hydrocarbon contamination or that cleaner residue remained. Therefore, the information obtained by white and UV light inspection was necessary to distinguish contaminants.

Materials compatibility testing included residue rinsability, effects on painted surfaces, total immersion corrosion, sandwich corrosion, and hydrogen re-embrittlement. Residue rinsability was performed per MIL-PRF-87937D Section 4.5.4 using pre-weighed, smooth-walled Al weighing dishes (5.71 cm outer diameter x 1.9 cm depth x 0.32 cm wall thickness) that were cleaned using Brulin Formula 815 GD at 10% dilution with distilled water, rinsed, and dried in an oven for one hour (hr) at 66°C. Cleaner test solutions then were diluted to 25% by volume using standard hard water, prepared by dissolving 0.40 grams (g) of reagent grade calcium acetate and 0.28 g of reagent grade magnesium sulfate in one liter of boiled, distilled water. Ten milliliter (mL) aliquots of 25% by volume cleaner or standard hard water (e.g., the control baseline) were placed into the dishes, were dried for 7.5 hrs in a convection oven at 66 °C, rinsed with distilled water for 1 min, re-dried for 7.5 hrs in an oven at 66°C, cooled to room temperature and placed in a dessicator. Dried dishes were weighed using an analytical balance (Dual Range Mettler-Torledo Model# PR5003) to determine the average weight change. Specimens then were subjected to a sash-type brush (2.5 cm diameter x 3.8 cm to 6.4 cm long) for 1 min under distilled water, rinsed for 30 s under running distilled water, dried in an oven for 7.5 hrs at 66°C, cooled to room temperature, and reweighed. The acceptance criteria for residue rinsability notes that weight change observed for the cleaner solutions should not be greater than that obtained with standard hard water tested under the same conditions.

The effects of the cleaners on unpainted surfaces were tested according to ASTM F 485 on 5.1 cm x 15.2 cm x 0.05 cm 4130 steel and 7075-T6 Al panels, with three panels being tested for each cleaner. Panels were wiped clean using methyl n-propyl ketone (MPK) and cotton, lintless cloths and visually examined for existing marks or scratches. One end of each panel was immersed for 5 min in sufficient cleaner maintained at room temperature to cover half of the panel. Removed panels were placed on a rack at a 45 degree (°) angle from the horizontal, dried in an oven at 66°C for 30 min, and cooled to room temperature for 15 min. Panels were rinsed for 1 min on both sides using tap water, rinsed again for 15 s using deionized water, placed again on racks at 45°, air dried for 20 min, and evaluated for etching, pitting or other adverse effects using visual inspection and SEM analysis (via Falcon Camscan Model# MV2300).

Total immersion testing was conducted per ASTM F 483 on three specimens (5.1 cm x 2.5 cm x 0.15 cm of the same alloy (i.e., pre-weighed SAE 1020 steel and 7075-T6 Al panels) for each cleaner. Three control panels of each alloy were pre-cleaned with a petroleum hydrocarbon-based mineral spirits stoddard solvent, type II per ASTM D 23, cleaned in MPK and dried and retained in desiccators. Three pre-weighed specimens of the same alloy were immersed vertically into 1000 mL glass beakers containing approximately 700 mL of each cleaner at 38°C for 7 days. Panels did not touch each other while immersed. Panels were removed, rinsed under hot tap water, rinsed with acetone, dried in an oven at 121°C until no moisture was visually observed on the panels, weighed with a Dual Range Mettler-Torledo Model # PR5003 and then returned to the cleaner. Panels repeated this cycle every 24 hrs to examine for weight loss.

Sandwich corrosion was evaluated on 4130 steel and 7075-T6 Al, 5.1 cm x 10.2 cm x 0.1 cm panels per ASTM F 1110. All panels were pre-cleaned using MPK. Eight sets of panels per alloy were configured with two panels each with one piece of 2.5 cm x 7.6 cm pieces filter paper, saturated in the respective cleaner or de-ionized (DI) water (e.g., control), placed between the panels. The sandwich was secured with 2.54 cm vinyl plating tape across the top of the sandwich and placed horizontally in an oven pre-heated to 38°C for a total of 8 hrs, followed by 16 hrs in a Singleton CCT 10P Chamber humidity chamber held at 38°C and 95% to 100% relative humidity. The cycle was repeated for 7 days after which the bottom panel was evaluated for corrosion using the severity rating system*** in ASTM F 1110-09.

Hydrogen re-embrittlement testing was conducted per ASTM F 519 immersion, sustained-load, as a service environment test using an Applied Test Systems Stress Rupture Machine Model# 2330 with Type IA 4340 steel specimens (0.64 cm outer diameter x 5.1 cm). Specimens were fully immersed in each cleaner for 155 hrs and loaded at 45% notched fracture strength (NFS) at temperatures between 20 and 30°C. If one specimen fractures, the remaining three specimens must be subjected to 2 hr, 5% incremental increase in stress to a maximum of 90% NFS.

EXPERIMENTAL RESULTS

Table 2 summarizes cleaning efficiency values, in comparison to the control formula **** on 4130 steel and 2024-T3 Al substrates, and the associated cleanliness inspections. For 2024-T3 Al, the control removed 97% of the grease, while 89% of the grease was removed from 4130 steel. It was desired that the alternative cleaner perform as well as or better than the control (i.e., obtaining 100% cleaning efficiency or better).

Table 2. Cleaning Efficiency Tests on 4130 Steel and 2024-T3 Al

Property	Material	Dilution Ratio	2ehl	EMIM acetate	EMIM ethylsulfate
Cleaning Efficiency	4130 Steel	Undiluted	100%	N/A	85%
		1:1 IL to water	101%	106%	92%
		1:3 IL to water	92%	97%	81%
		1:9 IL to water	84%	97%	95%
	2024-T3 Al	Undiluted	106%	N/A	103%
		1:1 IL to water	82%	101%	106%
		1:3 IL to water	100%	108%	101%
		1:9 IL to water	91%	93%	76%

None of the 1:9 cleaner dilutions performed satisfactorily in comparison to the control solution. The 1:3 cleaner dilutions did not perform satisfactorily on 4130 steel, but were satisfactory for Al. Additionally, concentrated EMIM acetate dissolved the cellulose sponge and could not be tested undiluted.

Undiluted 2ehl performed the best on 2024-T3 Al and similarly on 4130 steel, with a 1:1 dilution performing only slightly better. No distinct trend was observed with increasing dilution. A 1:1 dilution ratio of EMIM acetate performed best on 4130 steel, but the 1:3 dilution ratio performed slightly better on 2024-T3 Al. EMIM ethylsulfate did not perform as well as the control on 4130 steel in any concentration; however, all concentrations, except for a 1:9 dilution ratio, performed satisfactorily on 2024-T3 Al.

Compositional analysis, statistically comparing cleaned specimens to a bare substrate at a 95% confidence level, gave widely varying results on 4130 steel and 2024-T3 Al, with high statistical variability within process sets. Overall, 2ehl failed on both substrates, while EMIM ethylsulfate failed only on 4130 steel. Failures were indicated when the panel composition differed significantly from that of the bare substrate. EMIM ethylsulfate and EMIM acetate passed on the remaining substrates demonstrating (composition statistically similar to the bare panel). White light examination of panels tested with the 1:9 dilutions revealed minimal streaks and spotting on 4130 steel for all cleaners, while UV light only revealed several spots on panels treated with 2ehl. White light also showed residue, staining, and streaking on 2024-T3 Al specimens for all cleaners, but few to no spots when inspected using UV light. Essentially, more contamination remained on the surface, either residual cleaner or grease, than was noted in the control. Most panels also passed the water break evaluation with all tested cleaners, with only two 2024-T3 Al (one panel each cleaned using 2ehl or EMIM acetate) panels and one 4130 steel panel (cleaned using EMIM acetate) displaying water droplets. However, no correlation to residual grease could be made for the failing panels. Therefore, the failures were likely due to residual cleaner.

In residue rinsability testing, all three cleaners exhibited no weight changes within the reporting limit, thereby passing the test. This is interesting given that in the cleaning efficiency tests it was thought that the streaking evident on panels was due in part to cleaner residue. Residue rinsability data suggests that the streaking was more related to residual hydrocarbons.

When evaluating the effects of the cleaners on 4130 steel panels, EMIM ethylsulfate exhibited moderate chemical etching, as shown in Figure 1A. However, 2ehl and EMIM acetate only exhibited a surface cleaning effect or slight smoothing as is shown in Figure 1B and 1C.

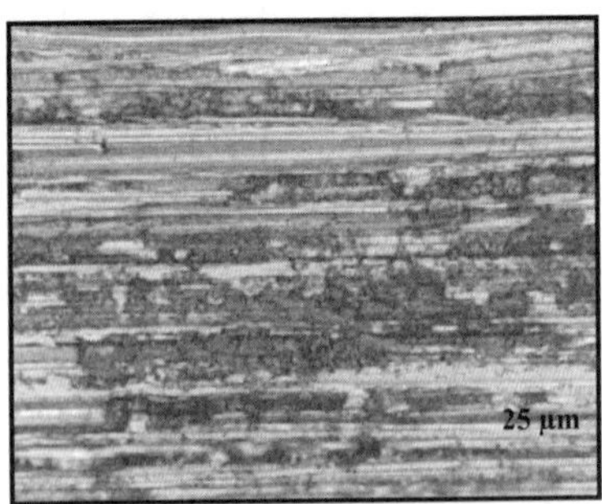

Figure 1A. 500X SEM image of 4130 steel etching by EMIM ethylsulfate

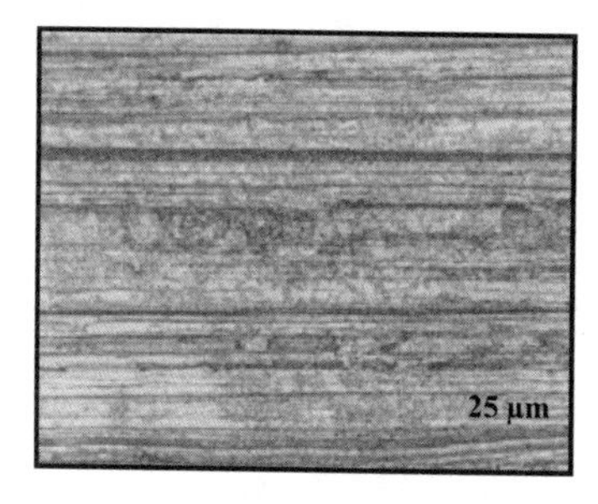

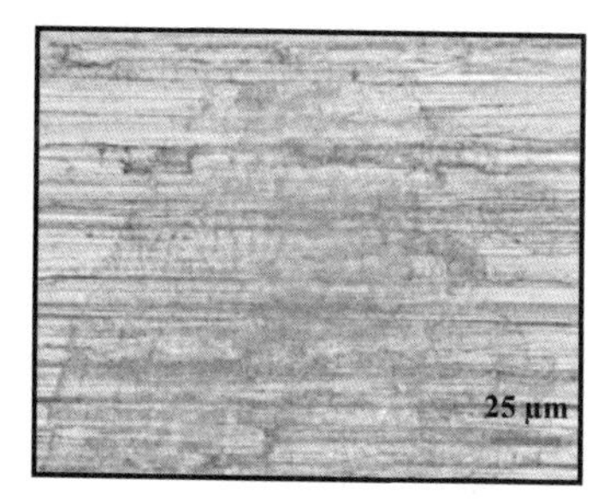

Figure 1B and 1C. 500X SEM images of 4130 steel smoothing by 2ehl and EMIM acetate, respectively

Others studies[17] have shown that EMIM ethylsulfate does not result in chemical attack of carbon steel substrates in a water-free environment generally. However, EMIM ethylsulfate, synthesized from methylimidazole and diethylsulfate, is not necessarily stable in the presence of water. It can hydrolyze to form EMIM hydrogen sulfate (HSO_4), super acid in the IL environment, and residual diethylsulfate could also hydrolyze into hydrogen sulfate or sulfuric acid, any of which could result in etching of the steel[18]. Although the cleaners were not diluted with water in this test, the panels were rinsed with water after immersion in the concentrated cleaners. In addition, the humidity of the environment was not controlled for the experiment.

Conversely, EMIM ethylsulfate showed only slight staining with no etching (see Figure 2C when 7075-T6 Al panels were immersed in solution. However, in 2ehl and EMIM acetate, small precipitates formed and/or etching aligned in the direction of the rolling bands (shown in Figure 3) was observed. Potential impurities from the production of EMIM acetate can result in residual chloride (as either EMIM Cl or hydrochloric acid) or acetic acid[18]. All forms could be corrosive to the Al substrate, especially in the IL environment. Regarding 2ehl, panel rinsing or humidity in the environment, could cause hydrolysis, resulting in lactic acid formation, again, resulting in etching of constituents of the Al and formation of precipitates.

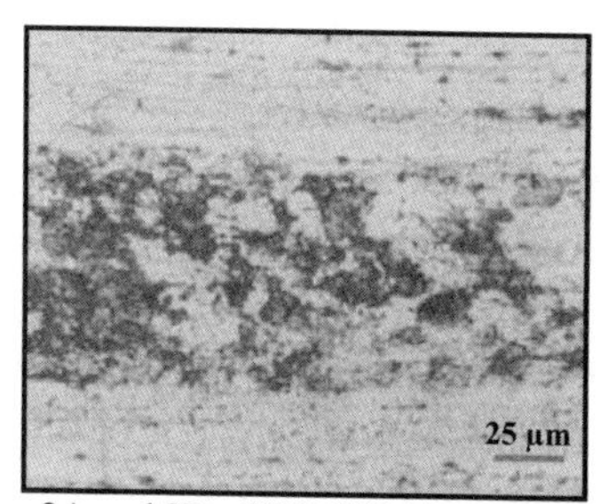

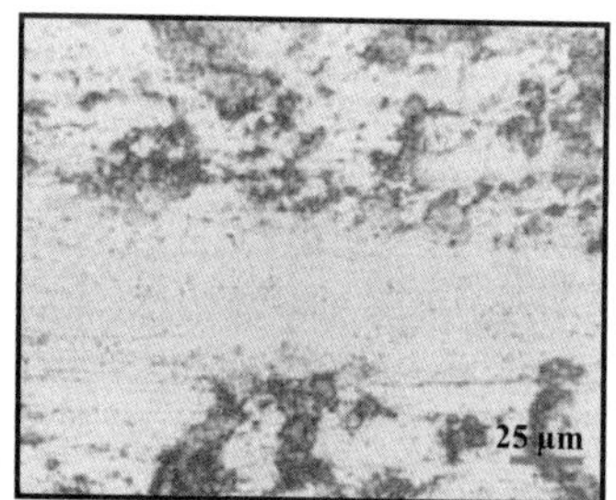

Figure 2A and 2B. 500X SEM images of precipitates and etching of 7075-T6 Al by 2ehl and EMIM acetate

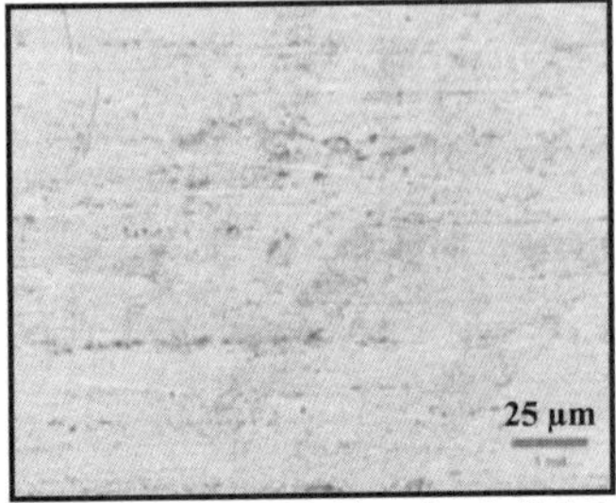

Figure 2C. 500X SEM image of 7075-T6 Al stained slightly by EMIM ethylsulfate

Although etching was noted on some specimens during the surface effects evaluation, total immersion corrosion testing did not reveal weight loss in excess of 0.25 mg per square centimeter (cm^2) for SAE 1020 steel and 0.15 mg/cm^2 for 7075-T6 Al exposure after 168 hrs exposure to any of the cleaners, i.e., the acceptance criteria for each alloy. All panels in each subset of each cleaner for both alloys exhibited an average weight loss across three panels that was than the reporting limit on both alloys, demonstrating weight loss that is less than the aforementioned criteria.

After seven days of sandwich corrosion testing, each panel was assigned a 1-4 rating evaluation per ASTM B 117. The control was rated with 4, with discoloration or corrosion being evident over more than 25% of its surface area. Although all panels passed sandwich corrosion testing when compared to the control (e.g., reagent water), there was still significant corrosion evident over more than 25% of the surface on both substrates. Corrosion ratings for tested panels are shown in Table 3 below.

Table 3. Sandwich Corrosion – Corrosion Rating

	Corrosion Rating	
	4130 Steel	7075-T6 Al
Control	4	4
2 ehl	4	3.75
EMIM acetate	3.75	4
EMIM ethylsulfate	4	4

Hydrogen re-embrittlement testing was conducted for all tested cleaners. During testing, concentrated cleaners were placed into Lexan containers in which the specimens would later be placed. During set up, 2ehl and EMIM acetate test solutions both dissolved the Lexan containers and the concentrated solutions leaked, unknowingly, onto the 4130 steel dummy bars used for the test set-up and remained there for 24 hrs. Upon discovery, the dummy bars were wiped clean, and test set up continued. During specimen loading, the dummy bars fractured, suggesting that the test solutions had embrittled the 4130 steel; however, the formal test could not be conducted to state the embrittling characteristics with certainty. Concentrated EMIM ethyl sulfate was found to be non-embrittling to 4340 steel, lasting 150 hrs at 173.8 kilopounds-force per square inch (ksi) [1197.5 megapascal (MPa)] applied stress, and had no effect on the Lexan container.

Because of the material compability issues between the Lexan containers and the 2-ehl and the EMIM acetate, a second hydrogen re-embrittlement test was attempted using a 1:1 ratio of each concentrated solution with water. Again, the Lexan container dissolved, and testing was discontinued. Testing was not further pursued with 2ehl or EMIM acetate.

CONCLUSIONS

Acceptable levels of cleanliness were achieved with 2ehl, and materials compatibility testing for this cleaner showed that the corrosivity of the chemistry was acceptable on both Al and 4130 steel. Likewise, minimal, if any, residue (as compared to that of standard hard water) remained on Al weighing dishes in residue rinsability testing of 2ehl. Its effects on unpainted 4130 steel surfaces were minimal (i.e., no permanent changes detected), but etching in the direction of the Al rolling bands was detected on 7075-T6 Al.

EMIM acetate performed similarly to 2ehl, with the only notable difference being a slight improvement in cleaning efficiency.

EMIM ethylsulfate demonstrated lower cleaning efficiency values than 2ehl and EMIM acetate on 4130 steel, but displayed similar results to the other tested cleaners on 2024-T3 Al. The corrosivity results were similar to the other tested cleaners as was residue rinsability. However, its effects on unpainted surfaces was very different from 2ehl and EMIM acetate in that it did not etch 7075-T6 Al and only produced minimal staining, but produced moderate chemical etching on 4130 steel. Likewise, EMIM ethylsulfate was the only cleaner that could be subjected to hydrogen re-embrittlement testing and produced passing results.

Because chemical etching was noted on 4130 steel or 7075-T6 Al for each cleaner tested, additional investigations should be conducted to determine whether the etching will adversely impact the mechanical properties of the substrates being cleaned or can be useful for advanced bonding applications. Therefore, it is suggested that additional work be conducted to observe surface chemistry changes, e.g., are metal sulfides formed and how intentional aqueous dilutions modify the corrosion rate. It may include modified hydrogen re-embrittlement testing using compatible test container materials and optimized dilution ratios. Additionally, axial fatigue testing will be required to determine effects of the cleaners on the integrity of sub-surface substrate properties. Additionally, chemical evaluations of the cleaners will be conducted to ensure that it meets military specification requirements.

Overall, the cleaners show potential ability for use in wiping applications, yet further evaluation is necessary to optimize cleaning performance for wiping applications, while ensuring material compatibility and conformance to military chemical compatibility requirements. In addition, use of the cleaners in immersion applications and their subsequent optimization or use with ancillary agitation equipment should be investigated.

FOOTNOTES

*HMIS hazard indicator: 4 = extreme; 3 = severe hazard; 2 = moderate; 1 = slight; 0 = none

**Xi indicates that substance is an irritant as identified by the European Union's Directive of Dangerous Substances, 2010

***0 = no visible corrosion or discoloration; 1 = very slight corrosion or discoloration (and/or up to 5% of area corroded); 2 = discoloration and/or up to 10% of area corroded; 3 = discoloration and/or up to 25% of area corroded; 4 = discoloration and/or more than 25% of area corroded and/or pitting present

****Control formula consisted of 60.0% by weight (wt.) distilled water, 30.0% by wt. D-limonene, 5.0% by wt. diethanolamine, and 5.0% by wt. Triton X-100.

REFERENCES

[1] Abedin, Sherif Zein El, and Endres, Frank, Ionic Liquids: The Link to High-Temperature Molten Salts, *Accounts of Chemical Research*, **40** (11), 1106-1113 (2007).
[2] Abbott, Andrew P. and McKenzie, Katy J. Application of Ionic Liquids to the Electrodeposition of Metals, *Phys. Chem. Chem. Phys.*, **8,** 4265-79 (2006).
[3] Holbrey, John, Turner, Megan B. and Rogers, Robin D., Chapter 1: Selection of Ionic Liquids for Green Chemical Applications, *Ionic Liquids as Green Solvents*, 2-12 (2003).
[4] Basionics, 1-Ethyl-3-Methylimidazolium Chloride (CAS No. 65039-09-0) Data Sheet, BASF Group, (2010).
[5] Basionics, 1-Butyl-3-Methylimidazolium Chloride (CAS No. 79917-90-1) Data Sheet, BASF Group (2010).
[6] Dr. Markus Wagner, MERCK; Dr. Megan O'Meara and Dr. Aurelie Alemany, BASF Corporation; Dr. Khalid Shukri and Dr. Andrew Abbott, Scionix, telephone correspondence, (November 2010).
[7] National Aeronautics and Space Administration (NASA) Final Report and Deliverables, "Precision Cleaning of Oxygen Systems and Components," NASA/CR-2009-214757, (2009).
[8] Technical Order TO 42C2-1-7, "Technical Manual: Process Instructions: Metal Treatments: Electrodeposition of Metals and Metal Surface Treatments to Meet Air Force Maintenance Requirements," (2006).
[9] Boeing Specification BAC-5408, "Vapor Degreasing, Revision M," (1998).
[10] Boeing Specification BAC-5765, "Cleaning and Deoxidizing of Aluminum Alloys,"(1995).
[11] Boeing Specification BAC-5555, "Phosphoric Acid Anodizing of Aluminum for Structural Bonding," (2001).
[12] Military Specification MIL-PRF-87937, "Performance Specification, Cleaning Compound, Aerospace Equipment, (2001).
[13] Military Specification MIL-PRF-680, "Performance Specification, Degreasing Solvent," (2006).
[14] Technical Order TO 1-1-691, "Cleaning and Corrosion Prevention and Control, Aerospace Equipment and Non-aerospace equipment, (2007).
[15] Technical Order TO 2-1-111, "Standard Maintenance Procedures, Navy and USAF, P&W Aircraft Engines." (2007)
[16] Technical Order TO 1-1-8, "Application and Removal of Organic Coatings, Aerospace and Non-aerospace Equipment," (2010).
[17] Uerdingen, Marc, Treber, Claudia, Balser, Martina, Schmitt, Gunter, and Werner, Christoph, Corrosion Behavior of Ionic Liquids, *Green Chem.,* **7**, 321-325 (2005).
[18] Dr. Tom Beyersdorff, IoLiTec, Inc., email correspondence, (July 2012).

EVALUATION OF PYCAL™ 94 AS AN ENVIRONMENTALLY FRIENDLY PLASTICIZER FOR POLYVINYL BUTYRAL FOR USE IN TAPE CASTING

Richard E. Mistler, Richard E. Mistler, Inc., Yardley, PA 19067
Ernest Bianchi, Maryland Ceramic & Steatite Co., Bel Air, MD 21014
William McNamee, Croda Inc., Griffin Innovation Center, New Castle, DE 19720

ABSTRACT

An environmentally friendly plasticizer for polyvinyl butyral has been evaluated and compared with two industry standards- butyl benzyl phthalate and dioctyl phthalate. Compatibility with a commonly used polyvinyl butyral, Butvar® B-98 is discussed. A direct comparison of this plasticizer is made with the industry standards in side-by-side tape casting runs.

INTRODUCTION

The object of this investigation was to evaluate a more environmentally friendly plasticizer for use with polyvinyl butyral in tape casting formulations. The new plasticizer, polyoxyethylene aryl ether (PYCAL™94), was compared directly with two of the most common plasticizers used in tape casting, butyl benzyl phthalate and dioctyl phthalate.

The paper is divided into two sections, one relating to the environmental and health considerations between the PYCAL™ plasticizer and the phthalate plasticizers and a comparison of the effect as an additive to the polyvinyl butyral binder and the second which relates to a direct comparison of tape formulations using the three different plasticizers.

The comparisons made included: the relative compatibility level for the most common grades of polyvinyl butyral, B-76 and B-98. The tape casting formulation comparisons included the following: slip viscosity, green bulk density (GBD), green oxide only density (GOOD), green tape thickness, tape drying characteristics, green tape character including flexibility and tackiness, and sintered properties such as fired bulk density (FBD) and shrinkage.

PART 1-PLASTICIZER COMPARISONS:

Environmental and Health Considerations

Butyl benzyl phthalate (BBP), commercially sold as Santicizer®, has two environmental and health issues: 1) reproduction toxicity and 2) dangerous for the environment. These issues impact mainly Europe, but spread in particular into Asia as producers there are becoming more and more concerned about imports into Europe. Specifically, BBP is classified according to its MSDS as "toxic to reproduction" Category 2 which is to be considered as toxic to reproduction to man. Another widely used phthalate plasticizer for tape casting, diethyhexyl phthalate (commonly called dioctyl phthalate), is also a Category 2 reprotoxic chemical. Phthalates, including BBP, are expected to be included on the "dangerous substances" or "substances to be avoided" industry chemical lists. BBP is also considered as very toxic to aquatic life and has a potential to bio-accumulate. This has less of an impact, but it compares poorly against the PYCAL™94 plasticizer. The data for PYCAL™94 indicate that it is classified as a non-hazardous substance according to Regulation (EC) No. 1272/2008. It passes all of the regulatory standards and is compliant in Europe, Asia, North America, and Australia. Therefore, based on the currently available data, PYCAL™94 has a better profile than BBP and other phthalate plasticizers.

Plasticizer Effects on Polyvinyl Butyral:

The effect of PYCAL™94 on the T_g (glass transition temperature) of polyvinyl butyral is still undetermined but it is known that the plasticizer does lower the value considerably and results in very plastic tape cast products at lower concentrations than the phthalates.

PYCAL™94 is found to have better compatibility and efficiency than traditional phthalate plasticizers in actual tape casting formulations. Compatibility of greater than 50% was found with the polyvinyl butyrals tested in this study (B-98 and B-76).

More data will be available with plots of T_g versus % plasticizer in the near future.

PART 2 –TAPE CASTING FORMULATION COMPARISONS:

EXPERIMENTAL PROCEDURE

Three tape casting batches were prepared with everything identical with the exception of the primary, Type I, plasticizer. The ceramic batch which was selected for the comparison was a 96% aluminum oxide formulation which is commonly utilized in thick film or multilayered ceramic packages. The formulations for the two control batches were as follows:

Control #1, Butyl Benzyl Phthalate:

Part 1:

Aluminum Oxide[1]	60.45 weight%
EPK Kaolin Clay[2]	0.63 weight%
Nytal 400 Talc[3]	1.89 weight%
Fish Oil, Z-3[4]	2.52 weight%
Xylenes	12.85 weight%
Ethyl Alcohol, 95% Denatured	12.85 weight%

Part 2:

Butyl Benzyl Phthalate, S-160[5]	3.78 weight%
Polyvinyl Butyral, B-98[6]	5.04 weight%

Control #2, Dioctyl Phthalate

Part 1:

Aluminum Oxide[1]	60.45 weight%
EPK Kaolin Clay[2]	0.63 weight%
Nytal 400 Talc[3]	1.89 weight%
Fish Oil, Z-3[4]	2.52 weight%

Xylenes	12.85 weight%
Ethyl Alcohol, 95% Denatured	12.85 weight%

Part 2:

Dioctyl Phthalate[7]	3.78 weight%
Polyvinyl Butyral, B-98[6]	5.04 weight%

The formulation for the experimental batch for comparison was:

Part 1:

Aluminum Oxide[1]	60.45 weight%
EPK Kaolin Clay[2]	0.63 weight%
Nytal 400 Talc[3]	1.89 weight%
Fish Oil, Z-3[4]	2.52 weight%
Xylenes	12.85 weight%
Ethyl Alcohol, 95% Denatured	12.85 weight%

Part 2:

Polyoxyethylene aryl ether[8]	3.78 weight%
Polyvinyl Butyral, B-98[6]	5.04 weight%

The fish oil is a dispersant for the inorganic components. The butyl benzyl phthalate, dioctyl phthalate, and the polyoxyethylene aryl ether are considered type I plasticizers for the polyvinyl butyral, i.e. they react chemically with the binder and change the T_g of the polymer and make it more flexible. The polyvinyl butyral is the binder for the system and it provides the strength and backbone for the tape.

Table 1 – Materials Sources

1. Aluminum Oxide, A-152 SG, Almatis, Leesdale, PA 15056
2. EPK Kaolin, Zemex Industrial Minerals, Inc., Atlanta, GA 30338
3. Nytal® 400 Talc, R.T. Vanderbilt Co., Inc., Norwalk, CT 06856
4. Blown Menhaden Fish Oil, Z-3, W.G. Smith, Inc., Cleveland, OH 44113
5. Santicizer® 160, Ferro Corporation, Bridgeport, NJ 08014
6. Butvar® B-98, Solutia,Inc., St. Louis, MO 63166
7. Dioctyl Phthalate, Sigma Aldridge Corp., St. Louis, MO, 6310
8. PYCAL™94, Croda, Inc., Edison, NJ, 08837

The procedure followed for each of the three batches was identical and was as follows:

1. Add one (1) Kg of ½" U.S. Stoneware Burundum milling media, 96% Alumina, to a Size 0 Roalox mill jar.
2. Dissolve the fish oil in xylenes and add to the mill jar.
3. Weigh and add the ethyl alcohol to the mill jar.
4. Add powders to the jar (A-152 dried in an oven at > 100°C for 24 Hours.)

5. Dispersion mill for 24 hours by rolling on jar rollers.
6. Weigh and add plasticizer to the mill jar.
7. Add the binder to the mill jar, stirring by hand to wet and mix the binder.
8. Mix for an additional 24 hours by rolling on jar rollers.
9. Pour slurry into an 80 oz. HDPE container.
10. Vacuum de-air for 8 minutes at 25" Hg pressure.
11. Measure viscosity and temperature of the slip.

At this point the batches were ready for tape casting using the following casting parameters:

1. Blade gap setting: 0.048"
2. Eight (8) inch wide single doctor blade.
3. Carrier: Silicone coated Mylar, G10JRM, 75 μm x 12 inch wide.
4. Casting speed: 20 inches per minute.
5. Air flow on lowest setting, no heat.
6. No underbed heat.

After drying the following measurements were made on the green tape:

1. Thickness
2. Green Bulk Density

Samples for sintering were punched from the control and experimental tapes. The punched pieces, which were 1" x 1.5", had the long axis oriented in the casting direction and the short axis oriented in the cross-casting direction. This provided a good basis for comparison of the shrinkage during sintering in the casting and cross-casting directions.

One sample from each batch was sintered in an electric furnace in an ambient atmosphere. It was found that sintering without cover plates resulted in warped samples, therefore a porous alumina cover plate was placed on top of the samples during sintering. This solved the warpage observed during the initial sintering run. The initial sintering run was made at a peak temperature of 1500°C for 2hours. Using this temperature resulted in incomplete sintering therefore the maximum temperature was increased to 1600°C and the hold time was increased to 3 hours. This yielded excellent sintered parts. The final sintering profile was as follows:

RT	to	500°C	@	3°C per minute
500°C	to	625°C	@	1°C per minute hold 1 hour
625°C	to	1500°C	@	5°C per minute
1500°C	to	1600°C	@	1°C per minute hold 3 hours
Furnace Cool to RT				

After sintering the samples were measured to determine the sintering shrinkage in the X, Y, and Z directions. Fired bulk density measurements were made using the Archimedes method by immersion in toluene.

RESULTS AND DISCUSSION

The viscosity measurements on the ready to cast slips were made using an RV-4 spindle at 20 RPM using a Brookfield Viscometer. The measurements were as follows:

Control Batch #1 (BBP)	5350 cP	at	22.9°C
Control Batch #2 (DOP)	5600 cP	at	21.4°C
Experimental Batch (PYCAL™)	5040 cP	at	21.7°C

Although the viscosity for the experimental batch was slightly lower than that of either of the control batches it was not statistically significant and was close enough to considered the same.

Table 2 includes the data and observations made on the green tapes which resulted from the casting runs.

Table 2 - Green Tape Results

Plasticizer	Thickness (in.)	Green Bulk Density (g/cc)	Green Oxide Only Density (g/cc)
BBP	0.0195-0.0221 +/- 0.0004	2.332	1.98
DOP	0.0197-0.0211 +/- 0.0004	2.333	1.98
Pycal™94	0.0193-0.0215 +/- 0.0005	2.307	1.95

The cast with BBP as the plasticizer had a yield of about 10 feet of strong, flexible tape. Some soft agglomerates or partially dissolved binder appeared during the pouring of the slurry. The tape was left in the casting machine overnight to dry fully. During this time, the tape curled upwards on each edge. Furthermore, some surface marks and cracks were seen on the initial section of the tape. This was probably due to the agglomerates mentioned above. The second half of the tape was free of major defects. The +/- 0.0004" thickness variation in Table 2 is the variation from side to side of the tape at the lead end of the cast.

The DOP control cast also had a yield of about 10 feet of strong, flexible tape, with some observable upwards edge curl during drying. The tape released easily from the carrier but was somewhat stiff. The cast was left in the machine overnight to dry completely. When the tape was rolled it appeared to be defect free. The thickness variation from edge to edge at the lead was +/- 0.0004".

The cast made with the Pycal™94 went well with a yield of about 10 feet of strong, flexible tape. The tape was left in the machine overnight to dry thoroughly. Slight edge curl was observed during the drying stage. The surface of the tape was smooth, and the tape released easily from the base film. The majority of the tape was free of defects.

When comparing and contrasting the tapes obtained from the three casts, the following generalizations can be made:

-The benzyl butyl phthalate batch appeared to be poorly mixed, the other two batches were very homogeneous.

-The viscosity measurements were similar for all of the batches, however the batch made with the Pycal™94 was the lowest. This can be an advantage, especially if you can lower the solvent content.
-The thickness uniformity was similar for all of the batches.
-The green bulk density was comparable for all of the batches, well within the experimental error limits.
-None of the casts was excessively tacky and all released from the base film easily on the day after the cast.
-The dioctyl phthalate tape was relatively stiff compared with the other two.
-The benzyl butyl phthalate and Pycal™ tape were both flexible, but the Pycal™ tape exhibited the highest degree of bending when hung over a rod to observe displacement.
-After about 2 weeks, tape samples of the Pycal™ batch became very difficult to remove from the silicone coated Mylar and left numerous pullouts and hundreds of tiny specs on the base film. This also created many small holes on the underside of the tape. There appears to be a reaction between the tape and the silicone coating over a period of time. The two control batches did not display this behavior. A reduction in the Pycal™ content was recommended to reduce the severity of bonding over time. The flexibility of the tape and the lower viscosity also point in this direction. Another experimental batch was therefore prepared to find the optimum concentration of the Pycal™ plasticizer. The results of that experiment are reviewed in the next section.

Experimental Batch to Optimize the Pycal™ Concentration:

Batch:

Part 1:

Aluminum Oxide[1]	61.22	weight %
EPK Kaolin Clay[2]	0.64	weight %
Nytal 400 Talc[3]	1.91	weight %
Fish Oil, Z-3[4]	2.55	weight %
Xylenes	13.01	weight %
Ethyl Alcohol, 95% Denatured	13.01	weight %

Part 2:

Pycal™94[8]	2.55	weight %
Polyvinyl Butyral, B-98[6]	5.10	weight %

The procedure for mixing the slurry and the parameters for the tape casting were exactly as described above for the first three batches.

RESULTS AND DISCUSSION FOR THE OPTIMIZATION RUN:

The batch given above is at 50 phr (parts per hundredweight resin). The batch previously described using the Pycal™94 was at 75 phr. The viscosity of the 50 phr batch measured using the same procedure was 6780 cP at a temperature of 18.0°C. This is considerably higher than the viscosity for the 75 phr batch which was 5040 cP. The temperature of the slip was lower for the present measurement but not enough to account for the difference of over 1000 cP. In all probability it is a combination of a decrease of total liquid components of 40 grams versus 60 grams and the plasticizing effect of the increased Pycal™94 in the previous batch. The viscosity is still in the acceptable range for tape casting.

Table 3-Comparison of Green Tape Results for 50 phr vs. 75 phr Pycal™94 Concentration

Pycal™94 Concentration (phr)	Tape thickness (in.)	Green bulk Density (g/cc)	GOOD (g/cc)
50	0.0210-0.0222 +/- 0.0006	2.258	1.95
75	0.0193-0.0215 +/-0.0005	2.307	1.95

As with the previous comparisons the +/- values given are the variation of thickness across the tape at the lead section. When comparing the tapes made using the different levels of plasticizer it can be stated:

- The GBD was higher for the 50 phr plasticized tape, however the green oxide only density was the same. The GOOD is a better predictor of the final sintered density.
- Both tapes were not tacky and released easily from the silicone coated Mylar carrier. However, as was described in the section above, the 75 phr tape exhibited sticking to the carrier after a period of two weeks and the 50 phr tape did not.
- Both tapes exhibited excellent flexibility when hung over a rod to observe displacement.
- The optimal Pycal™94 concentration for this 96% alumina formulation appears to be > 50 phr and < 75 phr.

The green bulk density values are averages of several measurements on the tape samples. The green oxide only density (GOOD) values are calculated from the GBD measurements and eliminate the organic content. The values obtained for the GOOD are good indications of the actual packing density of the tape.

The results for the sintering evaluation and comparison are presented in Table 4 as follows:

Table 4- Sintering Results

Property:	PVB/BBP	PVB/DOP	PVB/Pycal™94
Green Bulk Density	2.338 g/cc	2.308 g/cc	2.305 g/cc
Fired Bulk Density	3.763 g/cc	3.767 g/cc	3.751 g/cc
Shrinkage:			
Along Tape	20 – 21%	20 – 21%	20 – 21%
Across Tape	20%	20 – 21%	21%
Thickness	16%	13%	16%
FBD % of TD (3.836)	98.10%	98.20%	97.78%
Apparent Porosity	0.33%	0.36%	0.36%

Based upon these measurements it appears that all of the sintered alumina samples were very similar with the results within the experimental error limits, independent of the plasticizer used. All of the samples fired flat and were free of defects.

As an added test, green laminates were also fabricated from each of the tapes using four layers which were 3" x 3". The layers were stacked and laminated using a Carver press with heated platens. The samples were held for 10 minutes at a pressure of 900 – 1000 psi and a temperature of 90-105°C. All of the samples could be laminated without defects, though they were not sintered. Samples with dimensions of 1" x 1.5" were blanked from each laminate. The green bulk densities were comparable as shown in the next table.

Table 5 – Green Laminate Density

Sample	Green Laminate Density
PVB/BBP	2.631 g/cc
PVB/DOP	2.631 g/cc
PVB/Pycal™94	2.635 g/cc

SUMMARY AND CONCLUSIONS

Tape casting formulations for a standard 96% alumina ceramic have been evaluated using an environmentally friendly plasticizer, Pycal™94. The results were compared with industry standard plasticizers: benzyl butyl phthalate and dioctyl phthalate. The green tape properties including green bulk density, thickness uniformity, flexibility, and strength were all found to be equal to or better than the standards. During the experiments it was found that a lower amount of the Pycal™ can be used to yield the same properties. The sintered density, shrinkage during sintering, and fired part quality were equal to or better than the industry standard parts. In addition the tapes were compared in laminated parts with the Pycal™ samples once again equal to the industry standards with respect to lamination and laminated density. Based upon these results it has been determined that the Pycal™94 (Polyoxyethylene aryl ether) can be used as a direct substitute as a type 1 plasticizer for polyvinyl butyral in tape casting formulations.

Materials and Systems for Energy Applications

CONDUCTION PLANE GEOMETRY FACTORS FOR THE β"-ALUMINA STRUCTURE

Emma Kennedy and Dunbar P. Birnie, III
Rutgers University, MSE
New Brunswick, NJ, USA

ABSTRACT

The β"-alumina structure is important for sodium battery systems. Its structure is composed of layers of (111) oriented spinel-structure units separated by sodium containing conduction planes that are wide compared to the spinel units. Of critical interest is whether composition or structure modifications can be designed that would assist in speeding up the sodium conduction within these conduction planes. The present work looks at the geometric limitations to expansion or contraction of this conduction plane as we substitute different ions into the lattice in different sites. Clearly a wider or narrower plane could influence the conduction rate by influencing the sodium motion activation energy. We take a detailed look at the structure and specifically the oxygen atoms that "sandwich" the sodium conduction environment. These atoms (the O(3) and O(4) sites) are analyzed for different β"-alumina structures to understand the subtle but important changes that sodium ions may feel when diffusing in different compositions of this material.

INTRODUCTION

The structure of β"-alumina is based on a stacking of oxygen close-packed layers that builds an overall hexagonal structure[1]. The basic repeat unit consists of two parts: a four layer ABCA stack that is spinel-like in its structure and a conduction slab that has fewer oxygen atoms and more room for sodium ion motion. One full unit cell then requires three sets of this pairing. Figure 1 shows how the ABCA oxygen layer stacking sequence creates the spinel and conduction blocks of the structure. The distance "D" is a measure of the thickness of the spinel block. The distance "H" is the spacing between spinel blocks and contains the hexagonal honeycomb of sites that are important for sodium conduction. Note that the naming of the A, B, C type layers is somewhat arbitrary. Figure 1 shows this sequence increasing upward on the page.

The β"-alumina structure is identified as space group $R\overline{3}m$ (#166) from the international tables[2]; this is rhombohedral in symmetry and is usually represented by coordinates referenced to a hexagonal cell. There have been a number of detailed X-ray and neutron diffraction studies that have determined atom locations within the unit cell to high accuracy, which was discussed in a previous paper[3]. There are typically 5 unique oxygen locations, 4 unique aluminum sites, and usually one or two unique sites for sodium (or for ions that have replaced the sodium). For example the seminal structure determination by Bettman and Peters[4] gave atom locations as shown here in Table I. This table identifies the relative position these layers would occupy if they had perfect close-packing as found in the spinel structure and if this stacking continued through the conduction plane via the O(5) bridging atom. It is clear that the actual coordinates are close to their ideal hard-sphere packing spacing.

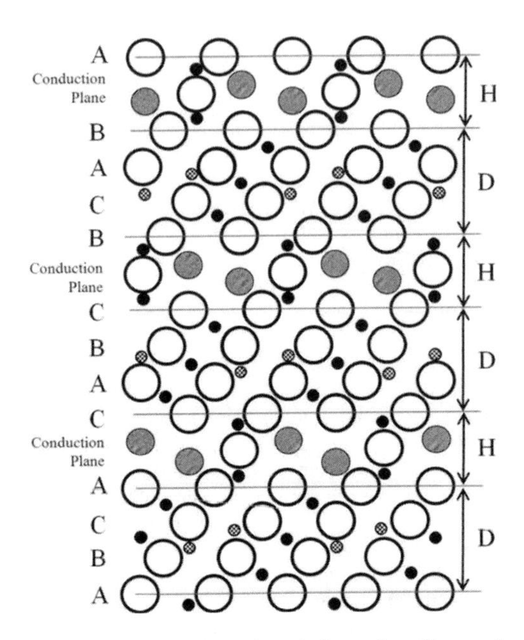

Figure 1. Crystal structure of β"-alumina viewed from the side emphasizing the stacking arrangement of close-packed oxygen layers that make up the structure. Spinel-related blocks are identified with D, and conduction slabs are identified by H. Large open circles are O^{2-}. The medium-sized shaded circles are Na^{+}. The small solid or checked circles are Al^{3+} sites.[3]

Table I. Unique Atom Site Locations for Bettman and Peters' Structure (in angstroms)[4]

Atom	# in Cell and Wyckoff position	X	Y	Z
Na	6c	0	0	0.1717
Al(1)	3a	0	0	0
Al(2)	6a	0	0	0.3501
Al(3)	18h	0.3362	0.1681	0.0708
Al(4)	6c	0	0	0.4498
O(1)	18h	0.1562	0.3124	0.0339
O(2)	6c	0	0	0.2955
O(3)	6c	0	0	0.0961
O(4)	18h	0.1657	0.3314	0.2357
O(5)	3b	0	0	0.5

STRUCTURE REVIEW

How much space is available for conduction and how is that space modified when composition adjustments are made? The conduction slab thickness calculated by Harbach[5] is based only on the external dimensions of the hexagonal unit cell ($\boldsymbol{a_o}$ and $\boldsymbol{c_o}$). However, the actual thickness of the conduction slab can be calculated directly for every case where a full crystal structure determination has been made.[3] If we look more closely at the atom identities and map them onto the side-view of the layer structure of the unit cell then we can label each of the oxygens as shown in Figure 2. Of course, the O(5) bridges adjacent spinel block layers and connects two tetrahedral Al(4) atoms pointing up and down. Atoms O(3) and O(4) are two oxygen locations that define the rest of the local environment for atoms within the conduction plane of the structure. Atoms O(1) and O(2) are internal to the spinel block and do not have direct bonds to any atoms in the conduction plane.

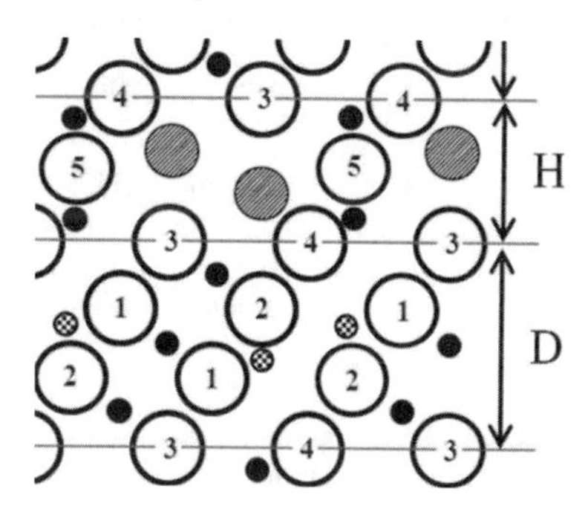

Figure 2. Crystal structure of β"-alumina viewed from the side emphasizing the stacking arrangement of close-packed oxygen layers that make up the structure. Specific oxygen crystallographic sites are numbered. O3 and O4 are used to calculate the conduction slab thickness.[3]

So, if we use the z-coordinate for atoms O(3) and O(4) then we can get an exact measure of the distances D or H. Using O(3) we get this conduction slab thickness:

$$H_3 = c_o\left\{\tfrac{1}{3} - 2z[O(3)]\right\} \tag{1}$$

And, using the z-coordinate of the O(4) oxygens then we get this measure of the conduction slab thickness:

$$H_4 = c_o\left\{2z[O(4)] - \tfrac{1}{3}\right\} \tag{2}$$

Edstrom et al.[6] used this metric when comparing conduction plane sizes for Ga^{3+}-Al^{3+} solid-solution formation, though found no trend even with large Ga^{3+} replacement. Both equations are based on the actual structure determinations, not on an assumed three-dimensional rigidity of the spinel-block that is implicit in Harbach's analysis.[5]

The H_3 and H_4 values were calculated and compared for several different β"-alumina structures containing different ions in the conduction plane[3]. Three β"-alumina structures were chosen because of their differing H values which can be seen in Table II. The first structure that was examined was the Bettman and Peters structure.[4] This structure is the basic structure which contains sodium in the conduction plane, and the H values for this structure fall in the middle of the range of values calculated. The next structure from Soetebier[7] comes from the smaller end of the H spectrum, in this structure the sodium atoms in the conduction plane are replaced with gadolinium atoms. The final structure that will be examined in this paper has a larger H value compared to the first two, and the sodium in the conduction plane has been replaced by indium this structure comes from Cetinkol.[8] As ions are conducted through each of these structures we would assume that it would more easily diffuse through the structures with larger H values.

Table II. H Calculations for Three Different β"-Alumina Structures in Angstroms.[3,4,7,8]

Atoms in the Condu Plane	Na	Gd	In
a_o	5.164	5.6067	5.604391
c_o	33.85	33.326	34.48072
z(O3)	0.0961	0.0977	0.093421
z(O4)	0.2357	0.2347	0.236835
H_3	4.77736	4.59677	5.05113
H_4	4.67356	4.53456	4.83891

SODIUM DIFFUSION GEOMETRY

Oxygen atoms in three different positions control the diffusion of sodium and other atoms through the conduction plane. These oxygen atoms can be seen in Figure 3, which is a 3D rendering of the Bettman and Peters β"-alumina structure[3] where every atom was removed except for the sodium atoms and the oxygen atoms that are within and surround the center conduction plane. The labels on the atoms in Figure 3 correspond to Tables III, IV, and V and Figure 4.

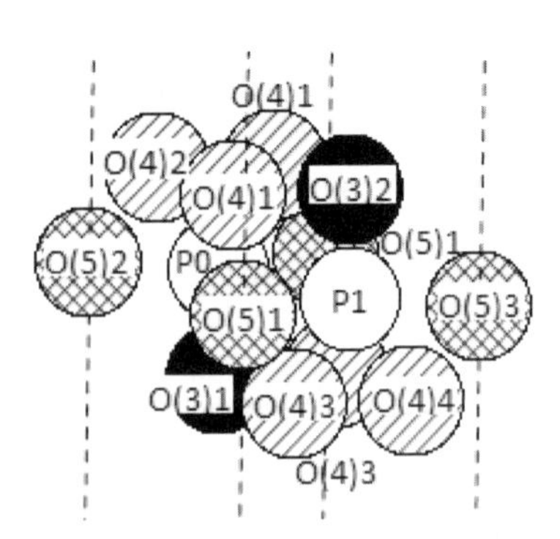

Figure 3. Center conduction plane of one unit cell where the Al, O1, O2, and O3 atoms have been removed. P0 and P1 correspond to Tables III-V, and are the initial and final positions of the sodium, gadolinium, indium, or other ions as they diffuse through the conduction plane.

Table III. Bond Distance Comparisons as Diffusing Na Ions Move From Position 0 to Position 1 Within the Conduction Plane.

Na Diffusion Position (P)	O(5)1	O(5)2	O(5)3	O(4)1	O(4)2	O(4)3	O(4)4	O(3)1	O(3)2
	Bond Distance (in angstroms)								
0	3.246	3.246	6.485	2.700	2.700	3.764	5.462	2.559	3.928
0.0625	3.148	3.447	6.282	2.665	2.843	3.621	5.272	2.545	3.775
0.125	3.061	3.649	6.079	2.644	2.992	3.484	5.084	2.548	3.628
0.25	2.923	4.052	5.673	2.649	3.307	3.233	4.712	2.603	3.349
0.375	2.836	4.457	5.267	2.717	3.641	3.016	4.346	2.718	3.100
0.5	2.807	4.862	4.862	2.841	3.988	2.841	3.988	2.887	2.887
0.625	2.836	5.267	4.457	3.016	4.346	2.717	3.641	3.100	2.718
0.75	2.923	5.673	4.052	3.233	4.712	2.649	3.307	3.349	2.603
0.875	3.061	6.079	3.649	3.484	5.084	2.644	2.992	3.628	2.548
0.9375	3.148	6.282	3.447	3.621	5.272	2.665	2.843	3.775	2.545
1	3.246	6.485	3.246	3.764	5.462	2.700	2.700	3.928	2.559

Table III shows how the bond distances between the diffusing atom (in this case sodium) and each type of oxygen atom changes as the atom diffuses from position zero (P0) to position one (P1). The bond distances that are the shortest are the bonds that will decrease the width of the conduction plane (H) and make it more difficult for ions to diffuse. At P0 it is clear that the bonds between the sodium atoms and the O(4)1, O(4)2, and O(3)1 atoms are making it more difficult for the ions to diffuse through the structure. At P0.5 diffusion is controlled by O(5)1 and then changes to O(4)3 and O(4)4 and O(3)2 as the sodium atom gets closer to P1. In Table IV we can see that the bonds between the diffusing ion and O(4)1 and O(4)2 are shorter at P0 compared to the Bettman structure and O(3)1 and 2 are shorter at positions close to P0.5. The rest of the bonds show some slight differences but for the most part they are the same. When we look at the bond distances for the Cetinkol structure the bonds between the diffusing atom and O(3)1 and 2 are noticeably longer, these bond lengths can be seen in Table V.

Table IV. Bond Distance Comparisons as Diffusing Gd or Na Ions Move From Position 0 to Position 1 Within the Conduction Plane.

Gd Diffusion Position (P)	O(5)1	O(5)2	O(5)3	O(4)1	O(4)2	O(4)3	O(4)4	O(3)1	O(3)2
	Bond Distance (in angstroms)								
0	3.255	3.255	6.483	2.501	2.501	3.829	5.491	2.639	3.783
0.0625	3.155	3.452	6.279	2.478	2.666	3.673	5.292	2.602	3.635
0.125	3.065	3.651	6.075	2.472	2.837	3.523	5.095	2.585	3.493
0.25	2.923	4.052	5.666	2.512	3.191	3.240	4.702	2.598	3.227
0.375	2.834	4.449	5.262	2.652	3.582	2.957	4.293	2.635	3.032
0.5	2.803	4.856	4.856	2.779	3.932	2.779	3.932	2.811	2.811
0.625	2.834	5.262	4.449	2.957	4.293	2.652	3.582	3.032	2.635
0.75	2.923	5.666	4.052	3.240	4.702	2.512	3.191	3.227	2.598
0.875	3.065	6.075	3.651	3.523	5.095	2.472	2.837	3.493	2.585
0.9375	3.155	6.279	3.452	3.673	5.292	2.478	2.666	3.635	2.602
1	3.255	3.255	6.483	3.829	5.491	2.501	2.501	3.783	2.639

Table V. Bond Distance Comparisons as Diffusing In or Na Ions Move From Position 0 to Position 1 Within the Conduction Plane.

In Diffusion Position (P)	O(5)1	O(5)2	O(5)3	O(4)1	O(4)2	O(4)3	O(4)4	O(3)1	O(3)2
	Bond Distance (in angstroms)								
0	3.267	3.257	6.483	2.596	2.596	3.953	5.575	2.895	3.889
0.063	3.156	3.453	6.277	2.578	2.759	3.799	5.378	2.857	3.748
0.125	3.066	3.651	6.073	2.579	2.929	3.648	5.180	2.830	3.616
0.25	2.922	4.049	5.665	2.625	3.279	3.369	4.789	2.830	3.710
0.375	2.833	4.450	5.259	2.736	3.644	3.117	4.402	2.886	3.164
0.5	2.802	4.854	4.854	2.903	4.020	2.903	4.020	2.999	2.999
0.625	2.833	5.259	4.450	3.117	4.402	2.736	3.644	3.164	2.886
0.75	2.922	5.665	4.049	3.369	4.789	2.625	3.279	3.710	2.830
0.875	3.066	6.073	3.651	3.648	5.180	2.579	2.929	3.616	2.830
0.938	3.156	6.277	3.453	3.799	5.378	2.578	2.759	3.748	2.857
1	3.257	3.257	6.482	3.953	5.575	2.596	2.596	3.889	2.895

DISCUSSION

As the atoms diffuse through the conduction plane they have to push through necks or short bonds in the structure, and the longer these bonds the easier it is for the atoms to diffuse. Figure 4 displays the average ion-oxygen distance in all three structures. The narrowest spot is between the ion at P0.125 and O(4)1, and the ion at P0.875 and O(4)3. This correlates with the H value that was calculated in Table II, and compared to the other two structures the width of the conduction plane is smaller, and therefore the bonds are shorter.

At this position in the structure (somewhere between P0.0625 and P0.25) the O(3)1 and O(4)1 are positioned in a slightly tighter triangle that the ion has to diffuse through, which can be seen in Figure 5 where the lines of this triangle have been drawn in. The bonds between this position and O(4)1 and O(3)1 for the Soetebier are around 2.47 and 2.59, the Bettman structure bonds are around 2.64 and 2.55, and the Cetinkol structure has bond distances around 2.58 and 2.83 respectively. This comparison shows that the Bettman structure has three bonds that are around the same length the O(3)1 and the two O(4)1 bonds, the Cetinkol structure has two bonds that are relatively similar to the Bettman bonds, the two O(4)1 bonds, but one bond that is significantly longer the O(3)1 bond. The O(5)1 atom also forms a neck when the diffusing atom is around P0.5. This neck bond distance is the same for the three different structures because the a-axis dimension is nearly the same for all three variants. Therefore this neck is not a determining factor when considering the ease of diffusion in different structures.

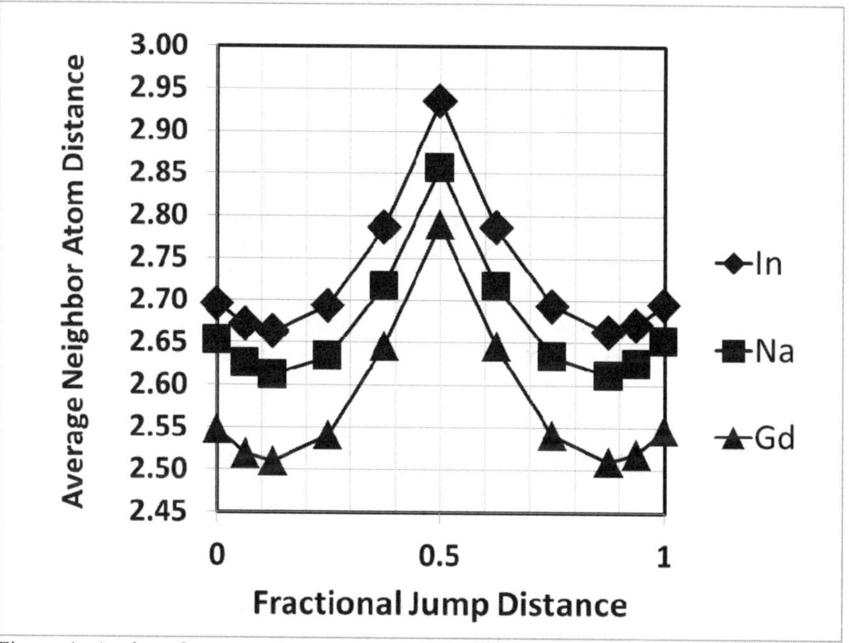

Figure 4. A plot of average Oxygen-ion separation as a function of diffusion jump as the diffusing atom moves from position 0 to identical position 1.

For all three of the structures the bonds are not so short that the diffusing atoms have difficulty making it through the structure. The point that is the tightest for all three structures is within the triangle that is formed by the O(3)1 and the two O(4)1 atoms and the same triangle that is formed from the O(3)2 and the O(4)3 atoms which is displayed in Figure 5. Although this is a tight point in the structure the atoms are able to diffuse without pushing a lot, this means the O3 and O4 atoms do not have to be that flexible to accommodate the diffusing atoms.

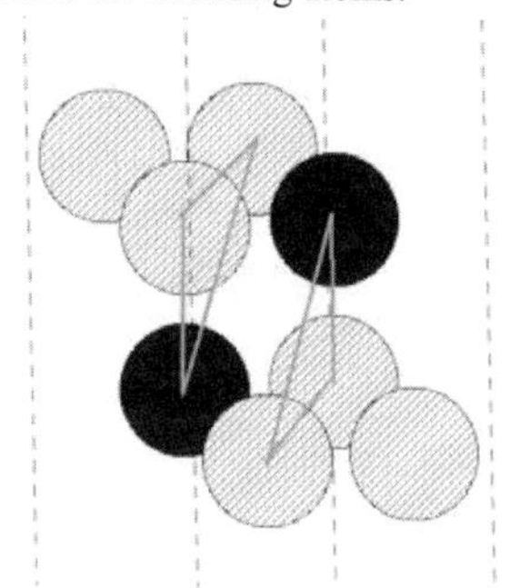

Figure 5. The Bettman and Peters' structure[3] with every atom removed except for the O(3) and O(4) atoms. The lines connecting O(3)1 and the two O(4)1 atoms, and O(3)2 and the two O(4)3 atoms show the tight neck region that the conduction ions have to diffuse through between P0.0625 and P0.25 and P0.75 and P0.938.

CONCLUSION

An analysis of the atoms within and surrounding the conduction plane of three β"-alumina structures was conducted to observe what happens in the to the surrounding oxygen atoms when the sodium ions are replaced with gadolinium or indium. The three oxygen atoms that affect diffusion are O(3), O(4), and O(5) however only the bonds with O(3) and O(4) and the diffusing ion differ in the three compared structures, and these are the two types of oxygen atoms that affect the conduction band width. Although the width of the conduction plane does change, the atoms aren't pushing very hard on the surrounding oxygen atoms, even when looking at structures with widths at the shorter end of the H value spectrum. The rate of diffusion is affected by the energy required to break shorter bonds of each structure and it would seem that the more open conduction planes would require less energy because although most of the bonds are very similar to the other structures that were looked at, there are some bonds that are long enough to make a difference in diffusion rate. In this case, the O(4)1 and 3 and O(3) atoms are the ones that would affect this rate because these bonds are the ones that change most with the type of diffusing ion.

ACKNOWLEDGMENTS

The authors are grateful for support from the HED Graduate Fellowship and the McLaren Chair Endowment.

REFERENCES

[1] J. L. Sudworth and A. R. Tilley The Sodium Sulfur Battery, New York: Chapman Hall, 477 (1985).
[2] T. Hahn, International Tables for Crystallography, 5th ed. Heidelberg: Springer (2005).
[3] D. P. Birnie III, On the Structural Integrity of the Spinel Block in the β"-Alumina Structure, *Acta Cryst.*, **B68**, 118-122 (2012).
[4] M. Bettman and C. R. Peters, The Crystal Structure of $Na_2O \cdot MgO \cdot 5Al_2O_3$, with Reference to $Na_2O \cdot 5A1_20_3$, and Other Isotypal Compounds, *J. Phys. Chem.*, **73**, 1774 (1969).
[5] F. Harbach, High Resolution X-ray Diffraction of Fully and Partially Magnesium Stabilized β"-Alumina Ceramics, *J. Mater. Sci.*, **18**, 2437–2452 (1983).
[6] K. Edstrom, T. A..Faltens, and B. Dunn, The Structure of Na^+ β"-Aluminogallate, $Na_{1+x}Mg_x(Al_{1-y}Ga_y)_{11-y}O_{17}$, x=0.67; y=0.23, 0.30 and 0.41, *Solid State Ionics*, **110**,137–144 (1998).
[7] Soetebier, F. PhD Hannover, 1-146 (2002).
[8] M. Cetinkol, P. L. Lee, and A. P. Wilkinson, Preparation and Characterization of In(I)-β"-Alumina, *Mater. Res. Bull.*, **42**, 713–719 (2007).

PHASE CHANGE MATERIALS AND THEIR IMPACT ON THE THERMAL PERFORMANCE OF BUILDINGS

Marsha S. Bischel, Ph. D.
William H. Frantz
Armstrong World Industries, Inc.
Lancaster, PA 17603

ABSTRACT

Globally, there is renewed interest in low energy strategies for thermally conditioning buildings. High thermal mass construction approaches offer one way to meet this objective via the passive storage and release of thermal energy. An alternative route is to use high energy density Phase Change Materials (PCM's). These are of interest due to their ability to store large amounts of thermal energy by means of a change in state at room temperature while using a comparatively small amount of material, and by providing a source of thermal mass to an existing building. However, along with thermal performance, these materials must meet a challenging host of other criteria, including fire performance and durability, while remaining financially viable and practical. This paper will explore recent developments in the field of phase change materials and their ability to impact the thermal performance of buildings, while meeting other important performance criteria.

PHASE CHANGE MATERIALS AND THEIR USES

Materials Used for Phase Change Properties

In their simplest terms, so-called phase change materials (PCM's) can be described as materials which have a change in latent heat (usually during the melting and solidification transitions) that is large enough to be exploited and used to regulate the thermal properties of the environment in which they are used. Ice is a classic example of such a material. The large latent heat associated with melting is exploited whenever it is used to cool a food or drink: during the melting phase, the ice absorbs large amounts of heat, helping to control the temperature of the immediate environment. When ice crystallizes, heat is released back into the environment. Because this effect is related to a fundamental material property, the ability to absorb and give off heat theoretically will not degrade over time, provided the material itself does not degrade.

This seemingly endless ability to absorb and release heat has long held the interest of those who design heat storage systems. In order to be useful for a long period of time, the material needs to be inert, and any liquid forms generally need to be contained. This has led to a variety of materials and systems using phase change technologies for various heating and cooling applications. These systems operate over a range of temperatures.

At least 500 different phase change materials have been identified;[1] in addition to having different chemistries, they differ in the temperature range at which the phase change occurs, and their thermal energy storage capacity. Among the more common materials used in commercial phase change systems are those based on crystalline alkane "paraffin" chemistry. By changing the length of the paraffin chain, the melting temperature of the wax can be "tuned" to the needed temperature, since the melting temperature increases linearly as molecular weight, or chain length, increases over the range of -50 to +80°C.[1,2] These systems also show relatively little super cooling.[2] Octadecane (a chain with 18 carbon atoms, or C-18) is a popular choice due to its low cost and ability to melt at a temperature that corresponds to room temperature, 25 – 26°C.[2,3,4] These octadecane materials can be manufactured from petroleum, or various bio-based sources (fats and oils from vegetable and animal sources, for example).

Hydrated inorganic salts, such as calcium chloride, calcium oxide and sodium nitrate, are also popular options for PCM's. These salts have a wide variety of melting temperatures and have large heat capacities, and are therefore useful in HVAC systems, solar energy storage systems and roofing systems; however, these can be toxic, corrosive, tend to exhibit super cooling, and segregate out. The performance of these materials may also degrade with time as a result of these issues.[5] However, there is current research in this area that would enhance the usefulness of these salts for these applications.

Other materials that have been used as PCM's include composites comprised of inorganic and organic components,[6] and non-paraffin based polymers such as butyl stearate,[7] polyhydric alcohols, polyethylene, polypropylene, decanes, dibasic esters, and others.

Uses of Phase Change Materials

To be useful in a phase change application, a candidate material should meet a number of basic criteria, including: having a high latent heat at the temperature of interest; undergo small changes in volume during phase transitions; be non-hygroscopic, non-corrosive and non-toxic; have little to no decomposition with repeated cycles and/or time; be stable with regards to mechanical actions; and show minimal super-cooling.[1,5] In addition, they must meet all of the other necessary performance attributes associated with the end-use.

Phase change materials have been investigated as additives in a wide variety of products, including: textiles, for use in sportswear, sleepwear, and military uniforms; hot/cold packs, for keeping food items, drugs, and blood at the appropriate temperature; and various building materials. In each case, the PCM has been selected in large part due to the temperature of its phase change. Thus, for packs designed to keep blood cool, a material changing at 22°C is used,[8] while for building materials, phase changes in the range of 18 – 26°C are common, depending on the specific application.

An analysis of the patent literature shows that phase change materials can be incorporated into other materials using a variety of methods, including: gross-encapsulation methods, such as placing a large quantity of the PCM in a bag, metal box or tube; macro-encapsulation within cementitious or polymeric shells; micro-encapsulation within a polymeric shell; via various coating or spray methods; incorporated into rubber-based or elastomeric compounds; added to extruded or melt spun pellets; or entrapped in non-woven fabrics.

A review of the patent literature suggests that much recent work has focused on the direct application of the micro-encapsulated paraffins in fabrics and materials such as gypsum and concrete and on creating novel methods of using macro-encapsulation techniques. In general, micro-encapsulated particles are in the range of 1– 1000 μm, while macro-encapsulated materials can be up to several millimeters in diameter.[1] However, since PCM's can undergo expansion and contraction during the phase changes,[9] the choice of shell material and particle size can be important, depending on the end use. Micro-encapsulation is often perceived as having certain advantages over the older macro-encapsulated technologies, including being easily integrated into a product matrix, being largely immune from accidental destruction, and by providing a large surface area for heat exchange, thus increase the heat transfer rate and the overall effectiveness of the PCM.[4] However, as particle size increases, the latent heat of the core material more closely matches that of the coated composite,[1] which has implications with regards to efficacy and cost effectiveness.

SPECIFIC USES OF PHASE CHANGE MATERIALS IN BUILDINGS

The use of phase changing materials in buildings and building products was first investigated in depth during the 1980's by the US Department of Energy. Scientists at ORNL demonstrated that the addition of PCM's to an indoor surface, such as wallboard, could change the climate within a space and be used to reduce the energy use of a building. Recent interest in these materials has coincided with the push for sustainable, energy-efficient buildings, such as those required by the US Green Building Council's LEED rating system, the European Directive of 2002[10,11] and the new California green building code, CALGreen.

The benefits of these PCM systems can include: reduction of the energy used to condition spaces; reduction of peak loads, reducing energy use and the size of needed equipment; improvements in occupant comfort by stabilizing temperatures and reducing fluctuations in ambient temperature; ease for use in retrofit and renovation projects; the ability to work with solar energy applications; and shift heating and cooling demands to off-peak periods.[5,11,12]

The basic principle behind the use of PCM's in buildings is that they provide a form of thermal mass. Traditional methods of building used materials that are inherently high in thermal mass, such as brick, stone, adobe and concrete. However, 20th century buildings moved away from these traditional materials, using lightweight, non-thermally massive materials such as steel and glass. These newer buildings do make use of high amounts of thermal insulation; while critical to the energy efficiency of a building, insulation performs differently than thermal mass.

Insulating materials, such as fiberglass, make use of low thermal conductivity to reduce the heat flow in or out of a space. Materials with high thermal mass typically have high specific heat coupled with high density, and will absorb heat when the surrounding environment is warm, and release this heat when the surrounding environment cools. The material will also appear to maintain a steady temperature for longer than does the surroundings. This is easily understood by considering a stone hearth in the interior of a home, which feels cooler for much longer on a hot day, or an exterior brick wall which is still warm well into the evening.

A wide variety of uses for phase change materials within buildings have been identified in the commercial, scientific and patent literature. These include, but are not limited to: drywall; plaster; rock wool ceilings; cementitious tiles; composite layered insulation structures with bulk quantities of macro-encapsulated PCM's; blown insulation; thermal control blankets; interior panels containing bulk quantities of macro-encapsulated PCM's; chilled ceilings; heat exchangers; under floor HVAC systems; wall coverings made of flexible sheet-goods; closed cell insulation; heat storage systems for solar cells; within a heat pump; open cell foam sheet; cementitious hollow core blocks; roof membranes; fire protection containers; insulation block for roofs; and agglomerate ceiling and floor tiles. The number of products which have been commercially viable is far smaller.

A key design requirement of any building material using PCM's to enhance thermal comfort and reduce energy usage is to determine the portion of the heating and cooling cycle that is to be impacted. Different melting temperatures are better for retaining and releasing heat versus assisting in cooling.[4] For cooling, the PCM should have a melting temperature that is at the upper level of what is acceptable for room temperature. Many researchers have used 22-26°C for cooling application, while melting temperatures of 18 - 22°C are used for heating needs.[13] This work will focusing on cooling systems.

Another key attribute of the system is the need for the PCM to completely change phase. For example, to use a PCM to assist in reducing the need for cooling, the PCM must be fully re-solidified during the night, so that it will be available for use the next day. This typically requires high air exchanges. If a space does not adequately cool down to a temperature below the freezing point of the PCM, the effectiveness of the system will be compromised.

While the thermal performance of a phase change material is key, other performance characteristics are also critical. Among these, the most important are related to life safety issues: smoke and fire performance, and seismic issues. Other important issues relate to the ease of installing and finishing the material, the ability to absorb or block sound, durability and longevity once installed, life cycle analyses, and the avoidance of any chemicals of concern. Finally, the cost-value proposition must be addressed if the product is to be a commercial success.

Table I shows a variety of commercially available building materials and prototype systems that contain phase change materials, and categorizes them according some of the performance attributes listed above.

Table I: Performance attributes of building materials containing phase change materials

Product Prototype Form	Storage per Unit Area [Wh/m^2]	Cost Per Unit Area [GBP/m^2]	Storage per Unit Cost [Wh/GBP]	Weight Per Unit Area [kg/m^2]	Storage Per Unit Weight [Wh/kg]
Ceiling panel with inorganic PCM (micro-encapsulated paraffin) filled insert	136	200.0	0.7	25.0	5.45
Salt hydrate in rigid plastic containers to be placed on the backs of metal ceiling panels	355	33.7	10.5	10.7	33.00
Magnesium oxide board with micro-encapsulated paraffin PCM	150	30.0	5.0	9.0	16.67
Paraffin PCM in a silicone material, foil faced board	88	45.5	1.9	4.5	19.44
Gypsum board with micro-encapsulated paraffin PCM	79	24.5	3.2	10.2	7.67
Salt hydrates in various plastic tubes and flasks	888	70.0	12.7	16.7	53.06
Palm oil and soy oil in plastic bubble mat	160	22.8	7.0	3.5	46.31
Encapsulated salt hydrate in plastic bubble mat	78	25.2	3.1	3.0	26.28
Cast structural concrete	12	1.5	7.7	24.0	0.49

It is apparent from the data in Table I that each product type had specific benefits and drawbacks, making it critical that an optimal design be selected to achieve the desired end product.

THERMAL TESTING OF BUILDING MATERIALS CONTAINING PHASE CHANGE MATERIALS

The thermal performance of phase change materials and building components which contain PCM can be evaluated experimentally on several different scales. Differential scanning calorimetry is commonly used to evaluate the "neat" and encapsulated forms of PCM. Typically these tests use very small samples of material on the order of 10 [mg] and volumes of 1 [cm^3].

Small scale thermal storage tests have been done using "hot/cold" boxes and relatively small samples of PCM-containing building components.[4,11,14,15] Such small scale tests might use 2000 [cm^3] of material and use the PCM in an actual product form. Larger scale tests have been conducted at various research universities, building research organizations, and at Armstrong. These tests occur in thermal test chambers having room-size dimensions and place the PCM-containing building material under realistic service conditions. Chamber areas can range from 10 - 20 [m^2] with heights of about 3 [m]. The amount of PCM-containing building material used can be of the order of 300,000 [cm^3]. Laboratory testing at this scale is critically important because it begins to involve key operational variables such as airflow velocities over building surfaces, thermal cooling loads, the fundamental arrangement of the air conditioning system, and other practical variables that greatly affect the actual performance of the PCM-containing building material.

Armstrong conducted independent, large scale thermal performance testing of PCM-filled ceiling products in an internal laboratory, and also commissioned testing by a third party organization involved with building performance research.

Finally, it should be noted that a specific PCM-filled ceiling product has been installed and monitored in five buildings located in the United Kingdom. The results from these tests will be published at a future date.

Chamber testing at Armstrong Facility

Large scale performance testing at Armstrong was based on the use of an air conditioning system known as "displacement ventilation," which uses low velocity air delivered low to the floor. The warm air moves up towards objects that give off heat (people, equipment, lights) and then progresses over the "room side" surface of the ceiling, transferring some of the heat to the PCM-containing ceiling panels. Typically, the warm air will then migrate towards open return air grills in the ceiling; once above the ceiling and in the "plenum space," it will continue flowing over the back surface of the ceiling panel toward the extract ductwork for the room.

Armstrong configured a heavily insulated, 9.7 [m^2] chamber to operate in this manner, with floor dimensions of approximately 2.6 x 3.7 [m] and a floor to roof height of 2.7 [m]. The walls, floor and roof were 330 [mm] thick with cork insulation having an estimated U value of 0.21 [W/m^2 K]. A wood timber frame structure was built inside the chamber so that the various ceiling panels and free hanging ceiling objects could be hung for testing. Typically the distance from floor to ceiling surface was 2.3 [m].

The heat given off by occupants, equipment, and lights was simulated in the chamber through the use of four "DIN man heaters."[16] Three 60 [W] incandescent light bulbs in each provided a heat load of up to 74 [W/m^2], equivalent to that generated by a range of activities from occupants performing sedentary office work to heavy physical exertion.

A simple 1465 [W] window air conditioner was used to remove heat from the chamber and simulate a "night purge" cycle. The supply and return grills were ducted so that cool supply air was delivered to floor level and warm return air was taken from high roof level, above the plane of the test ceiling. A precision thermostat replaced the built-in thermostat on the air conditioner.

Thin film, surface mount thermocouples were installed on the interior and exterior of the walls, floor and roof. They were used to help estimate uncontrolled heat loss or gains between the test room and the exterior. A stand inside the room held additional thermocouples at positions 100, 200, 500, 1500, and 1800 [mm] above the floor, and were used to record the temperature profile that developed. Additional thermocouples were mounted on the surfaces of test samples and, where possible, were also embedded within test pieces. Two power transducers were used to record the real time power usage of the DIN man heaters and the air conditioner.

A typical test cycle is described Table II, below. A test ceiling was installed in the chamber and allowed to come to equilibrium conditions for about 24 [h] prior to beginning a test cycle. At the start of a "work-day" a controlled thermal load was applied by the "DIN man heaters," mimicking the generation of heat in an occupied space during typical work hours. After this simulated occupancy period, the heaters were turned off, signally the end of the work day. A "purge period" was then started, during which the air conditioner was used to extract stored heat from the panels, allowing the PCM to re-solidify in preparation for use the next day. In practice, this heat would be removed by night-time ventilation or some other low energy, low cost method.

Table II: Test Cycle Description used at Armstrong Facility

Time	Thermal Load	AC Set point	Simulates	Status of PCM
Prior evening	Off	64 [F]	Night time purge. Conditions the test pieces to a standard starting point.	All PCM is solid.
8 AM	On	Increase to 75 [F]	Beginning of the workday. Heat sources are turned on to simulate occupation of space. PCM ceilings begin to store heat.	PCM begins to melt.
8 AM to 1 PM	On	75 [F]	PCM panels accumulate heat. Loads are not detected by the air conditioner and it remains off.	PCM continues to melt.
1 PM to 5 PM	On	75 [F]	PCM panels are "full" and can absorb no more energy. Loads begin to warm the air returning to the air conditioner. Air conditioner begins to operate to maintain comfort conditions.	PCM is fully melted.
5 PM	Off	Decrease to 64 [F]	End of the workday. Heat is turned off to simulate people leaving. Air conditioner set point is reduced to simulate a cool air night purge cycle.	PCM is fully melted.
5 PM to 11 PM	Off	64 [F]	Accumulated heat is removed from the ceiling panels and discharged outside of the building.	PCM solidifies.

The Armstrong facility was used to test approximately 65 different configurations of PCM-filled ceiling systems, including different systems, PCM materials, and energy storage densities. The materials tested included traditional lay-in ceiling tiles, commercially available panel systems, and various other prototypes, as described in Table I.

Chamber Testing at Third Party Facility

Additional large scale testing was performed in collaboration with a third party testing facility affiliated with a building research organization located in the United Kingdom. This chamber was approximately 16 [m^2] in area and was designed with a temperature controlled buffer space between the chamber and the exterior environment. This facility could also be reconfigured to simulate different air distribution approaches such as overhead air delivery with unducted air returns, overhead air delivery with ducted air returns, and displacement ventilation. A single design of PCM-filled ceiling tile system was used for all tests. Over a period of 20 days, a sequence of 12 different configurations was evaluated. This allowed the prototype product to be tested in various realistic configurations, and allowed the efficacy of the product to be more broadly evaluated. Figure 1 below shows some of the most important combinations of system type and PCM ceiling storage density. The third party testing was in general agreement with testing done by Armstrong for the comparable case of displacement ventilation. Evaluating alternative air distribution arrangements and operating conditions (different airflow rates, cooling loads, and purge air temperatures) provided valuable insight into how best to operate a PCM ceiling. These topics will be discussed in a future paper.

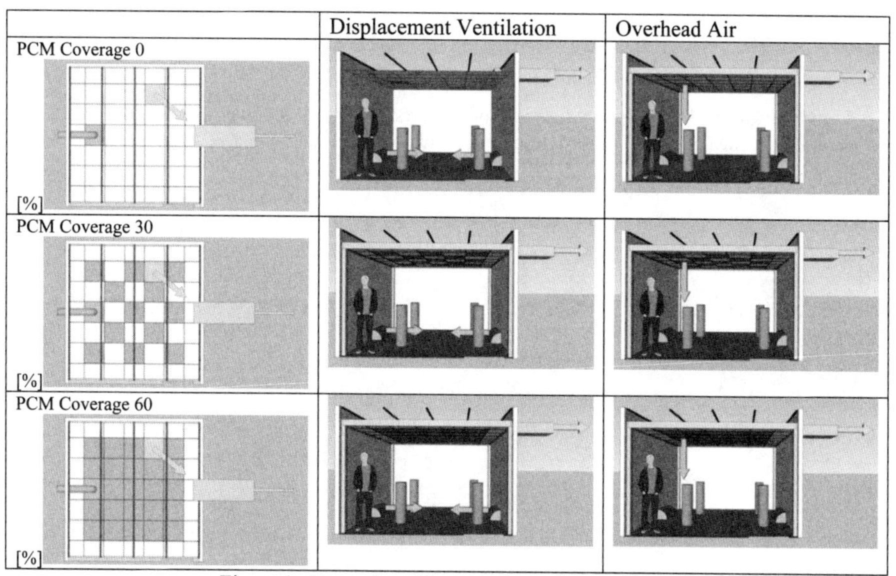

Figure 1: Examples of large scale testing configurations

RESULTS OF CHAMBER TESTING

In all cases, an empty metal pan ceiling was used as a reference case. Energy use for the test sample was then compared against that of the reference case.

With the Armstrong facility, it was observed that the addition of thermal mass typically reduced energy use over the 8 [h] of simulated occupancy by 5 - 50 [%], depending on details. Typical 24 [h] data traces are shown below, in Figures 2 - 3. Conditions surrounding the outside of the thermal test chamber were not controlled. Obviously, varying exterior conditions could increase or reduce the heat input to the room. To adjust for this variation, data from past reference cases were used to adjust energy use according to actual exterior temperatures monitored during a test.

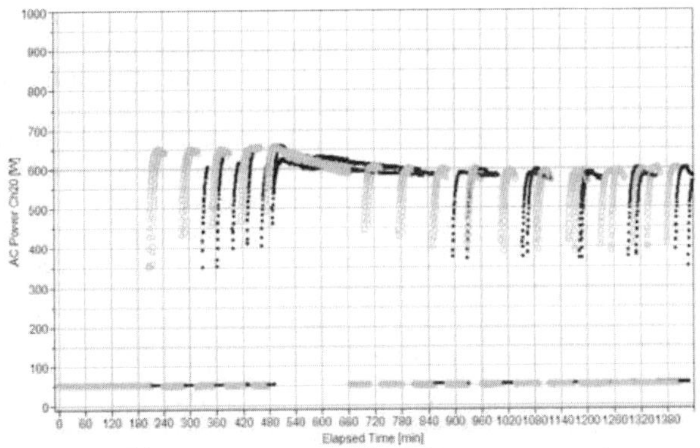

Figure 2. Air conditioner power usage. The light trace represents power usage with a standard metal reference ceiling. The dark trace represents power usage with a PCM ceiling installed.

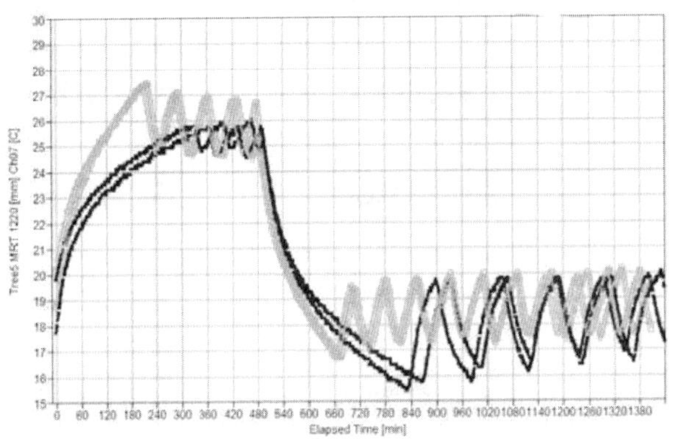

Figure 3. Mean radiant temperature ("globe temperature") in the test chamber. The light trace is for the metal control ceiling; the darker trace is for the PCM-filled ceiling.

Figure 2 shows the power used by the air conditioning unit. The reference metal ceiling and the PCM-filled ceiling are shown by the gray and dark traces, respectively. In this case, the metal reference ceiling used 1600 [Wh] of energy over the 8 [h] period of occupancy while the PCM-ceiling used only 750 [Wh] of energy over the same period. This represents a reduction of about 53 [%] of the energy needed for air conditioning.

The "mean radiant temperature" in the test chamber is shown in Figure 3. As can be seen by comparing the dark and light traces, the PCM-filled ceiling created a lower mean radiant temperature than the reference metal ceiling (25 - 27°C for the metal ceiling verses 25 - 26°C for the PCM ceiling). Overall, differences of 2 - 4°C were observed. In addition, there was less fluctuation in the trace for the PCM-filled ceiling, suggesting fewer changes in ambient temperature for an occupant in the room. These radiant cooling effects would likely result in greater thermal comfort for the occupants of a space.

The results of the large third party study will be published in detail at a later date. However, results from three cases using the same loading of PCM are shown in Table III. As these three cases show, the efficacy of the PCM-based system is dramatically affected by the type of ventilation used. It is therefore important to test such systems under a range of realistic service conditions if the full impact of the PCM-filled product is to be fully understood.

Table III: Results from Third Party testing

Type of Ventilation	Hours of projected Storage due to PCM
Displacement Ventilation	4
Overhead Air, Unducted Return	2 – 3
Overhead Air, Ducted Return	Approximately 1

RESULTS OF FIRE TESTING

With any building material, fire performance is a critical performance attribute. Fire performance is related to a number of factors, including fuel content, combustibility, smoke generation, and the formation of toxic by-products. In the case of organic phase change materials, this is important because the materials are inherent sources of fuel. With inorganic materials, combustibility is generally not an issue, but toxic by-products such as hydrochloric acid or ammonia gas may be generated, depending on the exact chemistry of the system. In addition, salt solutions kept in sealed containers could rupture in the event of a fire.

However, when reviewing the literature, it was apparent that most research into phase change materials ignored any issues related to life safety. Some researchers have looked at the incorporation of flame retardants into these systems; however, these add cost and complexity to the manufacturing processes, and sometimes used materials that are discouraged by the sustainable building standards, such as halogenated compounds.[2,11]

Because the fire performance of building products is strictly regulated, meeting the required performance levels was determined to be the first criteria when evaluating phase change materials and products using them for commercial applications.

Fire reaction testing of building products is a necessary part of the development of a practical construction material containing a phase change material. Specific testing requirements can vary depending upon the type of building material, the intended country of use, and the type of building and its occupancy. Requirements are different for insulation, construction board, floor surfaces, ceiling panels, articles of furniture, etc. Requirements can also vary dramatically from one country to the next. Building type and occupancy can affect design targets, with requirements being quite different for an office as compared to a hospital or school, for example. However, in most cases, fire performance relates to a few specific categories: flame spread, heat generation, smoke development, and the formation of any droplets.

This research focused on the European fire reaction tests for a ceiling product developed for use in the United Kingdom, typically in an office or educational building. The relevant testing protocol is detailed in EN 13823, and is known as the Single Burning Item test (SBI).[17] This is a "large scale" test that requires 15 finished ceiling panels for testing. Tests were run in an Armstrong fire test facility, and at a third party accredited laboratory in the UK.

Due to the large sample required for this test, a small "screening" test was also developed. This precursor test used a small sample of the product design placed in a horizontal metal pan. A thermocouple was placed between the sample and the metal pan and a propane torch was adjusted to generate a temperature of 430- 480°C on the sample surface. The torch was then left in place for 10 [min] and the behavior of the sample was observed. If the sample ignited, or if vapors from the sample could be ignited resulting in a sustained flame, then that sample was unlikely to perform well in the large scale EN 13823 test.

Use of the screening test allowed rapid evaluation of various phase change materials and prototype ceiling product designs. Materials or designs that performed poorly were eliminated from further consideration; those that looked promising were pursued further.

Ultimately, several different prototype products were evaluated using the two large scale SBI tests. Based on the results of these tests, it was shown that it is possible to engineer an appropriate system that meets the European fire requirements while using an otherwise combustible component. In fact, the fire performance achieved was better than for any other similar products that were commercially available.

CONSTRAINTS ASSOCIATED WITH COST AND INVESTMENT

For any building product to be commercially successful, it must have an attractive cost-value proposition. This needs to include not only the material costs, but the costs to install and maintain the product, as well as the expected life-time of the product. For products that are designed to reduce energy use or costs, the analysis will also need to include the expected cost of energy in the future, and a valid estimate of the savings associated with the product.

The cost analysis for building materials that contain phase change materials will be highly dependent on a number of factors, including: climate and the ability to use night-time ventilation; the cost and cost-structure of energy; the embodied energy in the product; and the type of heating and ventilation system used. The impact of varying the cooling system can be seen in Table III: a 75% reduction in energy storage for the third case as compared to the first makes a dramatic impact on the cost analysis.

Furthermore, the use of thermal mass can reduce the size of the air conditioning equipment, resulting in lower capital investments in equipment and lower maintenance costs. Thus, it may be difficult to develop a single, simple model that can be used to effectively predict energy savings for all cases.

In certain geographies that have enacted rigorous legislation designed to dramatically reduce energy use, the cost-benefit analysis may be altered by the need for compliance with the energy codes. For example, legislation in the European Union will require new structures to be "net zero" by 2021, and for countries to reduce primary energy use by 20% by 2020.[10,17] This regulatory pressure, combined with a climate that is conducive to effective natural night-time ventilation and high energy costs, has sparked interest in Europe in using phase change materials.

In general, it appears that energy costs in the United States are still too low to make "distributed" PCM-based thermal storage economically viable. In addition, large parts of the country do not have climates that are compatible with the use of natural ventilation for night time cooling, although other options are available for providing the needed cooling.

Thus, it would appear that making a valid cost analysis for any PCM-based building product is complex and is highly dependent on the particular details of the installation. Simplistic models presented in the literature may not adequately predict the ultimate cost-benefits of an installed PCM system.

CONCLUSIONS

Based on the results obtained during this broad-based assessment of the use of phase change materials in ceiling products, several conclusions may be drawn. First, it was demonstrated that the use of PCM-filled ceiling panels can reduce air conditioning energy use by 5 - 50%, depending upon how they are used. In addition, a significant radiant cooling effect was observed, with the PCM-filled ceilings staying 2 - 4°C cooler than a comparable metal ceiling, thus providing increased comfort to the occupants of the space.

Some operational constraints were also identified. The best energy and cost savings occur when night ventilation can be used to purge accumulated heat, and re-solidify the phase change material. Some cost savings is possible if off-peak power rates are available, and traditional air conditioning is used to remove the heat from the PCM.

In addition, it has been shown that the amount of energy storage ultimately associated with the PCM material is directly impacted by the choice of the HVAC system and the manner in which conditioned air is supplied to the space. By changing the HVAC system, the efficacy of the same PCM-filled product was dramatically changed, with some systems providing four times the benefit of others. This not only changes the cost-benefit analysis for a particular space, it demonstrates the need to perform full-scale testing of these systems, rather than simply make projections based on latent heat of fusion data.

Finally, it has been demonstrated that solutions meeting the requirements of the European fire requirements are possible through the careful choice of materials and design.

REFERENCES

[1]Mattila, H. R., ed., Intelligent textiles and Clothing, The Textile Institute, New York: CRC Press, 2006.
[2]Saylor, I. and A. Sircar, "Phase change materials for heating and cooling of residential buildings and other applications," *Proceedings of the 25th Intersociety Energy Conversion Engineering Conference*, **4** (1990) 236.
[3]Kissock, K., "Diurnal load reduction through phase-change building components," *ASHRAE Transactions*, Volume 112, Part 1, 2006.
[4]Schossig, P., H. M. Henning, S. Gschwander and T. Haussmann, "Micro-encapsulated phase-change materials integrated into construction materials," *Solar Materials & Solar Cells*, **89** (2005) 297.
[5]Khudhair, A. and M. Farid, "A review on energy conservation in building applications with thermal storage by latent heat using phase change materials," *Energy Conversion and Management*, **45** (2004) 263.
[6]Fang, X. and Z. Zhang, "A novel montmorillonite-based composite phase change material and its applications in thermal storage building materials," *Energy and Buildings*, **38** (2006) 377.
[7]Feldman, D. et al., "Obtaining an energy storing building material by direct incorporation of an organic phase change material in gypsum wallboard," *Solar Energy Materials*, **22** (1991) 231.
[8]Cryopack: http://www.cryopak.com/phase-22-insulated-shipping/
[9]Mulligan, J. C., et al., "Microencapsulated phase-change material suspensions for heat transfer in spacecraft thermal systems, *Journal of Spacecraft and Rockets*, **33** (1996) 278.
[10]European Union Energy Performance of Buildings Directive, 2002.
[11]Butala, V. and U. Stritih, "Cold storage with phase change material for building ventilation," *Journal of Ventilation,* **5** (2006) 189.
[12]Kosny, J. et al., "New PCM-enhanced cellulose insulation developed by the ORNL Research Team, Oak Ridge National Labs, 2006.
[13]Addington, M and D. Schodek, Smart materials and technologies for architecture and design professions, New York: Elsevier, 2005.
[14]Ahmad, M, et al, "Experimental investigation and computer simulation of thermal behavior of wallboards containing a phase change material," *Energy and Buildings*, **38** (2006) 357.
[15]Kondo, T. and T. Ibamoto, "Research on thermal storage using rock wool PCM ceiling board," *ASHRAE Transactions*, **112** (2006).
[16]EN 14240, "Ventilation for buildings; Chilled ceilings: Testing and rating"
[17]European Union Energy Efficiency Directive, 2012
[18]EN 13823, "Reaction to fire tests for building products. Building products excluding floorings exposed to the thermal attack by a single burning item"

HYDROGEN-EXPOSED WELDED SPECIMENS IN BENDING AND ROTATIONAL BENDING FATIGUE

Patrick Ferro
Gonzaga University
Spokane, WA, USA

Reza Miresmaeili
Tarbiat Modares University
Tehran, Iran

Rana Mitra
SE Louisiana State University
Hammond, LA, USA

Jason Ross
Gonzaga University

Will Tiedemann
Gonzaga University

Casey Hebert
Gonzaga University

Taylor Goade
Gonzaga University

Duncan Howard
Gonzaga University

Keith Davidson
Gonzaga University

ABSTRACT

Austenitic stainless steel specimens were fatigue tested under various conditions involving exposure to hydrogen. Some of the experimental parameters that were investigated include long-term exposure to high pressure hydogen (138 MPa at 300°C for 15 days), surface preparation (polishing compared to plasma-cut), and welding. The location of the weld on the welded samples was based on the fatigue crack initiating and propagating in the heat-affected zone (HAZ). Bending fatigue of Type 304 stainless sheet of 0.9 mm thickness is being investigated, and rotational bending fatigue is being explored for future testing. Friction stir welded samples are being preliminarily investigated.

INTRODUCTION

The challenges of preparing austenitic stainless steel specimens for hydrogen embrittlement fatigue testing have been described previously[1,2]. Earlier results have preliminarily shown that hydrogen precharged fatigue specimens may have shorter fatigue lives than control specimens.

In addition to studying the effect of hydrogen precharging on fatigue life of Type 304 austenitic stainless steel, the present investigation has begun to study the effect of welding on the fatigue life of the same types of samples. The investigation has attempted to determine if there is a synergistic effect of prior welding and hydrogen precharging on fatigue life. Some of the challenges have included locating the weld in the correct location on the specimen, and removing stress-concentrators from the weld prior to fatigue testing. The work has attempted to augment results as reported by Somerday[3]. In the work of Somerday et al., microcrack propagation at the austenite/weld ferrite interfaces was observed, as well as in the ferrite[3].

The investigation has used bending fatigue specimens that have been exposed to hydrogen under a range of exposure conditions including pressures up to 145 MPa at elevated temperature[1]. Earlier work has attempted to show the effect of intermittent exposure conditions, at 1 atm hydrogen exposure pressure[2]. Advantages of experimenting with austenitic stainless steel specimens are discussed, and it is shown how hydrogen embrittlement and fatigue experimentation can be performed without the necessity of in situ hydrogen testing apparatus[2].

Additional proposed work has investigated friction stir welded (FSW) samples. Samples of 6061 and 5086 grades of wrought aluminum are being investigated by milling bending fatigue samples from FSW sheets of 3.2 mm stock. The effect of the friction stir weld and hydrogen exposure has been proposed to be investigated. A flowchart of the testing for the overall reseach plan is shown in Figure 1.

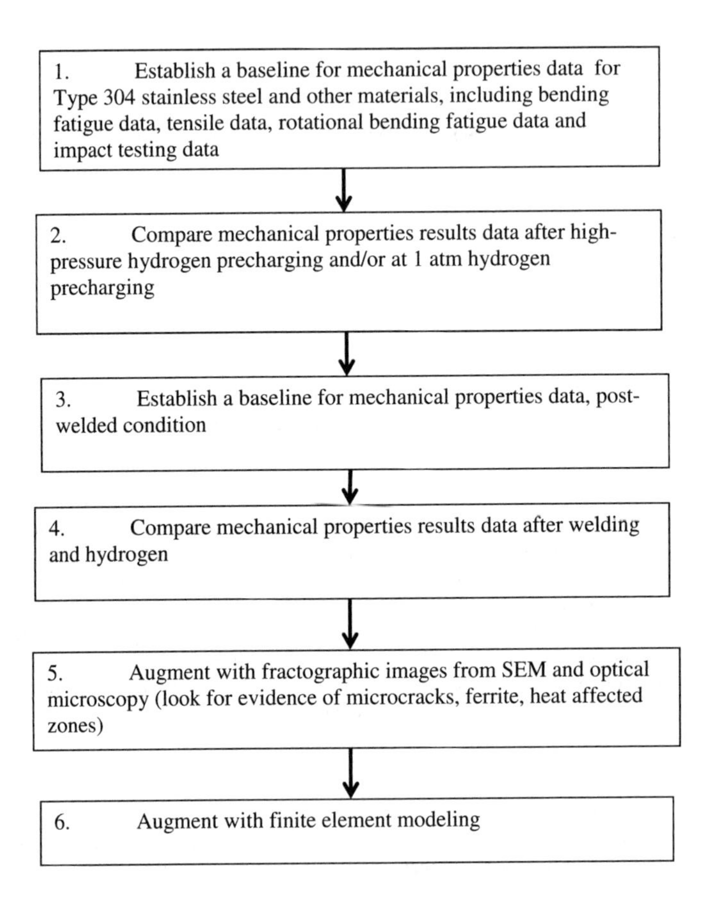

Fig. 1. Flow chart of testing plan for hydrogen embrittlement research at Gonzaga University.

PROCEDURE

The bending fatigue tests that were performed were using a VSS-40H bending fatigue testing machine from Fatigue Dynamics (Walled Lake MI). The cycling frequency was between 300 and 350 cycles per minute. After the specimen failed, it was removed and the location of the failure was identified by measuring the distance from the start of the specimen radius to the middle of the failure location on specimen. Some of the samples are analyzed with SEM. Further description of the experimental equipment and procedures may be found in previous papers[1,2].

Bending fatigue specimens were plasma cut from 0.9 mm thick sheets of Type 304 stainless steel, from Alcobra Metals (Spokane WA). The plasma-cut specimens were belt-ground along the periphery to remove slag. The specimens were hand sanded and polished down to 600 grit sand paper.

Some of the samples were highly precharged with hydrogen at Sandia National Laboratory prior to testing. These samples were sent to Sandia in the plasma-cut and edge belt ground condition, where they were exposed to hydrogen pressures up to 138 MPa (20000 psi), at temperatures as high as 300°C (575°F) for a period of time of up to 15 days[4]. The typical hydrogen concentration in 300 series stainless steels after similar conditions is approximately 140 ppm, by weight[4]. Other hydrogen charging protocols are described by Mine et al.[5]

Some of the bending fatigue samples were welded. An STT Lincoln automated MIG welding machine was used. The wire was 0.75 mm diameter 308 LSI. The voltage was maintained at 17V while the current ranged between 80 and 90 amps. Previous investigators have shown that the presence of ferrite in a post-welded microstructure correlates with higher levels of hydrogen embrittlement[4]. Higher levels of ferrite may be responsible for increased levels of hydrogen embrittlement sensitivity[4]. Welding parameters and filler metals may cause ferrite levels to change, and are thus a relevant area of research for the present investigation. The location of the weld centerline is 11 mm from the clamped end of the specimen (where the transition begins).

Rotational bending fatigue testing was preliminarily investigated. Figure 2 shows a drawing of the type of rotational bending fatigue specimens that were generated from 304 stainless for testing. An RBF 200 fatigue testing machine (Fatigue Dynamics, Walled Lake MI) was used. The machine has a slideable setting that allows for a range of bending stresses to be used in experimentation. The machine can be set between 0 and 22.6 N-m. For the preliminary experimentation reported, the machine was set at 7.3 N-m corresponding to a nominal longitudinal bending stress of 614 MPa. The rotational velocity ranged from 6000 to 10000 rpm in the testing that was performed. Figure 3 shows a photo of the RBF 200 rotational bending fatigue testing machine.

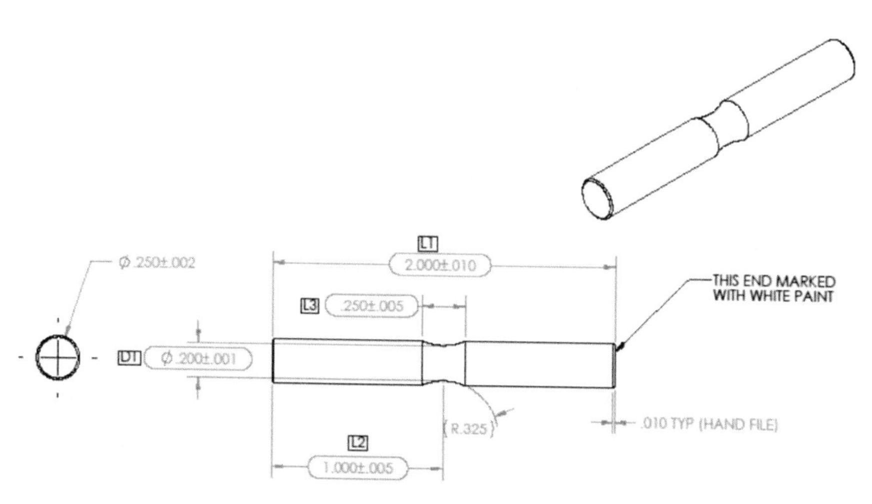

Fig. 2. Drawing of the rotational bending fatigue specimens that were machined from Type 304 stainless bar stock. Dimensions shown are in inches.

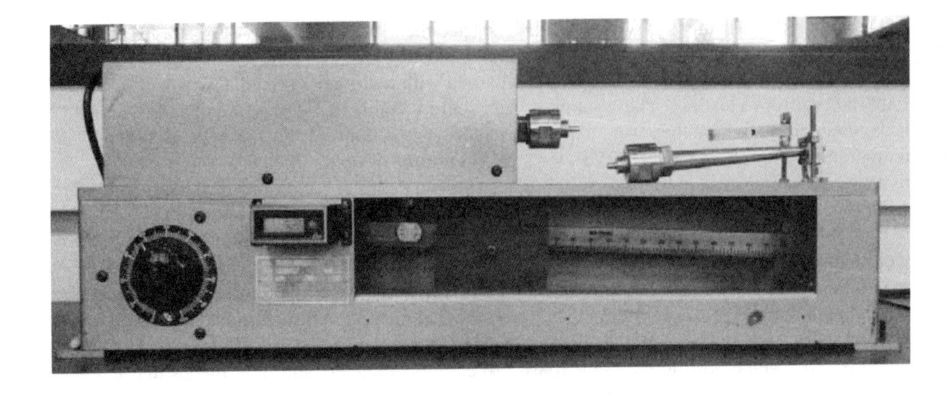

Figure 3. Photo of the RBF 200 from Fatigue Dynamics. The machine is a rotational bending fatigue tester, with a range of stresses between 0 and 22.6 N-m possible with a machine setting.

Friction stir welded samples were prepared by friction stir welding 3.2 mm thick plates of aluminum alloys. The plates were friction stir butt welded at rotation stir velocities of 2700 and 3200 rpm. The tool travel speed ranged between 14 mm min^{-1} and 107 mm min^{-1}. A cobalt-based multiphase alloy MP-159 made by Latrobe Specialty Metals was used as the tool material. Figure 4 shows a schematic of the friction stir welding configuration, from Somasekharan[6]. Table 1 summarizes the alloys and the friction stir weld parameters that were tested. In the Table 1, the offset was 2 mm on either the advancing side or on the retreating side or none (designated respectively as A, R or none).

Table 1. Summary data on FSW sample preparation

Al alloy	*Tool rpm*	*Travel speed*	*Offset*
5086-H116	2700 rpm	19.8 mm min^{-1}	A
5086-H116	3200 rpm	14.0 mm min^{-1}	none
5086-H116	3200 rpm	17.3 mm min^{-1}	R
6061-T6511	3200 rpm	106.7 mm min^{-1}	A
6061-T6511	3200 rpm	50.8 mm min^{-1}	A

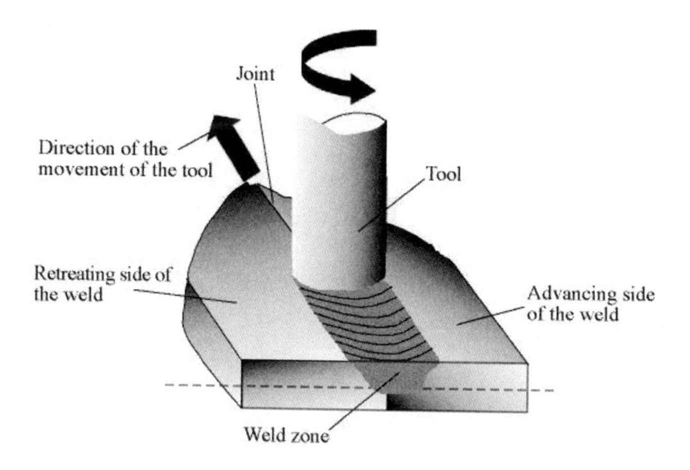

Figure 4. Schematic showing the friction stir welding configuration, from Somasekharan[6]. The tool is shown rotating counterclockwise, so that the advancing side falls on the right of the figure, and the retreating side on the left. In the experiments reported here, the "offset" refers to the displacement of the tool tip by 2 mm – either to the advancing side (referred to here as Offset A) or to the retreating side (offset R) or zero offset (none).[6]

To obtain bending fatigue specimens from the friction-stir welded plates, specimens were milled from the plates so that the weld centerline was the same as that used for the arc welded austenitic sheets (11mm from the transition). Milling was necessary to mill the relatively thick plates down to the 0.9 mm final thickness. Non-welded plates were also milled to excise specimens, for comparison with the welded plates.

RESULTS

Figure 5 shows bending fatigue data for 304 stainless specimens of 0.9 mm thickness. The maximum bending stress amplitude was calculated to be 277 MPa using the elasticity method of calculation, at a maximum deflection of 6.7 mm. Data for four difference specimen preparations are compared. The data appears to show that the effect of a weldment is to reduce the fatigue life of the specimens. The highest average fatigue life is for the unwelded specimens. The other three preparations are all variations of weld bead removal.

The work has attempted to determine the welded specimen preparation which may provide for a valid estimation of the effect on fatigue life for hydrogen-exposed, welded 304 stainless. One conclusion that may be drawn from Figure 5 is that the weldment changes the stress distribution to increase the stress concentration factor, effectively lowering the fatigue life. Because the bending fatigue specimens are tapered, the location of the maximum stress may be different for a welded specimen than for a non-welded specimen. Also, how the weld bead is removed affects the fatigue life and presumably the location of maximum stress.

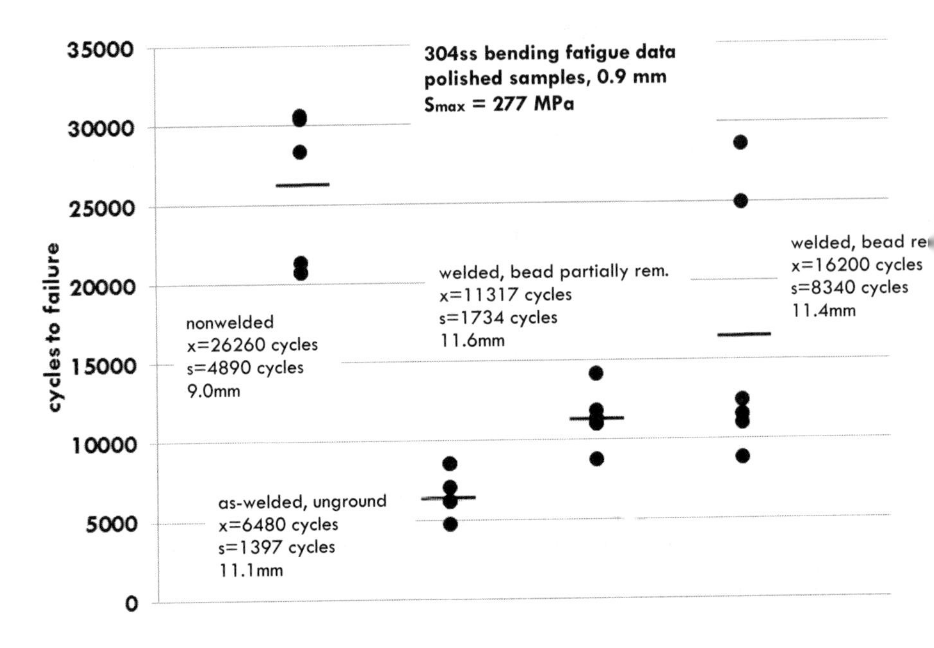

Figure 5. Bending fatigue results for 304 stainless specimens of 0.9 mm thickness. The results appear to indicate that not removing the weld bead lowers the average fatigue life. The maximum calculated applied bending stress is 277 MPa. The location of failure for the unwelded specimens was 9.0 mm from the transition at the clamped end. The location for the welded specimens was on average between 11.1 mm and 11.6 mm from the transition at the clamped end.

The location of the failure for the welded specimens was, on average, more than two millimeters further from the transition at the clamped end than the non-welded specimens. The typical failure location was in the weld metal. Since the nominal location of the weld center line was 11 mm from the clamped end, the average failure locations for each of the welded specimens was in the weld metal region of the microstructure.

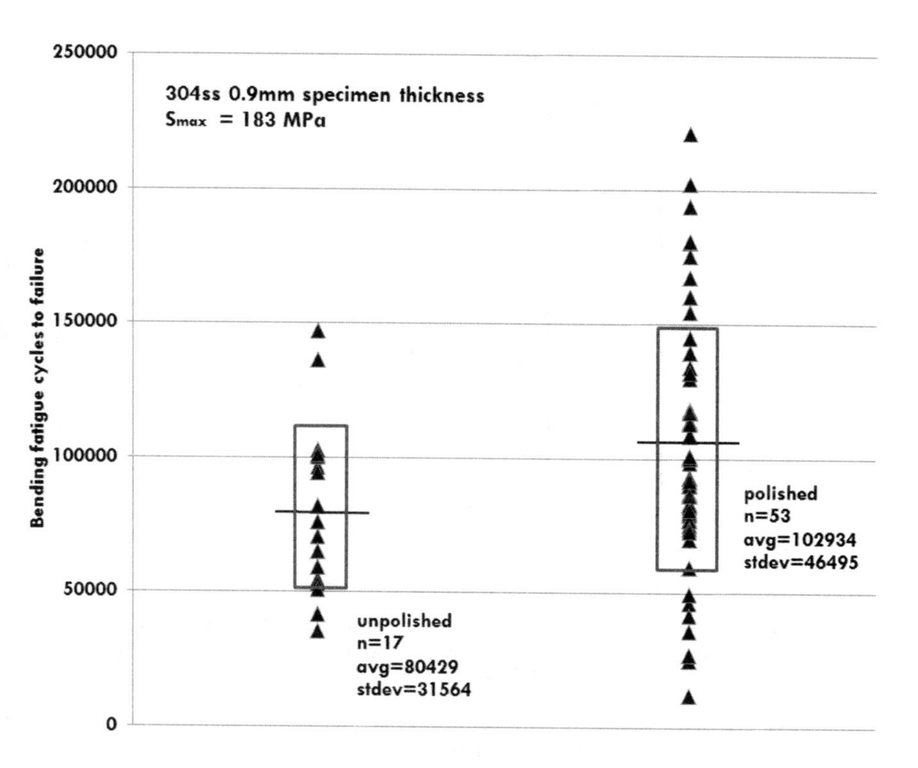

Figure 6. Bending fatigue data for 304 stainless specimens of 0.9 mm thickness. The data shows that polishing samples increases the bending fatigue life. A student's t-test probability of 0.01 was calculated, for unpaired heteroscedastic comparison, indicating that the two populations are different.

Figure 6 shows bending fatigue data for 304 stainless specimens of 0.9 mm thickness. The calculated maximum bending stress was 183 MPa. The data compares specimens that were polished to fineness achievable with 600 micron sand paper to specimens that were tested in the as-plasma cut condition. The effect of polishing appears to increase the bending fatigue life.

Figure 7 shows a probability density function plot of rotational bending fatigue data for 304 stainless samples of 6.35 mm diameter stock. The maximum rotational bending fatigue stress correlated with a machine setting of 7.3 N-m. The estimated maximum stress at that setting is 614 MPa. The data shown in fig. 7 represents 40 individual tests that were performed. The average cycles to failure was 86500 cycles, and the standard deviation was 39100 cycles. The dashed line in fig. 7 represents a Gaussian distribution of the data.

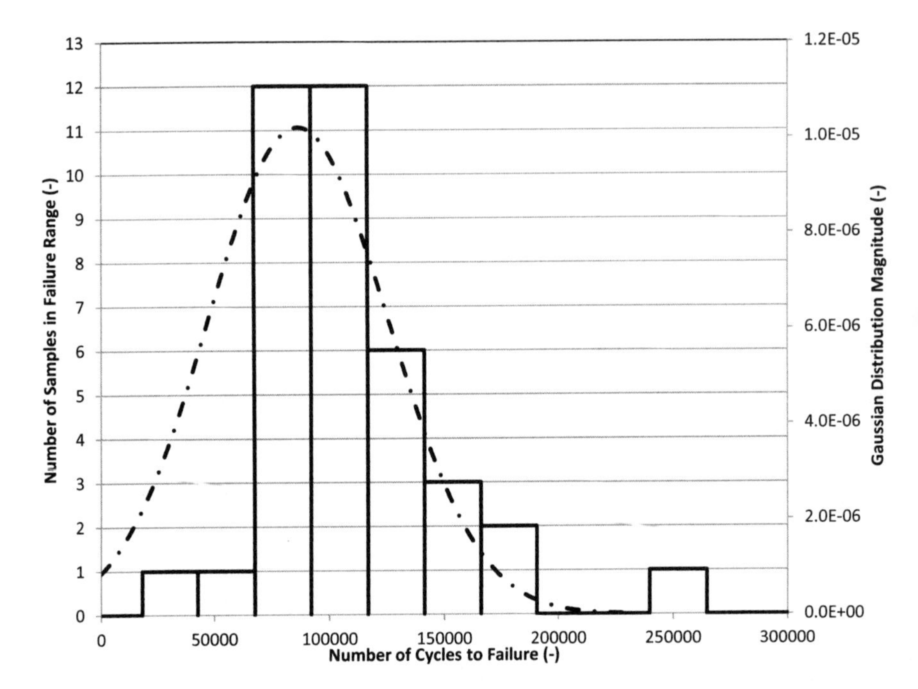

Figure 7. Probability density function for rotational bending fatigue data for 40 non-hydrogen exposed specimens. Data for 6.35 mm diameter 304 stainless specimens are shown. The square line represents the number of samples in a range. The average number of cycles to failure was 86500 cycles. The line represents a standard Gaussian distribution fit to the data points. The maximum bending stress correlated with a machine setting of 7.3 N-m, to give an estimated stress of 614 MPa.

The friction stir welded fatigue specimens failed at the weld during milling, due to lack of fusion. The control, non-welded specimens are currently being fatigue tested. Future friction stir welded specimens will be generated, to investigate the cause of lack of fusion and for comparison to the control specimens. Once the effect of the weld is understood on the fatigue life, the effect of hydrogen exposure will be compared.

CONCLUSIONS

Polished bending fatigue specimens of 304 stainless appear to have a longer fatigue life than that of as-plasma cut specimens. The effect of welding specimens is shown to cause failure at a nominally different location than that of an unwelded specimen. Further work is needed to investigate the effects of welding on the fatigue life, and to help elucidate the effects of welding parameters on the ferrite level of the microstructure and on the hydrogen embrittlement sensitivity.

Rotational bending fatigue data on austenitic stainless steel has been generated at a single rotational bending stress level of 614 MPa. The data will be compared with samples after exposure to hydrogen, at the same stress level.

Friction stir welding of aluminum alloys is being investigated. The proposed work is to compare the effect of friction stir welding and hydrogen exposure on fatigue life. Lack of fusion at the weld prevented testing of samples generated to date. Control samples (non-welded) are being prepared.

ACKNOWLEDGMENTS

The authors are grateful for the funding provided by Kyushu University, Tarbiat Modares University, Southeastern Louisiana University and Gonzaga University to conduct this work. The authors are also very grateful to Beau Grillo, James Moody, Steve Klemp and Verona Bravo in specimen preparation and equipment setup.

REFERENCES

[1]P. Ferro, Effect of Hydrogen on Bending Fatigue Life for Materials Used in Hydrogen Containment Systems, MS&T 2010 Conference, Houston TX, October 2010, Advances in Materials Science for Environmental and Nuclear Technology II, edited by S.K. Sundaram, T. Ohji, K. Fox, E. Hoffman, *Ceramic Transactions*, v. 227, pp. 39-49, American Ceramic Society Publication, Wiley ISBN 978-1-118-06000-1 (2011).

[2]P. Ferro et al., Fatigue Testing of Hydrogen-exposed Austenitic Stainless Steel, MS&T 2011 Conference, Columbus OH, October 2011, Columbus OH, Advances in Materials Science for Environmental and Energy Technologies, *Ceramic Volumes* (2012).

[3]B. Somerday et al., Hydrogen-assisted Crack Propagation in Austenitic Stainless Steel Fusion Welds, *Met. and Mat. Trans. A.*, **40A**, 2350-2362 (2009).

[4]C. San Marchi, B.P. Somerday, X. Tang, G.H. Schiroky, Effects of Alloy Composition and Strain Hardening on Tensile Fracture of Hydrogen-precharged Type 316 Stainless Steels, *Int'l J. Hydrogen Energy,* **33,** 889-904, (2008).

[5]Y. Mine, K. Tachibana, Z. Horita, Effect of High-Pressure Torsion Processing and Annealing on Hydrogen Embrittlement of Type 304 Metastable Austenitic Stainless Steel, *Met. and Mat. Trans. A*, **41A** (2010).

[6]A. Somasekharan, UTEP, El Paso TX.

3-D TIN-CARBON FIBER PAPER ELECTRODES FOR ELECTROCHEMICALLY CONVERTING CO_2 TO FORMATE/FORMIC ACID

Shan Guan; Arun Agarwal; Edward Rode; Davion Hill; and Narasi Sridhar
Research and Innovation, DET NORSKE VERITAS, INC.
Dublin OH, US

ABSTRACT

A plating process for fabricating three-dimensional tin-carbon fiber paper (CFP) electrodes has been developed. Major parameters such as electrolyte composition and pH have been optimized to provide a better tin coverage over carbon fiber compared to what we reported previously. A comparison to two commercial tin plating processes indicates that the new tin plating process is well suited for depositing tin onto porous substrates with a better uniformity and coverage. The Sn-CFP electrodes have been tested in a continuous flow cell for electrochemically converting CO_2 to formate or formic acid. Results show that using KCl-KOH electrolytes (Pt-Nb anode), at 3.75 V cell potential, the cathodic current density was 75 mA/cm^2 and faradic efficiency was 78% for converting CO_2 to formate. To convert CO_2 to formic acid, KCl-H_2SO_4 electrolytes (MMO anode) were chosen. The corresponding current density and faradic efficiency were 74 mA/cm^2 and 85% respectively at the same cell potential.

INTRODUCTION

There are essentially three strategies to reduce the accumulation of CO_2 in the atmosphere: reduce emissions by the use of energy efficient technologies and non-carbon fuels, long-term removal of CO_2 by storing it in stable geological media, and utilizing/converting the CO_2 into useful products that will otherwise require carbon sources [1]. It is likely that all these carbon capture and utilization (CCU) strategies will be employed in order to meet meaningful CO_2 reduction for future generations. CO_2 utilization technologies have the potential for immediate benefits to many industries, both to reduce CO_2 footprint and gain return on investment through product sales. Furthermore, CO_2 can be utilized as an alternative for hydrocarbon feedstocks for many chemicals [2-3]. Among different CCU technologies, electrochemical technology has attracted many attentions because it is modular, can be operated at ambient temperature, has the ability to make a variety of end products, and occupies a relatively small spatial footprint [4-5]. DNV has been involved in the in-depth development of electrochemical conversion of CO_2 to useful products such as formate/formic acid, which is referred to as ECFORM (Electrochemical Reduction of Carbon Dioxide to **Form**ate/**Form**ic Acid). Research interests are especially focused on developing electrochemical conversion of CO_2 to formic acid, where CO_2 can be utilized as a storage medium for renewables [6-7]. Formic acid is a perfect candidate to use as feedstock for electrofuels technology because it contains 53 g L^{-1} hydrogen (4.3 wt. % of H_2) at room temperature and atmospheric pressure, an equivalent density of 350 atm H_2. Furthermore, formic acid is in demand and can be used as energy storage medium, chemical feedstock, steel pickling, antibacterial agents, and deicing solutions [8-9].

To realize an economically viable electrochemical process, cathode is one of the most critical factors. The cathode can be either the catalyst or a substrate coated with a catalytic material on the surface of which the CO_2 conversion took place. An ideal electrode should possess at least the following aspects of technical advantages: high current densities (>100 mA/cm^2), high current efficiencies for target products (>70%), long catalyst life (>4000 hours), low cost, easy fabrication and be able to scale-up. These requirements will ensure the CO_2 conversion process to produce more useful products while consume less energy.

To electrochemically convert CO_2 to formate/formic acid, Tin (Sn) is a preferred catalyst due to its high current efficiency that was reported to be nearly 100% in some cases [10]. Depends on the design of reactors, Sn electrodes can be fabricated through different processes including electrochemical deposition (ECD), physical vapour deposition (PVD), impregnation or colloid method, and the modified gas diffusion electrodes (GDE) fabrication process. For each category, a variety of electrodes were reported that meet one or several aspects of the above requirements. PVD including sputtering and thermal evaporation can be applied to deposit thin films of Sn with high accuracy of film thickness and uniformity. In addition, when combined with photolithography technology, patterned thin films can be directly formed through PVD process. However, the cost for PVD is typically high because of the complexity of equipment and requirement for high vacuum systems. Furthermore, it is difficult to deposit thick film due to the low deposition rate and high residual stress associate with the PVD process [11]. Impregnation and colloid process have also been applied to fabricate Sn and other catalysts for battery and fuel cells. Sn compounds synthesized by these techniques require further reduction before they can be applied as catalysts. This was done either by in situ flow of hydrogen for several hours at a high temperature up to 600 ^{o}C, or was carried out in the solution by adding of excessive reducing agents [12-13]. These necessary post processes add the cost to the fabrication and make it difficult to scale-up. An alternative technique is to utilize the process similar to the fabrication of gas diffusion electrode for fuel cell application. To prepare such electrodes, suspended catalyst particles of Sn or Sn oxide were mixed up with a binder such as Nafion, and then, were applied to a substrate by hot pressing, brush painting or thermal spray. The current density of these electrodes is typically proportional to the loading of catalyst powders. Whipple et al. reported that a high current density of 100 mA/cm^2 for CO_2 conversion was obtained using a micro-fluidic reactor with an electrode area of 1 cm^2 [14]. However, health concerns remain for processing of Sn/Sn Oxide particles especially when nano-grainsized powers were used. Also, it has been a challenge to scale up such electrodes for a commercial electrochemical process; as most of the reports to date are dealing with small electrode with a surface area less than 5 cm^2.

Electrochemical deposition including electroplating and electroless plating is a mature technique for depositing metal and metal alloys. ECD process has a fast deposition rate that can be up to several μm/min. It does not require a vacuum system, is relatively low cost, and can deposit a variety of materials [15]. Furthermore, by incorporating with a suitable substrate such as porous materials of carbon paper, graphite or metallic foams, three-dimensional (3-D) electrodes can be achieved. Those electrodes have a real surface area that can be hundreds times more than the superficial area. As a result, they can generate large currents for electrochemical reaction, which is particularly well suited for the ECFORM application. For example, three-dimensionally ordered macro porous Ni-Sn structures fabricated through electroplating were reported [16]. They were designed to use as anodes for lithium batteries. During testing, these electrodes exhibited an initial discharge capacity of 455 $mAhg^{-1}$. However, due to the poor adhesion of the deposits, after 100 charge/discharge cycles, films delaminated from the current collector and caused the discharge capacity drop to 50 $mAhg^{-1}$. Electroplating has also been applied to fabricate proton exchange membrane (PEM) electrodes for fuel cells. Kim et al. investigated using pulse electroplating technique to directly load platinum catalysts onto carbon black substrates to form electrodes. The properties of catalyst layer can be controlled by optimizing deposition parameters such as duty cycle and peak current density. As reported, these electrodes yield a relatively high current density of 380 mA/cm^2 at a cell voltage of 0.8 V [17-18]. Unfortunately, platinum is not a preferred catalyst for converting CO_2 to formate/formic acid. To date, technique of Sn electroplating on porous substrates has not been successfully developed because achieving films with appropriate material properties has proven difficult due to a number of factors, including coating uniformity and control of grain-size. In this paper, we report a new electroplating process that was created for making 3-D Sn electrode using carbon fiber paper as the substrate.

EXPERIMENTAL

Substrate Preparation

The substrates selected for Sn plating is Carbon Fiber Paper (CFP) with a nominal thickness of 375 μm (Toray, TGP-H-120). Figure 1 is an SEM image of carbon fiber paper substrate with the fiber diameter varied from 6 to 10 μm. The CFP was cut into samples with a dimensional size of either 4 cm x 4 cm or 30 cm x 30 cm for Sn plating. They were first immersed in a 50% HCl solution at the room temperature for 30 min. After rinsing with deionized (DI) water, they were ultrasonically cleaned in DI water for 3 min. Because of the hydrophobic property of CFP, they need to be immersed in a 50/50 Ethanol to DI water solution for 5 min before transferring into a plating tank for processing. This step can help electrolytes penetrate into the interior of substrate. To electroplate 4 cm x 4 cm samples, a potentiostat (Princeton Applied Research, VMP3) was used as the power supply and platinum (Pt) wires were selected as the anode. For the larger sample of 30 cm x 30 cm, a DC power supply (BK Precision, 1710A) was used and Pt-Nd mesh was used for anode. Constant current mode was applied during plating, and the current density was controlled to be within 1.5 to 2.5 mA/cm^2. The electrolyte was heated using an automatic water bath and its temperature was kept within 58-65 °C during plating. The thickness of the catalyst was typically from 5 to 10 μm, and it was controlled by the plating time. After plating, the electrodes were rinsed with warm water, DI water, and then were ultrasonically cleaned for 3 min. If Sn deposits peel off from the CFP during ultroasonic clean, the electrode will not use for CO_2 conversion testing. Finally electrodes were blow-dried with N_2 gas.

Figure 1. Carbon fiber paper, TGP-H-120.

Electrodes Characterization and CO_2 Conversion Testing

The surface morphology of Sn electroplated electrodes before and after CO_2 conversion testing was examined using Scanning Electron Microscopy (SEM, JEOL JSM840-A). The material compositions including the surface oxygen percentage were determined by an Energy Dispersive Spectrum (EDS, PGT/Bruker), and the film thickness of tin oxide was estimated using an X-ray Photoelectron Spectroscopy (XPS, Surface Science Labs SSX-100). The ECFORM performance of 3-D Sn-CFP electrodes was tested using a continuous flow cell previously reported by DNV [6]. Figure 2 is a schematic drawing of the flow cell that shows each component. This configuration consists of three chambers: cathode, anode and CO_2 gas. The cathode (3-D Sn-CFP electrode) and anode were separated by an ion exchange membrane.

The CO_2 gas was flowing in from the back of the cathode and reaching the reactive interface of the porous cathode. The catholyte is 2M KCl saturated with CO_2, and the anolyte is either 1 M KOH or 0.5 M H_2SO_4 depends on the production routes of CO_2 conversion. During the testing, catholyte and anolyte were flowing into each own chamber continuously. A constant voltage generated by a potentiostat (Bio-Logic, SP-130) was applied across the anode and cathode, and the corresponding current was recorded. Products mixed up with catholyte were collected for Faraday Efficiency (FE) analysis. The concentration of the formate ions was determined using an Ion Chromatography (DIONEX, LC20). To calculate FE for ECFORM, the following equation was used:

$$FE = \frac{I_i}{I_{total}} \quad (1)$$

Where I_{total} denotes the total cell current and I_i is the current for generating formate determined by the IC method.

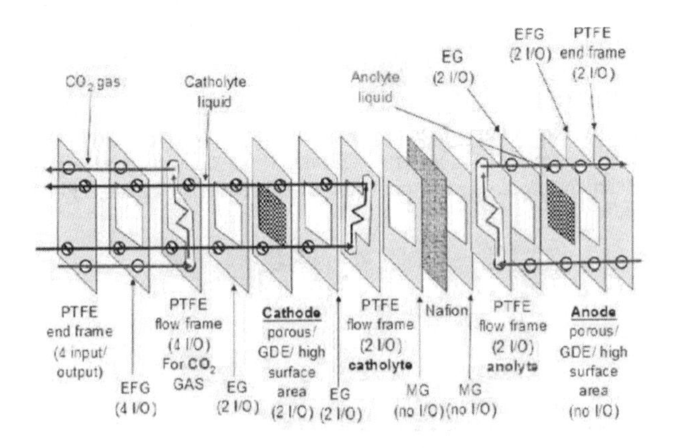

Figure 2. Schematic drawing of a flow cell for CO_2 conversion testing.

RESULTS AND DISCUSSIONS

Optimized Electrolyte and Plating Process

To maximize the CO_2 conversion capability of the catalyst and increase reactive current density, a uniform Sn coating that covers individual carbon fibers is preferred. Other coating properties are also important for the gas-liquid two-phase flow system such as the coating adhesion, surface oxide of the deposits, as well as the grain-size of electroplated Sn. In addition, the more carbon fibers plated with Sn, the more catalysts will be available for CO_2 conversion. This can not only increase current density but also improve FE of the ECFORM. It was found from our experiments that multiple factors can affect the critical properties of the electroplated Sn layers. Table 1 listed an optimized electrolyte composition and the plating process, as well as a comparison to the previous plating technique reported by DNV [6]. The optimized process was able to deposit a uniformed Sn layer that coats the carbon fiber paper to a thickness of about 50 μm on each side of the substrate. Also the Sn coating can easily pass the adhesion test, which was carried out by ultrasonically vibrating the samples at 200 W for 3 min or longer in the DI water.

Table 1. Plating process for making 3-D Sn-CFP electrodes[19]

	Optimized Process	Previous Reported Process [1]
$SnCl_2$•$2H_2O$ (g/l)	45	43
$K_4P_2O_7$ (g/l)	165	184
$C_2H_5NO_2$ (g/l)	15	12.5
$C_6H_8O_6$ (g/l)	10	
Polyoxyethylene (12) nonylphenyl ether (g/l)	0.1	
NH_4OH (g/l)		2.3
Electrolyte pH	4.5 (adjusted using diluted HCl)	8-9 (adjusted using diluted KOH)
Applied current density(mA/cm^2)	2.0	1.25
Electrolyte temperature (°C)	55-65	50-60

Figure 3A shows the uniformed Sn deposits on CFP obtained from the optimized process. This electrode is referred to as the "new electrode" for the rest of this paper. This is to distinguish it from the electrode we reported previously, which is described as the "old electrode". Figure 3B is an SEM image of an old electrode made using the previous reported process listed in Table 1. As can been seen from Figure 3B, the grain-size of Sn was larger compared to that of the new electrode, and Sn deposits aggregated to form some big crystals that poorly attached to the carbon fibers. Also some carbon fibers within the first layer of the substrate were exposed without Sn coatings. This will not only reduce the current density for CO_2 conversion, but also may affect the FE since carbon is not a good catalyst for producing formate or formic acid. For the optimized plating process, $SnCl_2$•$2H_2O$ is the source to provide Sn^{2+} for Sn plating. $C_6H_8O_6$ can prevent the oxidation of Sn^{2+} to Sn^{4+}, and thus increase the stability of the plating solution. $K_4P_2O_7$ and $C_2H_5NO_2$ act as complex agents in Sn plating to prevent the grain-size of Sn from growing too large. $C_2H_5NO_2$ can also reduce the surface tension of the carbon fiber paper, and help plating to take place inside the substrate. Non-ionic surfactant polyoxyethylene (12) nonylphenyl ether promotes the release of the hydrogen bubbles generated from the cathode process; as the result, a dense Sn layer free of pinholes demonstrated by Figure 4A can be obtained. During the technology development, we also tested several commercially available Sn plating processes and found none of them were suitable for making 3-D Sn-CFP electrodes. At our request, two independent plating companies (because of confidentiality agreements, their names will not be released) to make 3-D Sn-CFP electrodes using their own Sn plating processes and the CFP substrates provided by DNV. Both plating companies were using an acidic Sn plating bath at our request. Figure 4B and C show the surface morphology of 3-D Sn-CFP electrodes they delivered that was examined using SEM. As can be seen from the images, these plating processes formed large islands of Sn deposits on the carbon fiber paper. Some Sn deposits were big enough to block the CO_2 gas flow. Similar to DNV's old electrode, there were some carbon fibers on the top 1-2 layers without Sn coatings. This will cause similar problems for CO_2 conversion as we discussed above. And thus, it is safe to say that these commercial plating processes are not suitable for making 3-D Sn-CFP electrodes without major revisions.

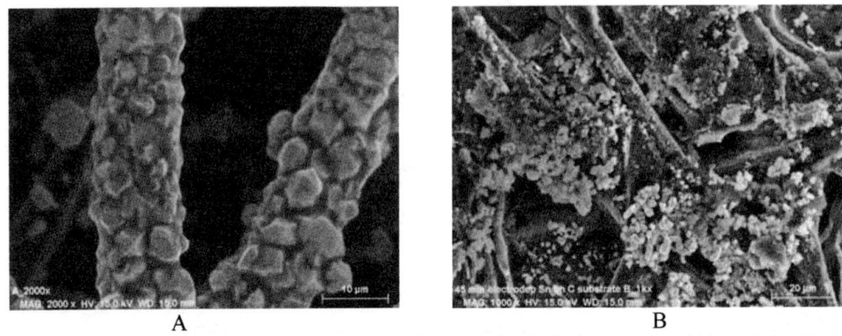

A B

Figure 3. 3-D electrodes fabricated using the optimized plating process (A), and previously reported plating process (B).

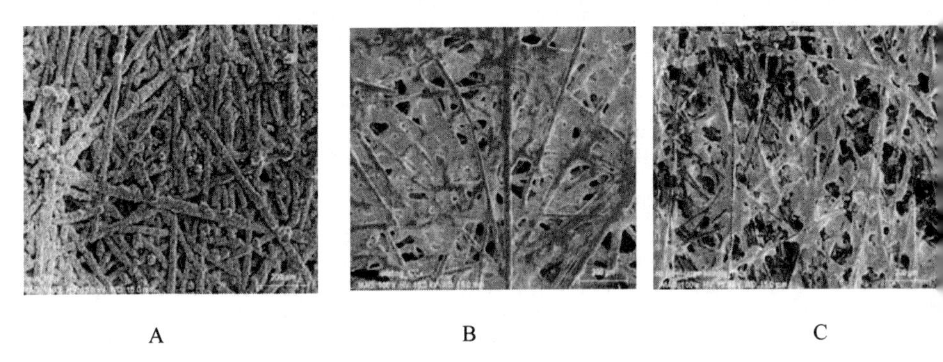

A B C

Figure 4. A comparison of electrode fabricated by the DNV optimized Sn plating process (A) to processes from two independent plating workshops (B and C).

Although metallic Sn is relatively stable in the air, it still tends to oxidize to form SnO_2 and SnO_4 [21], which are poor catalysts for CO_2 conversion. Sn deposits without oxide layer are highly preferred for our application. Fresh (hours after Sn plating) 3-D Sn-CFP electrode was analyzed using EDS, and results show that the Sn surface is free of oxide (Figure 5). In addition, electrodes placed in the air for more than 2 months were examined using an XPS that has a beam size around 600 μm. Both XPS survey scan and high resolution scan for the elements of interest (Figure 6a and 6b, respectively.) show that the surface of the new electrode is relatively clean with a very thin layer of oxide. The Sn 3d spectra of the new electrode found the presence of both Sn^{+4} (or Sn^{+2}) and Sn^0. This is an indication that the surface oxide is relatively thin. The thickness of the surface oxide layer can be estimated from the $Sn^{+2,+4}$ / Sn^0 using published electron mean free path values. It was estimated that the tin oxide thickness is ~5 nm.

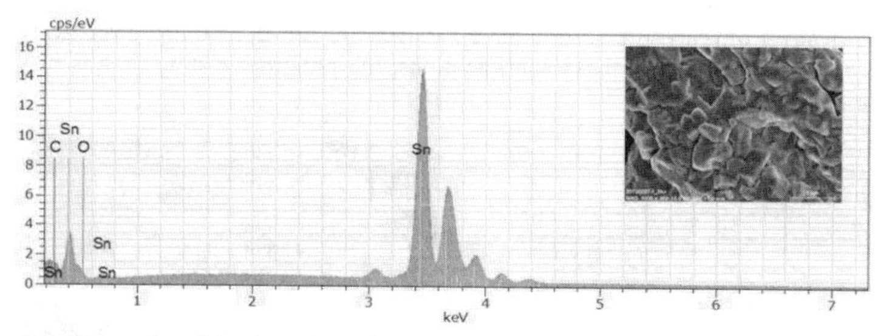

Figure 5. EDS results of Sn deposits as plated from the optimized Sn plating process show zero surface oxide (materials composition: 98.32% tin and 1.68% carbon).

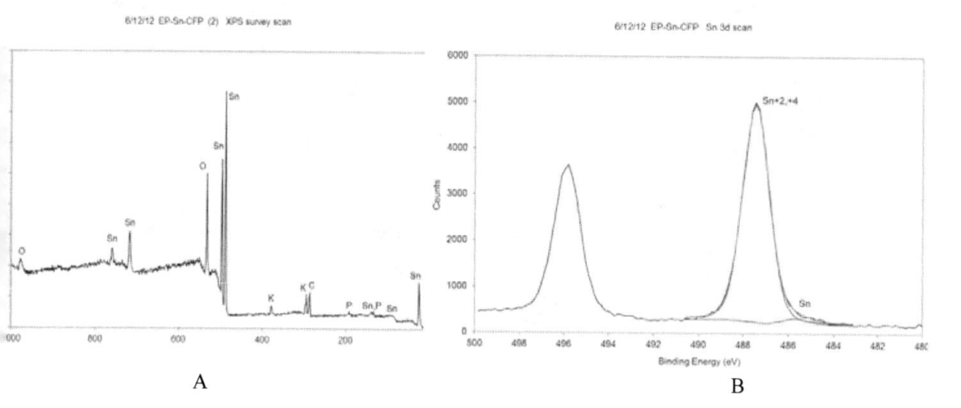

A B

Figure 6: XPS scan of "new electrode" that was place on shelf for 2 months: A)XPS survey scan, and B)Sn 3d high resolution scan.

It is important to control the grain-size of electroplated Sn in the process of fabricating 3-D Sn-CFP electrode. For the ECFORM application, for the same coating thickness, catalyst with a smaller grain-size is corresponding to a larger surface area, which can provide more surfaces for electrochemical reaction (CO_2 conversion). New electrode made using the optimized Sn plating process shows smaller grain-size compared to the old electrode, and this was confirmed using SEM. As can be concluded from Figure 3A, the Sn grain size is about 1-3 μm after plating for 60 min. Figure 7 demonstrated the surface morphology of deposits under different plating time intervals, i.e. 5, 10, 15, and 120 min. The Sn grain-size becomes larger as the plating time increased. Although smaller grain size is preferred for CO_2 conversion, the thickness of the deposits needs to be above 5 μm in order to increase the stiffness of the electrode for handling. Furthermore, the cathode chamber of the ECFORM reactor is a complex liquid-gas two-phase flow system. The pH of the electrolyte is acidic when 0.5 M H_2SO_4 was applied as the anolyte. Under these conditions, corrosion/erosion will occur on the surface of electrode although the rate is quite slow. Figure 8 are SEM images of electrode before and after CO_2 conversion testing. This electrode has a 600 cm^2 superficial surface area and was tested for CO_2 conversion for more than 4 hours. A quick comparison shows that there was Sn loss especially at the shaper edges of Sn crystals. Because the catalysts need to have a longer life above 4000 hrs, a relatively thicker Sn layer is necessary. Based on these considerations, the preferred plating time is between 60 and 120 min.

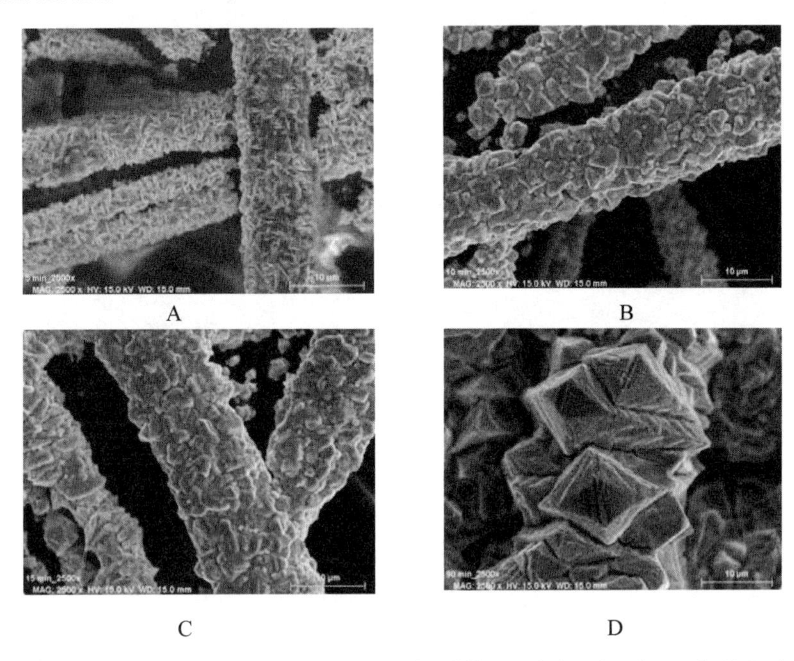

Figure 7. Surface morphology of electrode under different plating time interval, A) 5 min; B)10 min; C) 15 min; and D) 120 min.

Figure 8. Surface morphology of electrodes A) before, and B) after 4 hrs CO_2 conversion testing.

ECFORM Testing

Using Sn as a catalyst, CO_2 can be electrochemically converted to either formate or formic acid depending on the compositions of electrolyte (especially anolyte). When 2M KCl was applied as the catholyte and 1M KOH was used for anolyte, CO_2 can be converted to formate via the following electrochemical processes [6]:

Cathode reactions are:

$$CO_2\,(aq) + H^+ + 2e^- \rightarrow HCOO^-\,(aq) \quad (2)$$

$$CO_2\,(aq) + 2H^+ + 2e^- \rightarrow CO\,(g) + H_2O \quad (3)$$

$$2H^+ + 2e^- \rightarrow H_2\,(g) \quad (4)$$

And the anode reaction is:

$$4OH^- \rightarrow 2H_2O + O_2 + 4e^- \quad (5)$$

When anolyte was replaced by 0.5M H_2SO_4, the anode reaction becomes:

$$2H_2O \rightarrow 4H^+ + O_2 + 4e^- \quad (6)$$

The CO_2 conversion process using KCl-KOH electrolytes has been investigated extensively [6-7, 10, 21-22]. DNV had reported electroplated Sn electrodes for this application, which is referred to as "old electrode". These electrodes yield a cathodic current density of 45 mA/cm^2 and a faradic efficiency above 75% at a cell potential of 3.75 V. In the long term testing, the old electrodes kept an average FE above 60% while the current density was stable for 4 days [6]. Using the optimized Sn plating process, new electrodes of 3-D Sn-CFP were made. As a comparison, these new electrodes were also tested in the KCl-KOH electrolytes using the continuous flow cell described in the experimental section. Figure 9 is the performance of both new and old electrodes using KCl-KOH electrolytes for CO_2 conversion. Results show that new electrode exhibited superior properties in the KCl-KOH electrolytes system. When using Pt-Nd mesh for anode, the new electrode gave a current density of 75 mA/cm^2 at 3.75 V. While at the same cell voltage, the current density for old electrodes was about 45 mA/cm^2. This represents an increase of nearly 70%. It is worthwhile to note that due to the porosity of the electrode (3-dimentional); the current density discussed here was in fact the superficial current density, which is defined as the total current divided by the superficial electrode area. The improvement in the current density is attributed to the fact that new electrode has a uniform and dense Sn layer that completely covers each carbon fibers on the CFP surface. In addition, new electrodes show hydrophilic surface properties, which help the electrolyte to wet the surface, and as the result, a higher current density can be expected.

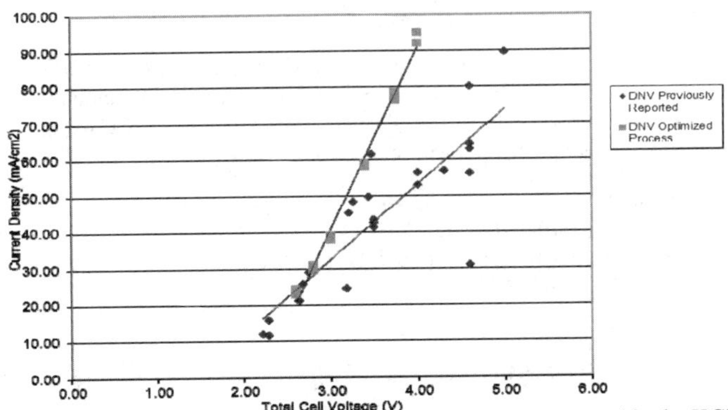

Figure 9. Improvement of current density for the optimized process tested in the KCl-KOH electrolytes.

Faradaic efficiency (FE) is an important parameter to the goal of an economically viable CO_2 conversion process. The higher the FE, the more current was utilized to convert CO_2 to useful products such as formate or formic acid. From the experiment, it was found that the new electrode shows a higher FE compared to the old electrode at the same cell voltage. For example, the FE at 3.75 V is about 90% while for the old electrode it is in the lower range of 80%. One reason is that new electrode has a better Sn coverage over the carbon fibers while for old electrode, some carbon were exposed without coating. Another explanation is that the catalyst of new electrode has fewer impurities such as tin oxide.

Using the improved plating process, we were able to fabricate 3-D Sn-CFP electrodes with a surface area of 600 cm^2 and above. Semi-scale up was carried out in the well-established KCl-KOH electrolytes. Testing results show that the current density is generally lower than those recorded in the small flow cell testing. At a cell voltage of 3.5 V, the recorded current density was between 30 to 35 mA/cm^2. This reduction in the current density is probably due to the CO_2 diffusion issue associated with the larger reactor. The KOH-KCl is a relatively stable electrolyte system that yields both high current density and FE for CO_2 conversion using the new electrodes. However, the disadvantage of this system is the high cost for chemicals. From reaction (5), it is easy to understand that KOH is a consumable chemical because OH^- was oxidized to oxygen and water. This cost barrier prevents it from becoming an economically viable process without considering any government subsidy through carbon tax. An alternative is to use acidic anolyte such as 0.5M H_2SO_4. In this case, the anodic reaction follows equation (6) which is the electrolysis of water. If anolyte is recirculated and water is added periodically, there will be no chemical consumption for the anodic process. This can greatly reduce the cost for ECFORM. Furthermore, the primary useful product for the KCl-KOH system is formate salt, which is generally applied for deicing (Sodium Formate), drilling fluids (Cesium formate) or chemical feedstocks. To produce energy storage chemicals such as formic acid, post processing steps are necessary which further increases the total cost. For the acidic process (KCl-H_2SO_4 electrolyte system), the useful cathodic product is formic acid, which is a well-known chemical for energy storage for fuel cell and battery. A major issue with the acidic system is potential corrosion of catalysts. The catholyte is strong acidic with a pH around zero. Sn catalysts can be etched in these solutions even under a high cathodic polarization. Another noteworthy problem for the acidic anolyte is the lower current density compared to alkaline anolyte at the same cell voltage, and hence, more energy is required to produce the same amount of products. Figure 10 plotted the flow cell testing results of KCl-H_2SO_4 system using Pt-Nd mesh as the anode. Compared to the alkaline system, under the same cell voltage, the current density is much lower. For example, at a cell voltage of 3.25 V, the current density for the acidic system is only 25 mA/cm^2, while it is over 70 mA/cm^2 for the alkaline system. Therefore, in order to save energy, new method to lower the total cell voltage must be explored. A straightforward approach is to reduce the anodic voltage drop in the ECFORM process, which can be realized by adopting a new anode material, MMO (Mixed Metal Oxide, Ti/Ta_2O_5/IrO_2, NMT Electrodes Pty Lta) to the system. This electrode material is stable in the 0.5M H_2SO_4 acidic anolyte, and can effectively reduce the voltage drop on the anode. As the result, a relatively high current density for CO_2 conversion can be obtained without further increasing the cell voltage. Figure 10 also recorded the current density of KCl-H_2SO_4 system using MMO anode, which shows a sharp increase compared to that using Pt-Nd anode. For example, after replacing anode with MMO, at a cell voltage of 3.5V, the current density increased from 20 mA/cm^2 to about 60 mA/cm^2. Further increasing the cell voltage to over 4.0 V, the current density will exceed 100 mA/cm^2. However a high cathodic polarization (corresponding to a high cell voltage) will promote the hydrogen evolution reaction (eq. 4) and as a result lower the FE. Figure 11 reveals the relationship between cell voltage and faradaic efficiency for both acidic and alkaline systems. For both systems, the FEs are in the higher range of 80% when the applied cell voltages are within the range of 3 to 3.75 V, which can be regarded as the preferred range for the ECFORM applications.

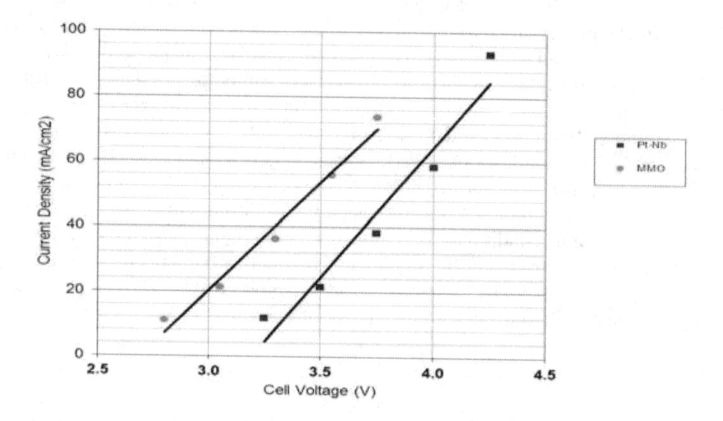

Figure 10. Current density vs. cell voltage with different anodes in the KCl-H_2SO_4 electrolytes.

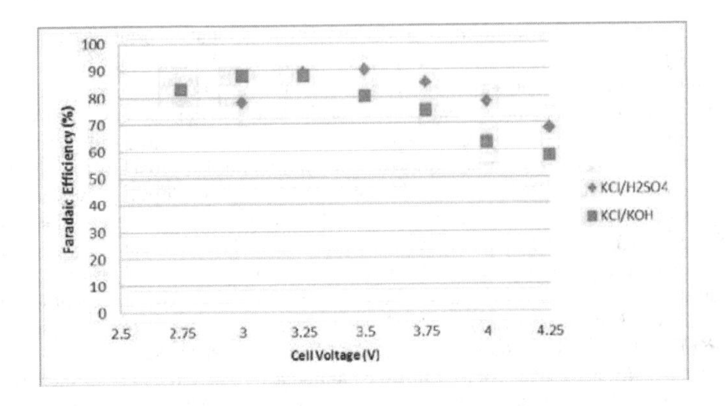

Figure 11. Cell voltage vs. FE for both KCl-H_2SO_4 and KCl-KOH electrolytes systems.

CONCLUSION

We created a Sn electroplating process that is well suited for making 3-D Sn-CFP electrodes for electrochemically converting CO_2 to useful products such as formate or formic acid. The new process shows advantages over commercially available plating processes evaluated by DNV for ECFORM application. Micro surface morphology of the electrodes made using the optimized Sn plating show a dense and uniform Sn layer that covers individual carbon fibers to a depth of about 50 μm. EDS and XPS results confirm that the surface of as-plated electrode is clean, and a very thin layer of Sn oxide formed after two-month shelf time. Flow cell testing show that these electrodes performed better than what we reported previously using KCl-KOH electrolytes. When tested in the KCl-H_2SO_4 electrolytes, the current density is generally lower compared to the alkaline conditions. This reduction can be compensated by utilizing MMO as the anode. As the result, a current density of about 75 mA/cm^2 was obtained using KCl-H_2SO_4 electrolytes at a cell potential of 3.75 V. Those electrodes exhibit fairly high faradic efficiencies in both electrolyte systems within the cell voltage range of 2.5 V to 3.75 V.

REFERENCES

1. M. Aresta, Carbon Dioxide: Utilization Options to Reduce its Accumulation in the Atmosphere, in *Carbon Dioxide as Chemical Feedstock* (eds M. Aresta), Wiley-VCH Verlag GmbH & Co. KGaA, Weinheim, Germany (2010).
2. M. Aresta, *Enzymatic and Model Carboxylation and Reduction Reactions for Carbon Dioxide Utilization* (eds M. Aresta and J.V. Schloss), Kluwer Academic Publishers , Dordrecht, The Netherlands (1990).
3. M. Aresta , *Carbon Dioxide Capture and Storage* (ed. M. Maroto -Valer), Woodhead Publishing Limited , Abington Hall, Granta Park, Cambridge, CB21 6AH, UK(2009).
4. L. J. J. Janssen, and L. Koene, The Role of Electrochemistry and Electrochemical Technology in Environmental Protection, *Chemical Engineering Journal*, **85**(2–3), 137-146 (2002).

5. N. Masuko, T. Ōsaka, and Y. Itō, *Electrochemical technology: Innovation and new developments*. Tokyo: Kodansha (1996).
6. A.S. Agarwal, Y. Zhai, D. Hill, and N. Sridhar, The Electrochemical Reduction of Carbon Dioxide to Formate/Formic Acid: Engineering and Economic Feasibility. *ChemSusChem*, **4**, 1301–1310(2011).
7. A. S. Agarwal, S. Guan, Y. Zhai, E. Rode, D. Hill, and N. Sridhar, Formic Acid and Formate Production through Electrochemical Reduction of CO_2 – An Assessment of Technology and Challenges, Abstract 1499, 220th ECS Meeting, Oct. Hawaii.
8. F. Joó, Breakthroughs in Hydrogen Storage – Formic Acid as a Sustainable Storage Material for Hydrogen, *ChemSusChem*, **1**, 805–808(2008).
9. P. G. Jessop, in *Handbook of Homogeneous Hydrogenation* (Eds.: J. G. de Vries, C. J. Elsevier), Wiley-VCH, Weinheim, Germany (2007).
10. C. Oloman, and H. Li, Electrochemical Processing of Carbon Dioxide, *ChemSusChem,* **1**, 385–391(2008).
11. S. Guan, and B. J. Nelson, Magnetic Composite Electroplating for Depositing Micromagnets, *J. MICROELECTROMECHANICAL SYS.*, **15**(2), 330-337 (2006).
12. H. Bönnemann, W. Brijoux, R. Brinkmann, T. Joußen, B. Korall, and E. Dinjus, Formation of Colloidal Transition Metals in Organic Phases and Their Application in Catalysis. *Angew. Chem. Int. Ed. Engl.*, **30**, 1312–1314(1991).
13. M. Götz, and H. Wendt, Binary and Ternary Anode Catalyst Formulations Including the Elements W, Sn and Mo for PEMFCs Operated on Methanol or Reformate Gas, *Electrochemica Acta*, **43**(24), 3637-3644(1998).
14. D. T. Whipple, E. C. Finke, and P. J. A. Kenis, Microfluidic Reactor for the Electrochemical Reduction of Carbon Dioxide: The Effect of pH, *Electrochemical and Solid-State Letters,* **13** (9), B109-111 (2010).
15. W. Ruythooren, K. Attenborough, S. Beerten, P. Merken, J. Fransaer, E. Beyne, C. Van Hoof, J. De Boeck, and J. P. Celis, Electrodeposition for the synthesis of microsystems, *J. Micromechanics and Microengineering,* 10(2), 101-107(2000).
16. K. Nishikawa, K. Dokko, K. Kinoshita, S. W. Woo, and K. Kanamura, Three-dimensionally Ordered Macroporous Ni–Sn Anode for Lithium Batteries, *J. Power Sources*, **189**(1), 726-729(2009).
17. H. Kim, N. P. Subramanian, and B. N. Popov, Preparation of PEM Fuel Cell Electrodes Using Pulse Electrodeposition, *J. Power Sources*, **138**(1-2), 14-24(2004).
18. H. Kim, and B.N. Popov, Development of Novel Method for Preparation of PEMFC Electrodes, *Electrochemical and Solid-State Letters*, **7**(4), A71-74 (2004).
19. Y. Zhai, S.Guan, N.Sridhar, and A. S. Agarwal, Method and Apparatus for the Electrochemical Reduction of Carbon Dioxide, WO2012040503.
20. M. Schwartz, Tin and Alloys, Properties, Encyclopedia *of Materials, Parts and Finishes* (2nd ed.), CRC Press (2002).
21. H. Li, and C. Oloman, Development of a Continuous Reactor for the Electro-reduction of Carbon Dioxide to Formate– Part 1: Process Variables, *J. APPL. ELECTROCHEM,* **36**(10), 1105-1115(2006).
22. H. Li, and C. Oloman, Development of a Continuous Reactor for the Electro-reduction of Carbon Dioxide to Formate – Part 2: Scale Up, *J. APPL. ELECTROCHEM,* **37**(10), 1107-1117(2007).

ANALYSIS OF THE THEORY OF FREQUENT CHARGE COLLAPSES IN A 25500KVA HERMETIC CALCIUM CARBIDE FURNACE

Hui SUN, Jian-liang ZHANG, Zheng-jian LIU, Ye-xiao CHEN, Ke-xin JIAO, Feng-guang LI
State Key Laboratory of Advanced Metallurgy, University of Science and Technology Beijing
Beijing 100083, China

ABSTRACT

The phenomena of charge collapses occur frequently during an investigation of the manufacturing process of a 25500KVA hermetic furnace in a calcium carbide plant, and the normal manufacturing works are affected seriously. Frequent charge collapses and large ones not only destroy the charge layer structure and dissipate a lot of enegy of reaction zone, but also result in a low quality of calcium carbide, a low reaction temperature and a hard discharge of the calcium carbide liquid. Based on the property of the charge during calcium carbide manufacturing, the paper discusses the mechanism of the charge collapses and analyses the causes of the frequent charge collapses in a hermetic calcium carbide furnace with the standpoint of the charge and electrodes operation, and in purpose of providing operable suggestions for preventing and reducing the occurrence incidence of the frequent charge collapses in practice.
KEY WORDS: hermetic calcium carbide furnace; frequent charge collapses; charge layer structure; calcium carbide liquid

1. INTRODUCTION

Hermetic calcium carbide furnace is used widely in the manufacturing of calclium carbide nowadays, and it can bring a good and safety environment compared to the semi-hermetic and opened ones[1]. What's more, the dust generated insides the furnace can be recovered effectively by extraction of the waste gas for the manufacturing of cement, glass and so on. After that, CO gas, as the exhaust gas, can be reused for heating the charge after combustion and can decrease the dissipation of energy from reaction zone as well as life of the furnace by the overflow of CO simultaneously.

Although hermetic calcium carbide furnace can realize the principle of safety and cleaning manufacturing of calcium carbide, it also has its own disadvantages in some occasions when the operation of electrodes inserted into the charge was introduced. During an investigation of the manufacturing process of a 25500KVA hermetic calcium carbide furnace in a plant, a frequent occurrence of the charge collapses was found due to the properties of the charge and the electodes operation. When the charge collapsed, on the one hand, a lot of gas burst into the top of the hermetic furnace from the cavity below which caused an increase of top pressure, on the other hand, some gas got out from the gap in the wall of it and burnt immediately can destroy the cooling facility and the electric device. Especially, a large-scale charge collapes will impact the electodes and make them soft broken or hardening broken to some extent[2]. Therefore, discovering the theory of the charge collapses and the causes of the frequent occurrence become more and more necessary.

2. ANALYSIS OF THE PROCESS OF CALCIUM CARBIDE MANUFACTURING

2.1 Process of calcium carbide manufacturing

As shown in Fig 1 below, the manufacturing of calcium carbide includes four independent processes, such as charge preparation, charge delivered, calcium carbide generation and calcium carbide liquid output.

From Fig 1 we can see that the charge for producing calcium carbide are consisted of carbon materials (or coke) and lime. Unlike traditional process which delived limesone into the hermetic furnace directly, there is a speclial equipment named lime kiln heated by the energy from a reaction of high-temperature gas from hermetic calcium carbide furnace with air which can produce a flame with a temperature of more than 1250 ℃ for generating lime by a chemical decomposition reaction $CaCO_3(s) \rightarrow CaO(s)+CO_2(g)$ at the temperature of 839℃. Many kinds of carbon materials besides coal and coke can be used in calcium carbide manufacturing, such as semi-coke, calcined coal, anthracite and so on, and the moisture content of semi-coke and anthracite needed to be decreased within the range of 2～4% by drying before weighing and mixing.

The size of the charge used plays an important role in practice, too large or too small (such as powder) will impact the permeability of the charge layer dramatically, therefore, the large ones should be broken into pieces within the size in the range of 10～40mm, and the small ones removed thoroughly. Then mix the carbon materials and lime at a ratio around 0.6 due to the difference of compositions and deliver them into the feeder storage room by herizontal conveyor belts and slope conveyor belts to annular feeder device successively after weighing by an electronic apparatus, and the charge drops into the hermetic calcium carbide furnace through the tubes inserted into the charge layer continuously.

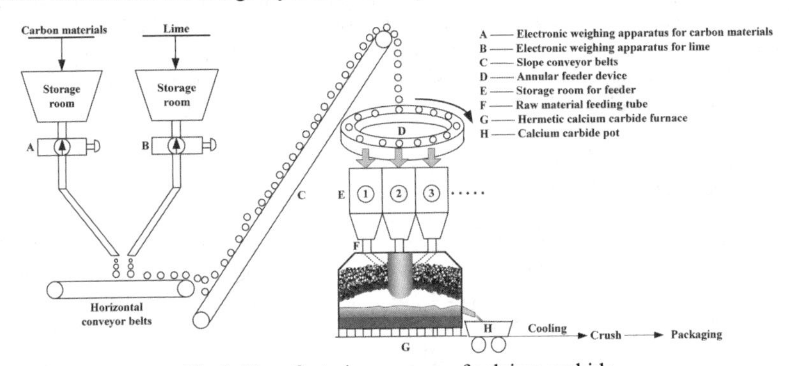

Fig 1. Manufacturing process of calcium carbide

When the charge are suplied, continuous manufacturing will do, and the red calcium carbide liquid are sliding down from the soft zone to the melted. The calcium carbide liquid will turn into block after cooling, and the final products will be used in industry by crush and packaging.

2.2 Theory of calcium carbide manufacturing

A main chemical reaction for calcium carbide generation takes place at a high temperature ranging from 1700℃ to 1820℃ due to the energy derived from the electric arc and resistance as shown in Eq. (1)[3-4], and many people take it as the major reaction for the manufacturing of calcium carbide[5-7].

$$CaO(s)+3C(s) \rightarrow CaC_2(s)+CO(g)+466kJ \quad (1)$$

But the traditional opinion insists that the main reaction includes two independent and continuous ones, which can be written as follows[8]:

$$CaO(s)+C(s)\rightarrow Ca(g)+CO(g)+523.79kJ \quad (2)$$

$$Ca(g)+2C(s)\rightarrow CaC_2(s)-59.41kJ \quad (3)$$

In reaction zone, the carbon materials react with lime and generate Ca vapor and CO gas, a small amount of Ca vapor and much of CO gas go up through the charge layer into the top room of the hermetic furnace,and most Ca vapor existed in the surface or the gap of the carbon materials for further reaction with them, and CaC_2 generated consequently.

Table 1. Additional Reactions in Reaction Zone

Reaction Number(k)	Additional Reactions	Energy Absorption and Loss(kJ)
1	$CaCO_3(s)\rightarrow CaO(s)+ CO_2(g)$	178.5
2	$CO_2(g)+C(s)\rightarrow 2CO(g)$	164.6
3	$SiO_2(s)+2C(s)\rightarrow Si(l)+2CO(g)$	573.6
4	$Fe_2O_3(s)+3C(s)\rightarrow 2Fe(l)+3CO(g)$	452.5
5	$Al_2O_3(s)+3C(s)\rightarrow 2Al(l)+3CO(g)$	1218.4
6	$MgO(s)+C(s)\rightarrow Mg(g)+CO(g)$	486.04
7	$H_2O(g)+C(s)\rightarrow H_2(g)+CO(g)$	46.75
8	$Ca(OH)_2(s)\rightarrow CaO+H_2O(g)$	108.94

At the same time, there is also some additional reactions during the process which can be listed in Table 1. Undoubtly, all the additional reactions are adopting energies which result in a large amount of energy loss additional. Several kinds of gases, such as H_2 gas, CO gas, CO_2 gas and other gases generated in reaction zone are of high temperature, and a lot of sensible heat loss when getting through the charge layer.

3. THEORY ANALYSIS OF CHARGE COLLAPSES

3.1 Properties of charge

In hermetic calcium carbide furnace where exists two material flows moving opposite direction, one is the burden flow downward from the surface of the charge, and the other one is the gas flowing upward through the charge layer from reaction zone, just the interaction of the two processes promotes the manufacturing of calcium carbide. An investigation are carried out for frequent charge collapses based on the properties and the mechanical analysis of the charge.

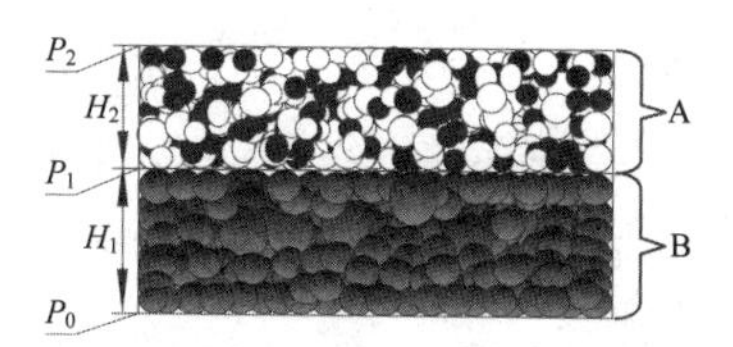

Fig 2. The layer structure of the charge in hermetic furnace

Fig 2 shows the entire charge layer in hermetic calcium carbide furnace where A is a mixture zone of the carbon materials with lime and B is a soft zone of the carboon materials with melted lime.

To calcium carbide manufacturing, a good permeability of the charge layer is of great significance to every supervisor. When the carbon materials and lime mixed and delivered into the hermetic furnace forms A, the permeability can be reflected by pressure drop which is calculated by Ergun equation listed in Eq.(4).

$$\frac{\Delta P}{H}=150\frac{\eta\omega(1-\varepsilon)^2}{(d_e\phi)^2\varepsilon^3}+1.75\frac{\rho\omega^2(1-\varepsilon)}{\phi d_e\varepsilon^3} \tag{4}$$

Where ΔP is the pressure drop of the gas flowing upward through the charge layer, ΔP=P_1-P_2, ω is the actual velocity of it under given temperature and pressure, ρ refers to the density of it, H and ε stand for height and porosity of A, and H=H_1, d_e and Φ are equivalent diameter and sphericity of the charge particles of A respectively.

As to Eq.(4), the first iterm can be neglected because of high velocity of the gas flowing upward through A from below, so the simplified Ergun equation can be written as:

$$\frac{\Delta P}{H}=1.75\frac{\rho\omega^2(1-\varepsilon)}{\phi d_e\varepsilon^3} \tag{5}$$

or

$$\frac{\Delta P}{H}=1.75\frac{1-\varepsilon}{\phi d_e\varepsilon^3}\rho\omega^2 \tag{6}$$

The main factor affecting the pressure drop can be divided into two parts, the first part $\frac{1-\varepsilon}{\phi\cdot d_e\cdot\varepsilon^3}$ stands for speciality of the charge particles of A, and $\rho\omega^2$ refers to the status of the gas, and the permeability index derived from the Eq.(6) is

$$K=\frac{\rho\omega^2}{\Delta P/H}=0.5714\frac{\varepsilon^3 d_e\phi}{1-\varepsilon} \tag{7}$$

Where, K is permeability index of A and a characterization of permeability.

From Eq.(5) and Eq.(7), a conclusion can be found that the pressure drop of the unit height of A is a negative correlation with the sphericity of the charge particles and the equivalent diameter of the charge of A as well as the permeability index of the charge layer of it. When the porosity increases, the pressure drop of the unit height of the charge layer of A decreases and the permeability index increases which can lead to a good operating environment with low abnormal conditions.

As to B, the assessment of permeability of the charge layer is different from A, and the pressure drop can be described in Eq.(8).

$$\frac{\Delta P}{H}=f_s\frac{\rho_s\omega^2}{2g\phi d_s}\frac{1-\varepsilon_s}{\varepsilon_s^3} \tag{8}$$

Where, ΔP is the pressure drop of the gas flowing upward through B, and ΔP=P_0-P_1, f_s is the drag coefficient, and $f_s=3.5+44S_r^{1.4}$, S_r, d_s are shrinkage and equivalent diameter of the carbon materials and melted lime in soft zone, and $S_r=1-\frac{H_s}{H_0}$, H_0 and H_s are the heights of B before and after shrinking respectively, and H=H_1=H_s, ε_s is the porosity of B, and $\varepsilon_s=1-\frac{\rho_s}{\rho_0}$, ρ_0 and ρ_s are densities before and after shrinking, g is acceleration of gravity. As shown in Eq.(8), the pressure drop of B is in agreement with the shrinkage of it.

Above all, the total pressure drop of the entire charge layer can be written as below.

$$\frac{\Delta P}{H}=\alpha f_s \frac{\rho_s \omega^2}{2g\phi d_s}\frac{1-\varepsilon_s}{\varepsilon_s^3}+1.75(1-\alpha)\frac{1-\varepsilon}{\phi d_e \varepsilon^3}\rho\omega^2 \tag{9}$$

Where

$$\alpha=\frac{H_1}{H_1+H_2} \tag{10}$$

According to an experience before, the pressure drop of B is hundreds of times that of A in general. Assume that the pressure drop per unit height of A is about 10 Pa/m, and that of B is 500 times of A, the relation between pressure drop and H1, H2 can be described in Fig 3.

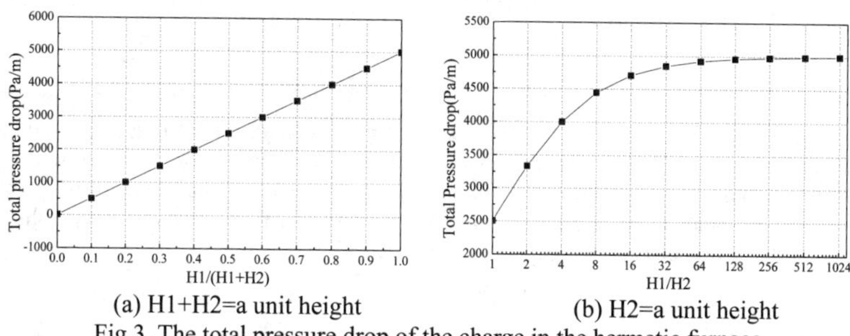

(a) H1+H2=a unit height (b) H2=a unit height

Fig 3. The total pressure drop of the charge in the hermetic furnace

As shown in Fig. 3(a), when the total height is a constant and H1 plus H2 equals to a unit height, the total pressure drop is linear to the proportion of H1 of the total height. It is to say that during the process of calcium carbide manufacturing where the height of the charge layer almost unchanged, the workers should make up their efforts to lower the temperature of reaction zone at a reasonable range so as to lower the height of the soft zone. In Fig. 3(b), H2 is treated as a unit height and the value of H1 to H2 is varied, a peak of the total pressure drop can be reached in pace with the increase of the ratio of H1 to H2, when the ratio is small, the total pressure drop varies greatly and it will slow down when the ratio increase continual, and an important phenomenon is that the pressure drop of B is the most of the total pressure drop of the entire charge layer.

3.2 Mechanical analysis of charge descending

In hermetic calcium carbide furnace, five essential conditions are required for charge descending: 1) the carbon materials and lime react well when heated by electric arc which can release a lot of energy; 2) when the charge moves down, the small size particles filled into the big ones lead to the volume shrinking; 3) a large cavity below reaction zone is necessary; 4) a large amount of gas generated in reaction zone can flow upward through the charge layer steady and uniformly; 5) the calcium carbide liquid should be discharged as soon as possible.

Fig 4 shows the mechanical analysis of the charge which bears four main forces that are gravity of the charge itself, supporting force by the gas in reaction zone, friction by the wall of the hermetic furnace and the interaction of the charge and the force given by electric arc, the balance analysis can be described as:

$$F=G-T-f-\Delta p\times A \tag{11}$$

Where, F is the force for descending, G is the gravity of the charge, T stands for the force given by electric arc, ΔP refers to the pressure drop of the gas coming from reaction zone, A is to the area of the surface below reaction zone, and f is the friction by the wall and the interaction of the charge.

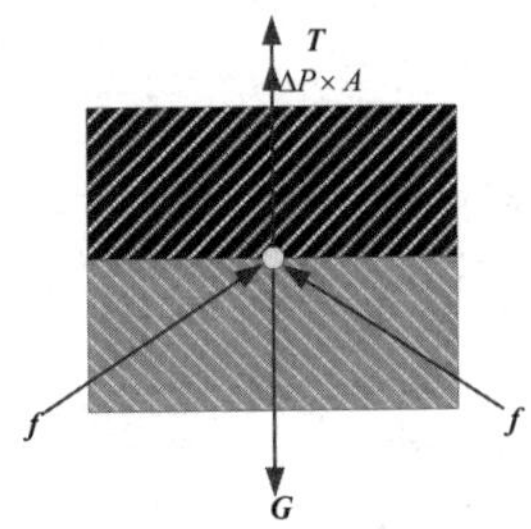

Fig 4. The mechanical analysis of the charge

It can be seen from Eq.(11) that the prerequisite for charge descending is that the value of F is greater than zero, otherwise the charge is suspended. Therefore, in order to ensure the charge descended smoothly, on the one hand, the charge filled into the hermetic furnace must be continual, on the other hand, the gas from reaction zone can flow upward through the charge layer well and a good permeability of the charge is necessary.

3.3 Theory of charge collapses

Based on the analysis above to discuss the theory of charge collapses, in general, the charge collapses are considered as a three-stage processes as shown in Fig 5:

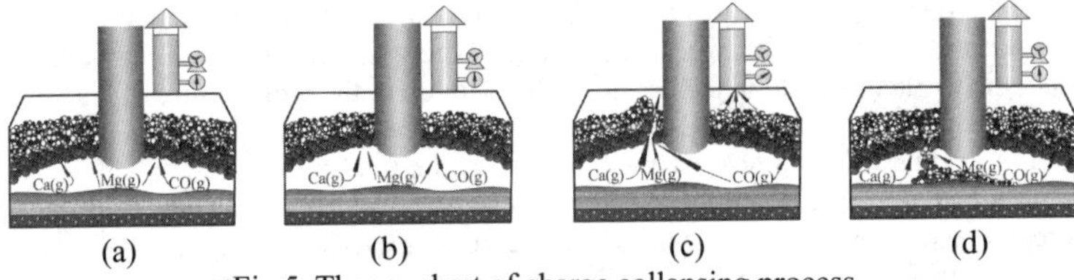

Fig 5. Thoery chart of charge collapsing process

The first stage is the charge layer suspended. Fig. 5(a) shows the charge layer before suspended, and the hermetic furnace is working well with the charge filled into it continually. If the size of the charge particles is not uniform and mixed with powder, what's worse, the carbon materials are of high moisture content, the permeability index of the charge will become more worse. As the manufacturing process goes on, the height of the charge layer is thinning, and a lot of gases, such as Ca vapor, Mg vapor and CO gas gather in the cavity below resulting in an increase of pressure drop. When the sum of T, f, $\Delta P \times A$ equals to G, the charge will be suspended which will be a reflection of unreasonable charge layer and an unstable environment for calcium carbide manufacturing in hermetic furnace that needs to take seriously as shown in Fig. 5(b).

The second stage is the charge moving. With the proceeding of the calcium carbide manufacturing, the height of the charge layer becomes more and more thinner, the width of the temperature belts narrows down, and the pressure drop increases because the gas of reaction zone is more which leads to the value of F tends to be positive. Fig. 5(c) shows the weak part of the charge layer which can't bear a huge impact by the gas below and move away with a big crack, a large number of gas run out and the pressure of the top furnace becomes larger immediately[9].

The third stage is the charge collapsing. At present, the pressure in the top of the hermetic furnace is always a little negative in order to prevent the gas from spirting and burning in the atmosphere. As shown in Fig. 5(d), after the gas all exits, the pressure in the cavity almost equals to the atmospheric pressure, and the charge collapses along the crack into the calcium carbide liquid near the electrodes which leads to a variation of the electrical parameters, and the final charge collapses are formed.

3.4 Characterization of charge collapses

Combined with the calcium carbide manufacturing process and found that a large number of gas outside of the hermetic calcium carbide furnace during the process of charge collapsing is burning mainly due to the reaction of CO gas and oxygen contacted in the atmosphere, which is the apparent phenomenon of the charge collapses.

In operation room, the workers can judge whether the charge collapses or not by the changes of three-phase currents, top pressure and temperature of the hermetic furnace and so on, and the intensity of the charge collapsing can be presented by them, else. Table 2. shows the electrical parameters before the charge collapsing, and the current values of the three electrodes A, B and C obtained by the calculation from Eq.(12) are 400.84A, 398.74A, 408.17A respectively.

$$I_0 = \frac{P_0}{\sqrt{3}U_0 \cos\varphi} \tag{12}$$

Where, I_0 is the calculated average current, P_0 is the active power, U_0 is the primary voltage and $\cos\varphi$ represents the power factor.

Table 2. Electrode parameters before charge collapsing

Electrode	Apparent Power (MVA)	Active Power (MVA)	Power Factor	Primary Voltage (KV)
A	8.49	6.81	0.802	36.69
B	8.50	7.00	0.824	36.90
C	8.71	7.12	0.817	36.98

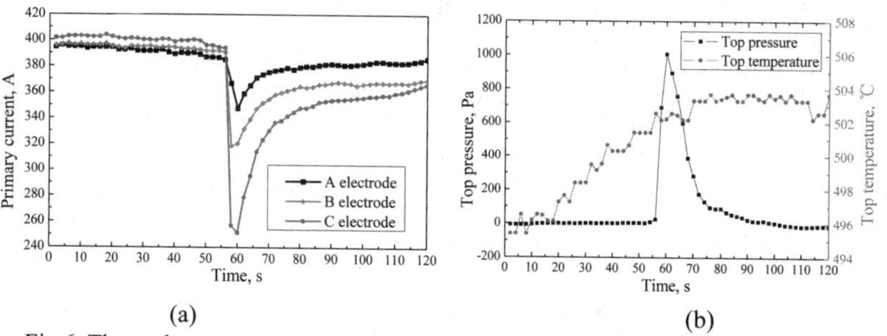

(a) (b)

Fig 6. Three-phase currents and top pressure, temperature during charge collapsing

Fig 6(a) shows the changes of the three-phase primary currents during the process of the charge collapsing, it can be seen that the currents drop sharply to the lowest at first and then rise up to the normal condition, the time is short with no more than two minutes. The reasons for charge collapsing attribute to two aspects, on the one hand, the increase of the electric arc resistance because the charge droped into the calcium carbide liquid that right below the electrodes, on the other hand, when charge collapses, there is a gap instantaneous existed in the charge layer which increases the resistance by-pass indirectly, and the entire resistance increases and the primary currents drop by the both. After then, the charge in the calcium carbide liquid reacts completed and the new charge filled the gap with time goes on to reach a recovery of the dropped currents to normal. It can also be seen that the currents of the three electrodes are not dropped simultaneously and the values are not uniform yet because the place where the charge collapsed is not the same.

As seen in Fig 6(b), a certain trend is found for the variations of the top pressure and the temperature during charge collapsing. When charge collapses happened, the top pressure goes up to the highest value which is about 1013 Pa immediately as analysed before and then lowers down gradually because the gas released thoroughly. Before the charge collapsing, the height of the charge layer reduced uniform and the temperature belts moved up, the surficial temperature rised and the top temperature did the same. The top temperature will decline and become stable after the charge collapsing, and the time always lags for a short time for top temperature reached the peak than the top pressure does.

3.5 Analysis of the causes of frequent charge collapses

The manufacturing process of calcium carbide is long so that any unreasonable operations can bring a bad influence to maintain the stability of it, and the causes of frequent charge collapses are dicussed based on the properties of the charge and the operation of electrodes.

The charge properties are mainly referring to particle size, quality of lime and moisture content of carbon materials, and the poor properties are the most impacts to frequent charge collapses. In a calcium carbide plant, the size of lime used is about 5~40mm, and that of the carbon materials is about 5~25mm, too large or too small and also the powders needed to treat again or remove. The charge should be used following the priciple of a reasonable particle size and a proper resistance of the charge in order to obtain a good permeability of the charge layer and to ensure a certain resistivity for a good operation of electodes. The moisture content of the carbon materials should be maintained less than 4% in order to prevent from reacting with lime for the powder of lime which can increase the specific surface area of the charge and cause the permeability of the charge layer worse and a bad environment for the calcium carbide manufacturing[10].

The content of MgO in the charge is essential, as required, it is not more than 1.0%, but it is always higher in practice as listed in Table 3. As we all know, MgO in the charge can be reduced by C of the carbon materials directly which generates a lot of Mg vapor and CO gas, and the reaction can be reversed under a certain temperature of 1843℃ as shown in Eq.(13). When the temperature is less than 1843℃, Mg vapor and CO gas react in the top of the hermetic furnace for particle mixture of MgO and C which will become blocks after hardening by the process of agglomeration and sintering[11-13].

$$MgO(s)+C(s) \rightleftharpoons Mg(g)+CO(g) \qquad \Delta_f G^0 = 613160-289.74T \qquad (13)$$

Table 3. Chemical composition of lime, %

Composition	CaO	MgO	SiO_2	Al_2O_3	Fe_2O_3	The Other
Content	86.9	1.60	2.78	0.73	0.257	7.733

Electrodes operation is the fundamental guarantee of the stability of calcium carbide manufacturing, and the core requirement is how to maintain the electrodes uniform and at the same level. However, due to the limited requirement of power, the electrodes can not transport the same power so much at the same time and then cause an unbalance with a high deviation of currents which lasts a long time, and the occurrence incidence of the frequent collapses will arise, as shown in Fig 7.

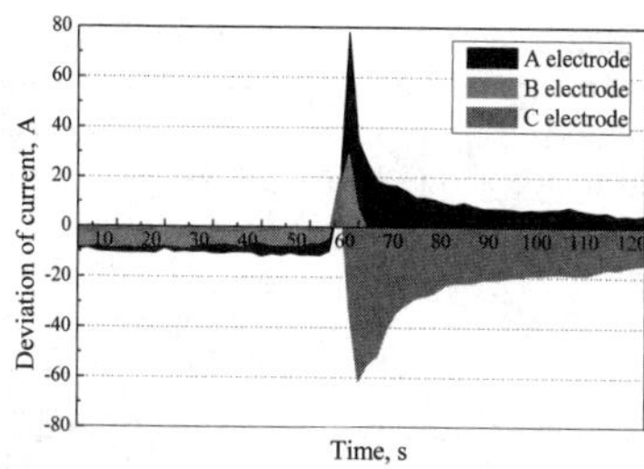

Fig 7. Tendency chart of the deviation of current during charge collapsing

3.6 The harms of frequent charge collapses

The frequent charge collapses have a greater impact to the stability of calcium carbide manufacturing and the productivity, and the main harms can be listed as follows.

(1) Destroy the charge layer structure, and the charge can't move downward smoothly. In hermetic calcium carbide furnace where the detailed structures from up to down are charge, red charge, soft charge, calcium carbide liquid and accumulating slags[14], when charge collapses, the gap generated by the charge moving are filled by new charge, which brings about a bad charge layer.

(2) The charge that dropped into the calcium carbide liquid can not react completely, the first thing is that the temperature in reaction zone declined and the calcium carbide liquid discharged difficult, what's worse, the impurities in the calcium carbide liquid increases to lower the quality of the calcium carbide product.

(3) Frequent charge collapses lead to the primary current, active power changed frequently, and many unnecessary operations of the electrodes is needed in order to maintain the balance of them.

(4) Frequent charge collapses result in an inceasing incidents of soft broken and hard broken of the electrodes.

(5) A lot of gases, such as CO gas, H_2 gas and so on burned outside the top of the hermetic furnace can damage the cooling devices easily and increase the costs for repairing.

4 PRECAUTIONS AND MEASURES FOR FREQUENT CHARGE COLLAPSES

Based on the analysis and the discussions of the theory and causes of the frequent charge collapses, some effective measures can be carried out as follows:

Firstly, improve the quality of the charge and reduce the moisture content, especialy extend the time for drying before filled into the furnace and remove the small particles or powder as soon as possible to increase the permeability of the charge layer.

Secondly, deal with the surface of the charge layer and remove the red and the hardened charge in order to make them not to affect the permeability of the charge layer.

Thirdly, make up efforts to adjust the three electrodes to an uniform level and to deal with the unbalance conditions in time.

5 CONCLUSIONS

Frequent charge collapses have a greater influence to maintain the stability of the calcium carbide manufacturing and to improve the productivity of the calcium carbide product. Based on the properties of the charge in the manufacturing process of calcium carbide, and make the electrodes operation as the essential requirment to improve the permeability of the charge layer and maintain the electrodes balance in order to obtain a reasonable and a uniform distribution of the charge layer. remove the particles of the charge whose size is too large or too small and also the powders. MgO in the manufacturing process of the calcium carbide is in cycle enrichment and will become the main substance of the hardened charge which impacts the permeability of the charge layer greatly and should be taken out as by-products for the procuction of cement and glass and so on.

6 ACKNOWLEDGEMENT

The present work was supported by National Science and Technology Support Program funded projects (No. 2011BAC01B02) and National Natural Science Foundation-funded project (No. 51174023).

7 REFERENCES

[1]Xiong Moyuan, Calcium Carbide Manufacturing and Deep Processing of Products, *M. Beijing: Chemical Industry Press*, (2010).

[2]Lu Changjie, Zhao Changqing, Analysis for Breaking Mechanism of Electrode for Silicon Smelting-furnace, *J. Carbon Techniques*, **3(27)**, 59-61(2008).

[3]Yan Zhenjun. The Cause and Precaution of Large Charge Collapses in Calcium CarbideFurnace, *J. Fujian chemical industry*, **4**,18-21(2005).

[4]M. S. Paizullakhanov, Sh. A. Faiziev. Calcium Carbide Synthesis Using a Solar Furnace, *J. Technical Physics Letters*, **32(3)**, 211-2(2006).

[5]Li Guodong, Fundamental research in the production of calcium carbide from fine biochars and fine calcium oxide, *D.* Beijing: Beijing University of Chemical Technology, 2011. (in Chinese)

[6]Tang Xubo, Ma Caixia, Liu Qingya, Li Guodong and Liu Zhenyu, Reactions of calcium compounds and coke for preparation of calcium carbide, *J. Journal of Fuel Chemistry and Technology*, **38(5)**, 539-43(2010).

[7]Liu Lu, Yang Pengyuan, Liu Hui, Thermodynamic analysis of calcium carbide synthesis and its thermal coupling with coke combustion, *J. Journal of Beijing University of Chemical Technology (Natural Science)*, **39(2)**, 1-6(2012).

[8]C. W. Zhu, G. Y. Zhao and V. Hlavacek, A d.c. plasma-fluidized bed reactor for the manufacturing of calcium carbide, *J. Technical Physics Letters*, **30(9)**, 2412-19(1995).

[9]Huang Haisheng, Zeng Daobiao, Discussion on Top Eruption Explosion and Practice of Increasing Manufacturing and Reducing Consumption of 16.5 MVA Electric Furnace, *J. Ferro-alloys*, **2(217)**, 8-11(2011).

[10]Li Guangsen, Wang Zhihua, Teng Fei, On-Line Determining Device of Permeability for Sinter Mixture, *J. Journal of Iron and Steel Research*, **24(1)**, 59-62(2011).

[11]Guo Jingkun, Feng Chude. The recent progress nano-ceramics, *J. CHINESE JOURNAL OF MATERIAL RESEARCH*, **9(5)**, 412-9(1995).

[12]Amit Ron, Nick Fishelson, Nathan Croitoriu, Dafna Benayahu, Yosi Shacham-Diamand, Theoretical examination of aggregation effect on the dielectric characteristics of spherical

cellular suspension, *J. Biophysical Chemistry*, **140(1-3)**, 39-50(2009).
[13]A.S.A. Chinelatto, R. Tomasi, Influence of processing atmosphere on the microstructural evolution of submicron alumina powder during sintering, *J. Ceramics International*, **35(7)**, 2915-20(2009).
[14]Chu Shaojun, Zeng Shilin, Huang Zucheng, Cause of The Accident on The Slag Surge and The Interior Explosion of The Ferroylloy Furnace, *J. Ferro-alloys*, **2(205)**, 15(2009).

CORROSION BEHAVIOR OF AISI 304L STAINLESS STEEL FOR APPLICATIONS IN NUCLEAR WASTE REPROCESSING EQUIPMENT

Negin Jahangiri; A. G. Raraz; J. E. Indacochea
Civil and Materials Engineering Department, University of Illinois at Chicago

S. McDeavitt
Nuclear Engineering Department, Texas A&M University

ABSTRACT

Developed by the Advanced Fuel Cycle Initiative of the Department of Energy, UREX+ is an advanced solvent extraction process to separate uranium from nuclear waste. The material selection for constructing the centrifugal contactor that is utilized in the operation of this process is very critical in order to ensure its reliable performance. Our work focuses on assessing the corrosion of 304L stainless steel in acidic aqueous solutions mimicking the environments present in a centrifugal contactor as the key equipment of UREX+ process during its operation. Corrosion tests are conducted in the following aqueous solutions: 5.0M HNO_3; 5.0M HNO_3 + 0.1M HF; and 5.0M HNO_3 + 0.1M HF + 0.1M Zr^{4+}. Electrochemical impedance spectroscopy (EIS) tests are considered to assess the corrosion mechanisms. X-ray photoelectron spectroscopy is also used to study the surface analysis. The results show that 5.0M HNO_3 + 0.1M HF solution is the most corrosive environment and addition of Zr^{4+} ions reduces the corrosion caused by HF to levels similar to those found in HNO_3 solutions.

INTRODUCTION

With increasing demand for nuclear power as an alternative source of energy and with the goal of extensive decrease in the dependency to fossil fuels, fuel reprocessing technologies capable of recycling fissile materials from nuclear waste have been developed. Amongst them, UREX+, Advanced Uranium Extraction process, is part of the Department of Energy's Advanced Fuel Cycle Initiative (AFCI) for reprocessing the spent nuclear fuels and recycling the uranium. The objective of this process is to improve waste management by means of enhancing energy recovery and proliferation resistance as well as the repository capacity[1-3].

UREX+ has been demonstrated in the lab-scale at Argonne National Laboratory (ANL)[4]. The centrifugal contactor is the key element in effective performance of this process. The contactor, which works on the foundation of solvent extraction processes, is exposed to acidic media (aqueous phase) containing nitric and hydrofluoric acids. Nitric acid is the major acid used as the dissolvent medium in the spent nuclear fuel reprocessing systems[5-6]. In addition to nitric acid, varying amounts of fluoride is added to the aqueous phase solution in order to dissolve the remnant materials which are insoluble in nitric acid[7-8].

304L stainless steel has been selected as the constructing material for centrifugal contactors based on its corrosion resistance when exposed to an oxidizing medium such as nitric acid. Corrosion resistance of stainless steel comes from its ability to form a passive film enriched with Cr, however, the corrosion rate of passive metals changes with the presence of aggressive ions such as fluoride ions[9-12].

In order to study the corrosion behavior of the centrifugal contactor after exposure to the aqueous phase present in UREX+ process, 304L stainless steel corrosion coupons were exposed to three different solutions: 5.0M HNO_3, as the benchmark for our study; 5.0M HNO_3+0.1M HF to study the effect of fluoride ions; and 5.0M HNO_3+0.1M HF+0.1M Zr^{4+} in order to investigate the impact of presence of zirconium ions since zirconium is part of spent nuclear fuels as fission product[13]. These corrosion studies are performed by means of electrochemical impedance spectroscopy (EIS), and X-ray photoelectron spectroscopy (XPS) is used in order to investigate the chemical changes of the surface of stainless steel coupons when exposed to the three different solutions.

EXPERIMENTAL PROCEDURE

Testing material

The composition of AISI 304L stainless steel corrosion coupons used in this investigation has the following chemical composition: 0.019 wt.% C, 1.76 wt.% Mn, 0.45 wt.% Si, 0.3 wt.% Mo, 8 wt.% Ni, and 18.0 wt.% Cr. Samples were annealed at 400°C for two hours in order to remove any residual mechanical stresses. The surface of the samples were ground from 120 to 1200 grit using silicon carbide paper and an automatic grinder/polisher machine prior to corrosion testing and cleaned using de-ionized water before changing the abrasive paper grit. The samples were then polished to a 1 µm surface finish using diamond paste. After that they were cleaned thoroughly with acetone, dried and then inserted in Polytetrafluoroethylene (PTFE) specimen holders.

Electrochemical Impedance Spectroscopy

The electrochemical cell setup is a three-electrode cell arrangement with flat 304L SS specimens as the working electrode, two glassy graphite rods as the counter electrodes and, due to the aggressive environment towards glasses by the presence of HF, an Accumet® gel-filled Calomel reference electrode with epoxy body and porous polymer junction was used as the reference electrode. The working electrodes are flat disks of 3 mm in thickness and 1.5 cm in diameter. The specimen holder allows exposure of 1 cm^2 of the surface of sample to the solution. A polyethylene Nalgene® beaker which could hold 300 mL of solution was used as part of the electrochemical cell since the original glass container of the electrochemical cell could not be used due to the presence of hydrofluoric acid. The top part of the original glassy cell which is designed to hold the electrodes was cut and used with the polyethylene beaker. By this arrangement, the solution was in contact with the polyethylene beaker during the test and the electrodes could be held in place by the glass top. 200 mL of each acidic solution was prepared for each experiment. All the acidic solutions were prepared with distilled and deionized water and HNO_3 and HF acids of pure analytic grade. Zr^{4+} ions were added as zirconyl(IV) nitrate. The solutions were made 2 hours prior to the tests, and were naturally aerated by ambient exposure.

The Electrochemical impedance spectroscopy experiments were carried out after the samples were immersed in the solution for 2 hour, using a computer-controlled Gamry PCI 4 potentiostat-galvanostat unit. These tests were conducted at the frequency range of 10^5 to 10^{-2} Hz and a 10 mV amplitude sine wave at open circuit potential and acquisition rate of 10 points per decade. The Zsimpwin 3.20 software was employed to fit the electrochemical data in order to obtain proper electrical models and perform the data analysis.

Surface Analysis

For surface analysis, XPS measures the energy spectra of electrons emitted after exposure to X-rays and can be used to determine the chemical state of near-surface elements, which makes it a practical technique to evaluate the corrosion resistance of stainless steels by identifying the compounds formed at its surface[10,14]. After taking out the samples from the solution, they were rinse with deionized water and dried and immediately introduced to the analysis chamber for XPS testing. Since XPS is a very sensitive technique, the samples were transformed to the XPS chamber using Sample-Storr™ vacuum containers to avoid exposure to the air. The XPS analyses were performed with a Kratos Axis-165 XPS instrument, using a monochromatized Al K-alpha x-ray source ($h \cdot v$=1486.7eV). Survey scans were conducted using pass energy of 160 eV, while narrow scans were conducted using pass energy of 20 eV. The binding energy scale was adjusted against the C(1s) peak at 284.5eV. Spectra of C, Fe and Cr were recorded. The spectra obtained were fitted, after background subtraction, following the Shirley procedure with Gaussian-Lorentzian curves using XPSPeak 4.1 software. Iterative fitting has been used to obtain the minimum difference between the experimental and the fitted data. NIST Standard Reference Database 20[15] is used to match the peaks in the XPS spectra.

RESULTS AND DISCUSSIONS

The results of EIS tests are presented as Nyquist plots including the chracteristic frequencies. The Nyquist plots presented in Figure 1, which correspond to the 304L stainless steel samples immersed in 5.0M HNO_3 and 5.0M HNO_3 + 0.1M HF + 0.1M Zr^{4+} solutions for 2 hours are semi-circular arcs, indicating the presence of a protective passive layer. It can be observed that in both of the solutions, the diameter of the semicircles are about the same, resulting in the same polarization resistances and consequently, similar resistance to corrosion.

The experimental data presented in Figure 1 are fitted to the Randles equivalent circuit, shown as an insert in Figure 1, which has the arrangement of $(R_s(CPE \| R_p))$, where R_s is the ohmic resistance of the electrolyte, R_p is the polarization resistance, and Q is the constant phase element also known as CPE which is the double layer capacitance (C_{dl}) dependent on frequency[16]. Replacing the capacitor by CPE is a common practice in electrochemical studies and improves the fitting of the in experimental EIS curves. The relationship between capacitor (C) and CPE (Q) for the classical depressed semicircles is given by the following equation:

$$Q = \frac{C}{(\omega_{max})^{\pi - 1}} \qquad \text{Equation (1)}$$

where ω_{max} is the frequency in which the imaginary part (−Z”) reaches a maximum and n shows CPE’s deviation from the ideal capacitance behavior in which n=1[17-18].

The Nyquist plots of 304L SS coupons tested in 5.0M HNO_3 + 0.1M HF solutions are shown in Figure 2, including experimental and fitted data. This plot presents contrasting difference in shape and scale magnitudes of the axes compared to the plots presented in Figure 1. The Nyquist curve shown in Figure 2 has a capacitive loop along with a tail turning away from the Z_{real} axis at lower frequency range. The occurrence of this tail which is called the Warburg impedance indicates a diffusion-controlled corrosion mechanism[19]. Warburg impedance is arranged in series with the polarization resistance in the modeled electric circuit shown as an insert in Figure 2. While corrosion of 304L stainless steel in 5.0M HNO_3 + 0.1M HF solutions can be related to the corrosion processes controlled by mass transfer, the information obtained by EIS shows that charge transfer is the controlling mechanism for corrosion in 5.0M HNO_3 and 5.0M HNO_3 + 0.1M HF + 0.1M Zr^{4+} solutions.

The similar behavior of the two solutions, 5.0M HNO_3 and 5.0M HNO_3 + 0.1M HF + 0.1M Zr^{4+}, is related to the similarity of the passive layers formed on the surface of specimens after immersion in these solutions. In HNO_3 solutions the stainless steel samples form a passive layer that is mainly constituted of Cr_2O_3. Zr ions behave as corrosion inhibitors by complexing the aggressive fluoride ions into zirconium-fluoride, preserving the protective passive film. The passive film of the sample exposed to 5.0M HNO_3 + 0.1M HF + 0.1M Zr^{4+} was analyzed via XPS. The XPS spectra revealed that the passive film was mainly chromium oxide, the same as the sample just exposed to 5.0M HNO_3 solutions.

Calculated values of the polarization resistances corresponding to each tested solution are given in Table 1. The polarization resistance, R_p, which is equal to the semicircle radii of the

Nyquist plots and is inversely proportional to the corrosion rate values, is one order of magnitude smaller for the 5.0M HNO_3+0.1M HF solution, compared to the 5.0M HNO_3 and 5.0 M HNO_3+0.1M HF+0.1M Zr^{4+} solutions. The corrosion rates of 304L stainless steel in the three solutions are obtained based on the method suggested by Silverman[20] and are presented in Table 1. Similar values of corrosion rates for 304L stainless steel samples exposed to 5.0M HNO_3 and 5.0 M HNO_3+0.1M HF+0.1M Zr^{4+} solutions are in contrast with the corrosion rate values of the sample immersed in 5.0 M HNO_3+0.1M HF solution, which is more than 16 times higher. Due to high stability of Zr-F complexes, Zr^{4+} ions are expected to be effective inhibitors[21]. The inhibitor efficiency for zirconium ions can be obtained by the following equation in order to quantify the corrosion inhibitor characteristic of Zr^{4+} ions[22]:

$$P=(w_0-w)/w\times 100 \qquad \text{Equation (2)}$$

where, w_o is the corrosion rate in absence of the inhibitor, corresponing to the corrosion rate in 5.0M HNO_3 + 0.1M HF solutions, and w is the corrosion rate in the same environment with the added inhibitor, referring to the corrosion rates in 5.0M HNO_3 + 0.1M HF + 0.1M Zr^{4+} solution. Based on this equation, the calculated inhibitor efficiency of Zr^{4+} ions is 93.9%, confirming the stability of the Zr-F complexes.

The surface chemistry of corrosion coupons was assessed by X-ray photoelectron spectroscopy (XPS). The peaks of the XPS spectra were matched with the data present in the NIST Standard Reference Database[15]. Figure 3 shows the XPS spectra for Cr on the surface of (a) the as polished sample and corrosion coupons after corrosion tests in: (b) 5.0M HNO_3; (c) 5.0M HNO_3 + 0.1M HF; and (d) 5.0M HNO_3 + 0.1M HF + 0.1M Zr^{4+} solutions. This figure shows the enrichment of the chromium oxide passive film after corrosion test in 5.0M HNO_3 and 5.0M HNO_3 + 0.1M HF + 0.1M Zr^{4+} solutions compared to the as polished sample, as the Cr(ox)/Cr(metal) peak area ratio for the $Cr2p_{3/2}$ peak is increased from 5.67 for the as polished sample to 10.18 and 9.48 for the samples immersed in 5.0M HNO_3 and 5.0M HNO_3 + 0.1M HF + 0.1M Zr^{4+} solutions, respectively. These results confirm the inhibition efficiency of zirconium by showing the similar behavior of 304L stainless steel specimens in both 5.0M HNO_3 and 5.0M HNO_3 + 0.1M HF + 0.1M Zr^{4+} solutions regarding the formation of a chromium oxide layer on the surface of the samples. However, for the 5.0M HNO_3 + 0.1M HF solution, no peak related to chromium oxide was detected, suggesting the dissolution of chromium oxide layer by hydrofluoric acid, and instead, the peak corresponding to the chromium fluoride with higher binding energies compared to the chromium oxide peak, appeared in the spectra.

In Figure 4, the XPS spectra for Fe on the surface of corrosion coupons is presented, showing the decrease in the ratio of Fe(ox)/Fe(metal) peak areas for the $Fe2p_{3/2}$ spectra of the samples immersed in three solutions compared to the as-polished sample. The Fe(ox)/Fe(metal) peak area ratio for the as-polished sample is 4.39 which reduces to 1.37, 1.40 and 1.56 after immersion in 5.0M HNO_3, 5.0M HNO_3 + 0.1M HF + 0.1M Zr^{4+} and 5.0M HNO_3 + 0.1M HF solutions, respectively. These results show the selective dissolution of Fe due to its faster diffusion of compared to chromium[23]. The passive layer formed on the surface of 304L stainless steel samples after immersion in 5.0M HNO_3 and 5.0M HNO_3 + 0.1M HF + 0.1M Zr^{4+} solutions is rich in chromium oxide which is the result of iron oxide dissolution from surface of the specimens[10,24]. XPS results confirm the presence of Cr_2O_3 in both 5.0M HNO_3 and 5.0M HNO_3 + 0.1M HF + 0.1M Zr^{4+} solutions while in 5.0M HNO_3 + 0.1M HF solutions, the presence of fluoride ions does not allow the formation of the protective passive film.

CONCLUSION

Addition of 0.1M Zr^{4+} ions to a solution containing 5.0M HNO_3 + 0.1M HF increases the polarization resistance obtained by EIS to the levels found in 5.0M HNO_3 solution. This is due to the formation of zirconium fluoride complexes as a result of the reaction of Zr ion with hydrofluoric acid. The high efficiency of zirconium in forming these complexes explains the reduction of corrosion rate of 304L stainless steel samples in 5.0M HNO_3 + 0.1M HF+0.1M Zr^{4+} solutions compared to 5.0M HNO_3 + 0.1M HF solution.

XPS results showed the formation of the passive film on the surface of 304L SS samples exposed to 5.0M HNO_3 and 5.0M HNO_3 + 0.1M HF+0.1M Zr^{4+} solutions, while for the samples immersed in 5.0M HNO_3 + 0.1M HF solutions, the spectra showed the peak corresponding to chromium fluoride, with no trace of chromium oxide peak present. XPS results confirm the information obtained by the EIS experiments and explain the corrosion rate reduction of the samples exposed to 5.0M HNO_3 + 0.1M HF+0.1M Zr^{4+} solution compared to 5.0M HNO_3 + 0.1M HF solution, which is due to the reaction of zirconium and fluoride ions and formation of Zr-F complexes which allows the 304L stainless steel samples to form the chromium oxide passive layer that is also present on the surface of the samples immersed in nitric acid solution.

REFERENCES

[1]G. F. Vandegrift, M. C. Regalbuto, S. Aase, A. Bakel, T. J. Battisti, D. Bowers, J. P. Byrnes, M. A. Clark, J. W. Emery, J. R. Falkenberg, A. V. Gelis, C. Pereira, L. Hafenrichter, Y. Tsai, K. J. Quigley, M. H. Vander Pol, Designing and demonstration of the UREX+ process using spent nuclear fuel, ATALANTE 2004, paper no. 012-01 (Nimes, France: Advances for Future Nuclear Fuel Cycles, June 21-24 2004).

[2] M.C. Regalbuto, J. M. Copple, R. Leonard, C. Pereira, and G. F. Vandegrift, Solvent Extraction Process Development for Partitioning and Transmutation of Spent Fuel, Las Vegas, NV, November 9-11 (2004). Eighth Information Exchange Meeting on Actinide and Fission Product Partitioning and Transmutation.

[3]U.S. Department of Energy, Office of Nuclear Energy, Science, and Technology, “Advanced Fuel Cycle Initiative: Status Report for FY 2005,” Report to Congress. February 2006.

[4]C. Pereira, G. F. Vandegrift, M. C. Regalbuto, A. Bakel, D. Bowers, A.V. Gelis, A. S. Hebden, L. E. Maggos, D. Stepinski, Y. Tsai, J. Laidler, LAB-SCALE DEMONSTRATION OF THE UREX+1a PROCESS USING SPENT FUEL, WM’07 Symposium, February 25, – March 1, 2007, Tucson, AZ .

[5]W. H. Smith, G. M. Purdy, Chromium in Aqueous Nitrate Plutonium Process Streams: Corrosion of 316 Stainless Steel and Chromium Speciation, Waste Manage., **15**, 477-484 (1995).

[6]F. Balbaud, G. Sanchez, P. Fauvet, G. Santarin, G. Picard, Mechanism of corrosion of AISI 304L stainless steel in the presence of nitric acid condensates, Corros. Sci., **42**, 1685-1707 (2000).

[7]W. E. Clark, R. E. Blanco, Dissolution of LMFBR Fuels: Survey of the Corrosion of Selected Alloys in HN03-HF Solutions, Oak Ridge National Laboratory, Oak Ridge, TN, 1971. USAEX Report ORNL- 4745.

[8]R. S. Ondrejcin, B. D. Mc Laughlin, Corrosion of High Ni-Cr Alloys and Type 304L Stainless Steel in HNO_3 – HF, Savannah River Laboratory, Report DP-1550, (1980).

[9]C. C. Seastrom, Passivity of Chromium and Stainless Steel in Hydrofluoric Acid. Corrosion, **20**, 179t-183t (1964).

[10]S. J. Kerber, J. Tverberg, Stainless Steel Surface Analysis, Adv. Mater. Processes., **158**, 33-36 (2000).

[11]W. Lai, W. Zhao, F. Wang, C. Qi, J. Zhang, EIS study on passive films of AISI 304 stainless steel in oxygenous sulfuric acid solution, Surf. Interface Anal., **41**, 531–539 (2009).

[12]R. R. Maller, Passivation of stainless steel, Trends Food Sci. Technol., **9**, 28-32 (1998).

[13]B. S. Matteson, P. Tkac, A. Paulenova, Complexation Chemistry of zirconium(IV), uranium(VI), and iron(III) with Acetohydroxamic acid, Separ. Sci. Technol. **45**, 1733-1742 (2010).

[14]W. P. Yang, D. Costa, P. Marcus., Resistance to Pitting and Chemical Composition of Passive Films of a Fe-17%Cr Alloy in Chloride-Containing Acid Solution, J. Electrochem. Soc., **141**, 2669-2676 (1994).

[15]NIST X-ray Photoelectron Spectroscopy Database, NIST Standard Reference Database 20, Version 3.5, http://srdata.nist.gov/xps.

[16]M. Sluyters-Rehbach, Impdances of Electrochemical Systems: Terminology, Nomenclature, and Representation: I. Cells with Metal Electrodes and Liquid Solutions, Pure & Appl. Chem., **66**, 1831-1891 (1994).

[17]C.H. Hsu, F. Mansfeld, Technical Note: Concerning the Conversion of the Constant Phase Element Parameter Yo into a Capacitance, Corrosion, **57**, 747-748 (2001).

[18]M. E. Orazem, B. Tribollet, Electrochemical Impedance Spectroscopy, (Hoboken, NJ: John Wiley & Sons, 2008), p. 171.

[19]E. Barsoukov, J. R. Macdonald, Impedance Spectroscopy: Theory, Experiment, and Applications, 2nd Ed., John Wiley & Sons, NJ, USA, 2005.

[20]D. C. Silverman, Practical Corrosion Prediction Using Electrochemical Techniques, Uhlig Corrosion Handbook, 2nd Edition, John Wiley and Sons, NY, (2000).

[21]A. E. Sillén, L. G., Martell, J. Bjerrum, Stability Constants of Metal-Ion Complexes, Special Publication, No. 17, Chemical Society, London (1964).

[22]R. W. Revie, Uhlig's Corrosion Handbook, 2nd ed. (New York: NY: John Wiley & Sons, Inc., 2000).

[23]S. Ningshen, U. Kamachi Mudali, G. Amarendra, Baldev Raj, Corrosion Assessment of Nitric Acid Grade Austenitic Stainless Steels, Corros. Sci., **51**, 322–329 (2009).

[24]G. Henkel, B. Henkel, Information on Passivation Layer Phenomena for Austenitic Stainless Steel Alloys, Technical Bulletin, Pickling and Electropolishing. Essay No. 45/Rev. 00 (2003).

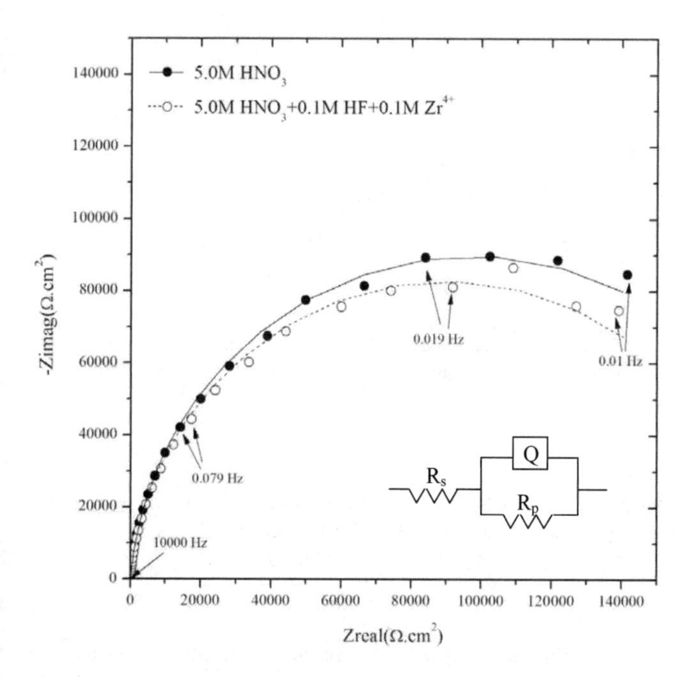

Figure 1. Nyquist plots for 304L Stainless Steel corrosion coupons in 5.0M HNO_3 and 5.0M HNO_3 + 0.1M HF + 0.1M Zr^{4+} solutions after 2 hours of immersion. Insert is the equivalent electrical circuit. Symbols: experimental data, Lines: fitted data.

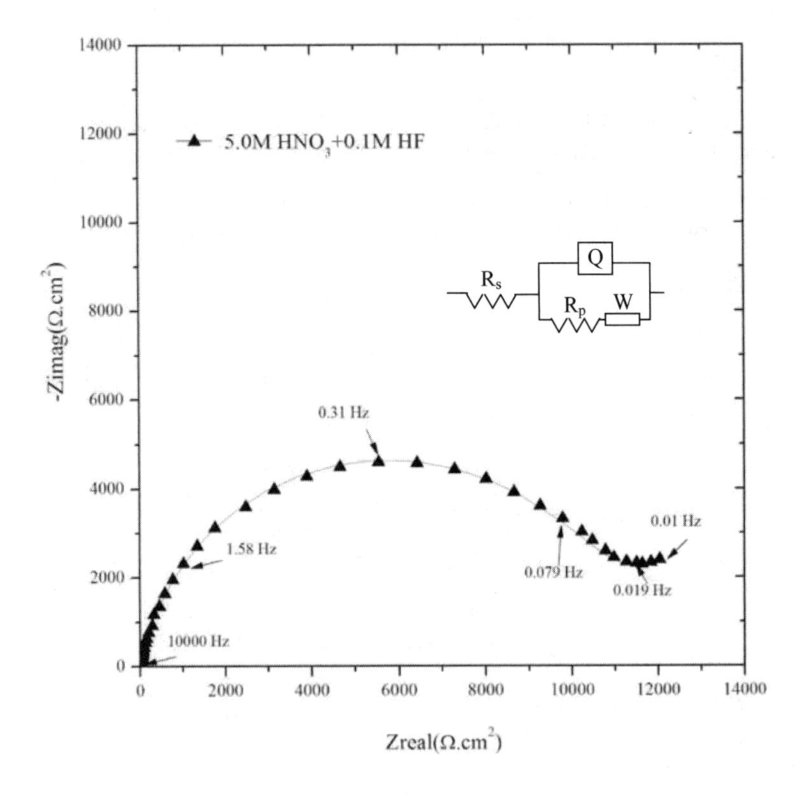

Figure 2. Nyquist plots for 304L Stainless Steel corrosion coupons in 5.0M HNO_3 + 0.1M HF solutions after 2 hours of immersion. Insert is the equivalent electrical circuit. Symbols: experimental data, Lines: fitted data.

Table 1. Fitting parameters for the EIS data obtained for immersion of the 304L SS samples in different solutions.

Solution	**R_s (Ω)**	**Q ($S.s^n$)**	**n**	**R_p (Ω)**	**Warburg, Y_0 ($S.s^{0.5}$)**	**Corrosion rate (mm/yr)**
5.0M HNO_3	0.6173	4.223×10^{-5}	0.9462	1.952×10^5		0.0013
5.0M HNO_3 + 0.1M HF	2.474	4.902×10^{-5}	0.8765	1.082×10^4	0.0016	0.023
5.0M HNO_3 + 0.1M HF + 0.1M Zr^{4+}	0.5982	3.806×10^{-5}	0.9456	1.802×10^5		0.0014

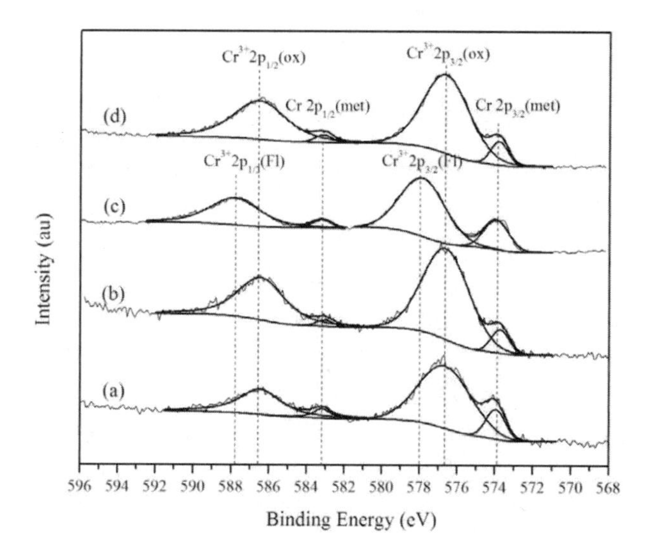

Figure 3. Cr 2p XPS spectra recorded for: (a) polished sample and samples immersed in (b) 5.0M HNO_3; (c) 5.0M HNO_3 + 0.1M HF; and (d) 5.0M HNO_3 + 0.1M HF + 0.1M Zr^{4+}.

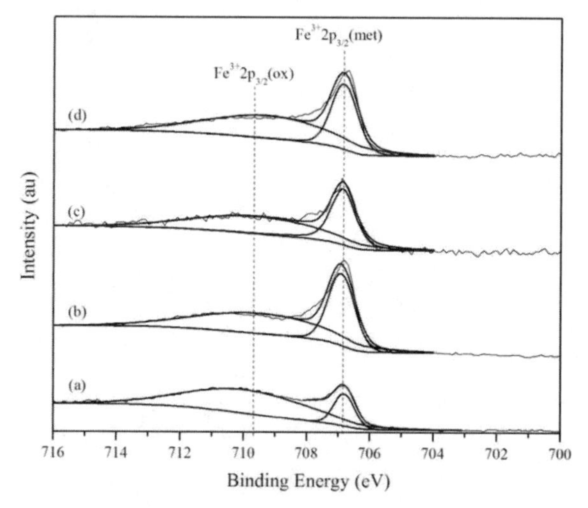

Figure 4. Fe 2p XPS spectra recorded for: (a) polished sample and samples immersed in (b) 5.0M HNO_3; (c) 5.0M HNO_3 + 0.1M HF; and (d) 5.0M HNO_3 + 0.1M HF + 0.1M Zr^{4+}.

Author Index